RF and Digital Signal Processing for Software-Defined Radio

RF and Digital Signal Processing for Software-Defined Radio

A Multi-Standard Multi-Mode Approach

Tony J. Rouphael

AMSTERDAM • BOSTON • HEIDELBERG • LONDON • NEW YORK • OXFORD
PARIS • SAN DIEGO • SAN FRANCISCO • SINGAPORE • SYDNEY • TOKYO

Newnes is an imprint of Elsevier

Newnes is an imprint of Elsevier
30 Corporate Drive, Suite 400, Burlington, MA 01803, USA
Linacre House, Jordan Hill, Oxford OX2 8DP, UK

Library of Congress Cataloging-in-Publication Data
Application submitted.

British Library Cataloguing-in-Publication Data
A catalogue record for this book is available from the British Library.

ISBN 978-0-7506-8210-7

For information on all Newnes publications
visit our Web site at www.elsevierdirect.com

Working together to grow
libraries in developing countries

www.elsevier.com | www.bookaid.org | www.sabre.org

ELSEVIER   BOOK AID   Sabre Foundation
             International

Contents

For my son, John Joseph Rouphael,
You are the wind beneath my wings.

Now I am the last to keep Vigil,
like a gleaner after the vintage;
Since by the Lord's blessing I have made progress
till like a vintager I have filled my winepress,
I would inform you that not for myself only have I toiled
but for every seeker after wisdom.

The book of Sirach 33:16-18

Acknowledgments

I am deeply indebted to many individuals without whom this book would not have been a reality. Special thanks go to Harry Helms, Rachel Roumeliotis, Heather Scherer, Rajashree Satheesh Kumar, and Anne McGee from the editorial staff at Elsevier for all their help and guidance all throughout the writing of this book. I also would like to express my most sincere gratitude to the various reviewers who were instrumental in preparing the final manuscript. I especially would like to thank David Southcombe, Farbod Kamgar, Scott Griffith, Sumit Verma, Ricke Clark, Hui Liu, Mariam Motamed, and Kevin Shelby. Finally I would like to thank my wife Tabitha for all her love, support, and constant encouragement. Without her this book would not have been completed. Toby you are my guardian angel!

Introduction

The aim of this book is to present the key underlying signal-processing principles used in software-defined radio (SDR) analysis and design. The various chapters span topics ranging from analog and digital modulation to radio frequency (RF) and digital signal processing and data conversion. Although the intent is to cover material relevant to the signal processing used in SDR, the same material could be applied to study more traditional radio architectures.

1.1 The Need for Software-Defined Radio

The SDR forum[1] defines SDR as a "*radio in which some or all of the physical layer functions are software defined.*" This implies that the architecture is flexible such that the radio may be configured, occasionally in real time, to adapt to various air standards and waveforms, frequency bands, bandwidths, and modes of operation. That is, the SDR is a multifunctional, programmable, and easy to upgrade radio that can support a variety of services and standards while at the same time provide a low-cost power-efficient solution. This is definitely true compared to, say, a Velcro approach where the design calls for the use of multiple radios employing multiple chipsets and platforms to support the various applications. The Velcro approach is almost inevitably more expensive and more power hungry and not in the least compact. For a device that supports multi-mode multi-standard functionality, SDR offers an overall attractive solution to a very complex problem. In a digital battlefield scenario, for example, a military radio is required to provide an airborne platform with voice, data, and video capability over a wide spectrum with full

---

[1]See www.sdrforum.org

interoperability across the joint battle-space. A traditional legacy Velcro approach in this case would not be feasible for the reasons listed above. In order to create a more adaptable and easily upgradeable solution, the Department of Defense and NATO created the Joint Tactical Radio System (JTRS) initiative to develop a family of software-defined and cognitive radios that would support interoperability across frequency bands and waveforms as well as satisfy the ease of upgradeability and configurability mandated by modern warfare.

In the commercial world, the need to adopt SDR principles is becoming more and more apparent due to recent developments in multi-mode multi-standard radios and the various complex applications that govern them. The flexibility of SDR is ideally suited for the various quality-of-service (QoS) requirements mandated by the numerous data, voice, and multimedia applications. Today, many base-station designs employ SDR architecture or at least some technology based on SDR principles. On the other hand, surely but slowly, chipset providers are adopting SDR principles in the design of multi-mode multi-standard radios destined for small form-fit devices such as cellphones and laptops [1].

1.2 The Software-Defined Radio Concept

The SDR architecture is a flexible, versatile architecture that utilizes general-purpose hardware that can be programmed or configured in software [2]. Compared to traditional architectures that employ quadrature sampling, SDR radios that employ intermediate frequency (IF) sampling tend to do more signal processing in the digital domain.
This particular radio architecture, shown conceptually in Figure 1.1, moves the data

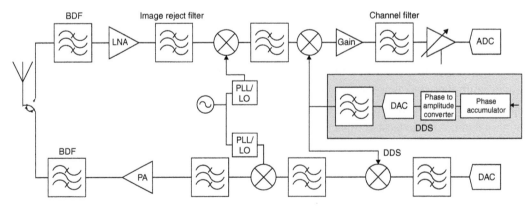

Figure 1.1 Ideal IF-sampling transceiver (digital section not shown) for a TDD system

conversion closer to the antenna. However, more digital signal processing should by no means imply that the radio is any easier to design or that IF-sampling is superior to quadrature sampling. All it means is that IF-sampling architecture could *possibly* afford more flexibility in supporting multiple waveforms and services at a reasonable cost and complexity. Furthermore, IF-sampling architecture does not imply that the ensuing design becomes any less complicated or any more robust!

As a matter of fact, the added flexibility of the radio serves to add to its complexity, especially when dealing with coexistence scenarios and interference mitigation. However, the advantage of SDR over traditional solutions stems from its adaptability to its environment and the number and type of application that it can support. This can be accomplished since the hardware itself and its operation is abstracted completely from the software via a middleware layer known as the *hardware abstraction layer* (HAL). The purpose of the HAL is to enable the portability and the reusability of software [3]. This common framework enables the development and deployment of various applications without any dependencies on the particular radio hardware.

The SDR concept further assumes certain smarts in the antenna, the RF, and the DSP. In the case of the antenna, this is manifested in terms of its flexibility to tuning to various bands, and its adaptability in terms of beamforming (self-adapting and self-aligning), MIMO operations, and interference rejection (self-healing). Obviously these aims, when it comes to the antenna, are much easier said than done. In terms of analog and digital signal processing, the SDR manages the various stages of the RF to make it more adaptable to its ever-changing environment in order to achieve the desired QoS. Likewise, this flexibility is extended to the data conversion mixed-signal blocks that are configured to receive the desired signal. This entails changing the sampling rate of the converter and its resolution depending on the environment (e.g., blockers, interferers, IF frequency, etc.) and the signaling scheme in a scalable, power-efficient manner that guarantees adequate performance. The digital signal processing is also configured and programmed to complement the RF in terms of filtering, for example, and to perform the various modem functionalities. The modem is designed to support multiple waveforms employing various data rates and modulation schemes (e.g., spread spectrum, OFDM, etc.). In most applications, this is done using a *field programmable gate array* (FPGA) with application specific capabilities. This endows the radio with real-time reconfiguration and reprogramming capacity, which allows it to roam over various networks in different geographical regions and environments supporting multiple heterogeneous applications.

The efficient implementation of such a radio relies on robust common hardware platform architectures and design principles as defined and outlined in the Software Communication Architecture (SCA) standard supported by the SDR forum and adopted by the Joint Program Office (JPO).

1.3 Software Requirements and Reconfigurability

Current SDR implementations mostly rely on reconfigurable hardware to support a particular standard or waveform while the algorithms and the various setups for other waveforms are stored in memory. Although application-specific integrated circuits lead to the most efficient implementation of a single-standard radio, the same cannot be said when addressing a multi-mode multi-standard device. For a device with such versatility, the ASIC could prove to be very complex in terms of implementation and inefficient in terms of cost and power consumption. For this reason, SDR designers have turned to FPGAs to provide a flexible and reconfigurable hardware that can support complex and computationally intensive algorithms used in a multitude of voice, data, and multimedia applications.

In this case, one or more FPGAs are typically used in conjunction with one or more DSPs, and one or more general-purpose processors (GPPs) and/or microcontrollers to simultaneously support multiple applications. FPGAs are typically configured at startup and with certain new architectures can be partially configured in real time. Despite all of their advantages, however, the industry still has not adopted FPGA platforms for use in handheld and portable devices due to their size, cost, and power consumption. Similarly, DSPs and GPPs can also be programmed in real time from memory, allowing for a completely flexible radio solution. Unlike dedicated DSPs used in single standard solutions, SDR employs programmable DSPs that use the same computational kernels for all the baseband and control algorithms.

1.4 Aim and Organization of the Book

The aim of this book is to provide the analog and digital signal processing tools necessary to design and analyze an SDR. The key to understanding and deriving the high-level requirements that drive the radio specifications are the various waveforms the radio intends to support. To do so, a thorough understanding of common modulation waveforms and their key characteristics is necessary. These high level requirements are then disseminated onto the various radio blocks, namely the antenna, the RF, the mixed-signal data conversion (ADC and DAC) blocks, and the digital signal processing block. Of course there are other key components related to the various processors and controllers, and the software and middleware. These topics are beyond the scope of this book.

The book is divided into four sections presented in ten chapters. Section 1 is comprised of Chapters 2 and 3. Chapter 2 deals with analog modulation techniques that are still relevant and widely in use today. Spectral shaping functions and their implementations are also discussed in detail. Chapter 3 addresses digital modulation schemes ranging from simple *M*-PSK methods to spread spectrum and OFDM. In both chapters figures of merit and performance measures such as SNR and BER under various channel conditions are provided.

Section 2 deals mainly with the RF and analog baseband. This section is divided into three chapters. Chapter 4 addresses the basics of noise and link budget analysis. Chapter 5 deals with nonlinearity specifications and analysis of memoryless systems. Design parameters such as second and third order input-referred intercept points, intermodulation products, harmonics, cross-modulation, and adjacent channel linearity specifications are a few of the topics discussed. Similarly, Chapter 6 further addresses RF and analog design principles and analysis techniques with special emphasis on performance and figures of merit. Topics include receiver selectivity and dynamic range, degradation due to AM/AM and AM/PM, frequency accuracy and tuning, EVM and waveform quality factor and adjacent channel leakage ratio are among the topics discussed.

Section 3 addresses sampling and data conversion. This section is comprised of Chapters 7, 8, and 9. In Chapter 7, the basic principles of baseband and bandpass sampling are studied in detail. The resolution of the data converter as related to sampling rate, effective number of bits, peak to average power ratio, and bandwidth are derived. The chapter concludes with an in-depth analysis of the automatic gain control (AGC) algorithm. In Chapter 8, the various Nyquist sampling converter architectures are examined in detail. Particular attention is paid to the pros and cons of each architecture as well as the essential design principles and performance of each. Chapter 9 discusses the principles of oversampled data converters. The two main architectures, namely continuous-time and discrete-time $\Delta\Sigma$-modulators, are examined in detail. Again, the various pros and cons of each method are presented.

Section 4 consists of Chapter 10. Chapter 10 introduces the basics of multirate signal processing techniques such as interpolation and decimation. A detailed study of the various filtering structures used to perform these basic operations is provided. Topics such as polyphase filtering structures, half-band and M-band filters, and cascaded integrator-comb filters are considered. Irrational sampling rate conversion techniques that employ interpolating polynomials such as the Lagrange polynomial are also discussed. An effective structure that enables the implementation of such polynomial interpolators, namely the Farrow filter, is also discussed.

References

[1] Srikanteswara S, et al. An overview of configurable computing machines for software radio handsets. IEEE Comm. Magazine July 2003;134–41.
[2] Kenington PB. Emerging technologies for software radio. Electronics & Communication Engineering Journal April 1999;69–73.
[3] Munro A. Mobile middleware for the reconfigurable software radio. IEEE Communications Magazine August 2008;38(8):152–61.

CHAPTER 2

Common Analog Modulation and Pulse-Shaping Methods

The choice of a modulation method and pulse-shaping scheme depends greatly on the availability of the spectrum, the desired data rate, and the radio channel itself. In turn, the modulation method influences the design complexity of the modem. In several SDR applications, the designer is faced with supporting multiple waveforms each with its own modulation and pulse-shaping technique, bandwidth, transmit mask and transmit power, etc.

A modulation scheme is influenced by various factors. The most commonly desired features of any waveform are:

- *Spectrally efficient modulation* that occupies the least amount of spectrum and minimally interferes with adjacent and neighboring channels

- *Robust performance* under certain channel conditions, such as multipath fading and heavy shadowing, Doppler and frequency drifts, interference and intentional jamming, etc.

- Achievement of a low bit error rate performance at low SNR and minimal transmit power

- Allowance for cost effective, low power, small size and weight solutions

Analog modulation is still widely used in one form or another in both military and civilian applications. These applications still rely on legacy waveforms that use amplitude modulation (AM), phase modulation (PM), and frequency modulation (FM) for their modulation scheme. In today's state of the art JTRS-SDRs, for example, these waveforms are generated digitally and not in the analog domain. Therefore, when discussing analog modulation in the context of SDR, the reader must keep in mind that, despite the classifications of these waveforms as analog, they are often generated digitally.

The aim of this chapter is to present an overview of common analog modulation and pulse-shaping techniques and to highlight the parameters that affect the SDR design. Section 2.1 discusses AM modulation schemes. Section 2.2 discusses FM and PM modulation schemes. Common pulse-shaping functions are presented in Section 2.3.

2.1 Amplitude Modulation

Amplitude modulation is still widely used in legacy DoD radios and supported by JTRS. In amplitude modulation, the baseband signal $x(t)$ is modulated on a frequency carrier f_c

$$y(t) = A_c \left\{1 + x(t)\right\} \cos(2\pi f_c t) \tag{2.1}$$

where A_c is the carrier amplitude. When the modulating signal $x(t)$ is sinusoidal, the relation in (2.1) becomes

$$y(t) = A_c \left\{1 + \kappa \cos\left(2\pi f_m t\right)\right\} \cos(2\pi f_c t) \tag{2.2}$$

where κ is the modulation index, and

$$\kappa = \frac{A_m}{A_c} \tag{2.3}$$

where A_m is the maximum magnitude of the message signal. In order to recover the baseband signal the modulation index must be $|\kappa| \leq 1$. The modulation index is an indication of the power efficiency of the AM transmit system. In (2.2), it is obvious that the carrier does not provide any information and that all the information is conveyed in the message signal.

The spectrum of (2.1) is divided into an upper sideband (USB) and lower sideband (LSB) as depicted in Figure 2.1 and can be expressed as

$$Y(f) = \frac{A_c}{2}\left\{1 + X(f)\right\} * \left\{\delta(f - f_c) + \delta(f + f_c)\right\} =$$

$$Y(f) = \frac{A_c}{2}\left\{X(f - f_c) + X(f + f_c) + \delta(f - f_c) + \delta(f + f_c)\right\} \tag{2.4}$$

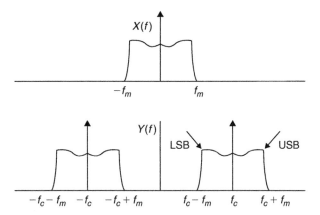

Figure 2.1 Spectrum of a double sideband AM signal with carrier

It is important to note that the spectrum of a DSB AM signal is limited by the bandwidth of the message signal $x(t)$[1]

$$B = 2f_m = \frac{\omega_m}{\pi} \tag{2.5}$$

where f_m is the highest frequency contained in the message signal $x(t)$. For AM modulation, it is important to retain the shape of the signal as much as possible since the signal amplitude carries the wanted information. The average power of an AM signal is

$$P_y = \int y^2(t)\, dt = \int \left\{ A_c \left\{ 1 + x(t) \right\} \cos\left(\omega_c t\right) \right\}^2 dt$$

$$= \int A_c^2 \left\{ 1 + 2x(t) + x^2(t) \right\} \frac{\left\{ 1 + \cos\left(2\omega_c t\right) \right\}}{2}\, dt$$

$$P_y = \frac{A_c^2}{2} \left\{ 1 + E\left\{ x(t) \right\} + E\left\{ x^2(t) \right\} \right\}$$

$$= P_c \left\{ 1 + E\left\{ x(t) \right\} + E\left\{ x^2(t) \right\} \right\} \quad \text{where } P_c = \frac{A_c^2}{2} \tag{2.6}$$

[1]Note that a single-sideband (SSB) AM signal occupies only half the bandwidth of a DSB AM signal.

where $\omega_c = 2\pi f_c$. For $x(t) = \kappa \cos(\omega_m t)$, the average power of the AM signal becomes

$$P_y = P_c\left(1 + \frac{\kappa^2}{2}\right) \tag{2.7}$$

where P_c is known as the carrier power. Generally, for random like signals, the peak to average power ratio of an AM signal is similar to that of white noise and is approximately 10 dB.

While the SNR of the received signal is an important figure of merit, various receiver architectures and implementations can impact the post-detection SNR. For AM receivers employing square-law detectors, the postdetection SNR[2] could be related to the predetection SNR, or carrier-to-noise ratio (CNR) in this case, as [1]:

$$SNR_{\text{postdetection}} = \frac{P_c \kappa^2}{2N_0 B} = \left(\frac{P_c}{N_0 B_{IF}}\right)\left(\frac{B_{IF}}{B}\right)\left(\frac{\kappa^2}{2}\right) = \frac{B_{IF} \kappa^2}{2B} SNR_{\text{predetection}} \tag{2.8}$$

Note that in SDR IF-sampling architecture, the sampling takes place at IF, and hence the bandwidth of the selectivity filter at IF determines the amount of noise and interference that the analog-to-digital converter (ADC) sees.

The AM signal described in (2.1) is a double sideband signal with carrier (DSB+C). For completeness, we list the various flavors of AM modulation, namely:

- Double sideband (DSB)
- Single sideband (SSB)
- Vestigial sideband (VSB)

DSB can be classified into two different categories; DSB+C and DSB with suppressed carrier (DSB-SC). In either case, the upper band in Figure 2.1 is known as the upper sideband (USB), whereas the lower band is known as lower sideband (LSB). USB and LSB are mirror images of each other with opposite phases. Both spectra contain sufficient information to demodulate the signal. The main difference between DSB+C and DSB-SC is the carrier term itself.

---

[2]Postdetection SNR is the SNR of the envelope signal after the detector, whereas predetection SNR implies the SNR of the signal before the detector.

Define efficiency as ratio of energy-containing information to the total transmitted energy:

$$\varepsilon = \frac{\kappa^2}{2 + \kappa^2} \times 100\% \qquad (2.9)$$

The maximum efficiency of 33.33% occurs when $\kappa = 1$. A lower modulation index implies a lower efficiency. Figure 2.2 shows a DSB+C signal with modulation indices of 100%, 50%, and 150%. Note that in the latter case the envelope is distorted and may not be recovered. The relation in (2.9) implies that DSB-SC is a 100% efficient AM modulation scheme. Traditionally, the real benefit behind DSB+C modulation is its simplicity. DSB+C can be generated without a mixer using a nonlinear device such as a diode, for instance. Using two DSB+C modulators with inputs having opposite signs,

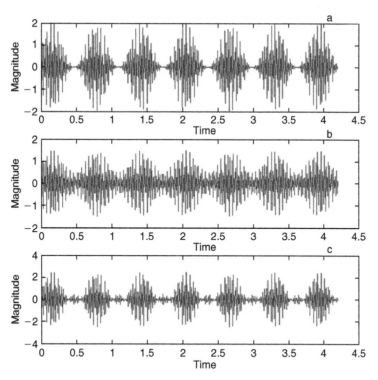

Figure 2.2 DSB+C signal with modulation index of: a-100%, b-50%, and c-150%

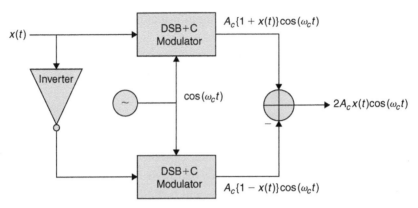

Figure 2.3 A DSB-SC modulator constructed in a balanced topology using two DSB+C modulators

one can construct a cheap DSB-SC modulator using a balanced mixer topology as shown in Figure 2.3. Another advantage of DSB+C is the ability to demodulate coherently using envelope detection techniques. In either case, the information transmitted on the USB and LSB is identical and, consequently, twice the bandwidth necessary is used to convey the same information. To use only the necessary bandwidth needed to transmit the signal, a single-sideband (SSB) modulation scheme is used which employs either the USB or LSB portion of the DSB spectrum.

SSB modulation can be easily obtained from DSB-SC via bandpass filtering before amplification and transmission. A key disadvantage of SSB is obviously the bandpass filter, which is required to have significant rolloff and rejection characteristics. To ease the problem, SDR designers who choose to generate SSB AM via filtered DSB-SC tend to use a low IF frequency in order to simplify design of the bandpass filter. An alternative and more common method to generate SSB uses the Hilbert transform. The choice of USB or LSB depends largely on the frequency plan of the radio. SSB-SC requires the same transmit power as DSB-SC in order to achieve the same SNR despite the fact that DSB-SC utilizes twice the bandwidth. The reason is that DSB-SC receivers utilize both bands to demodulate the signal, and hence no energy is wasted.

Vestigial sideband (VSB) AM modulation, on the other hand, offers a reasonably good frequency response at low frequencies while maintaining bandwidth efficiency. VSB modulation constitutes a compromise between DSB and SSB. SSB modulation methods exhibit a poor frequency response at low frequencies due to the imperfections in the sidebands of the filter, or due to the 90° phase shifting technique. In SDR, however, the 90° phase shift is generated digitally since quadrature modulation is performed before the DAC, which is a luxury that legacy AM radios did not have. DSB modulation methods, on the other hand, display good frequency response at low frequencies but are bandwidth inefficient. In VSB, a filter similar to that used for DSB is used to retain the desired band along with a portion of the rejected sideband.

2.2 Frequency and Phase Modulation

Frequency and phase modulation are a form of angle modulation:

$$y(t) = A_c \cos(2\pi f_c t + \phi(t) + \phi_0) \tag{2.10}$$

where f_c is the carrier frequency, ϕ_0 is the initial phase, and $\phi(t)$ is a baseband controlled function defined as

$$\phi(t) = \begin{cases} 2\pi\Gamma_f \int\limits_{-\infty}^{t} m(\tau)d\tau & \text{for FM} \\ \Gamma_p m(t) & \text{for FM} \end{cases} \tag{2.11}$$

where Γ_f and Γ_p are the frequency and phase deviation constants, respectively. In FM, in the event where the modulating signal is a sinusoid with amplitude K and frequency f_m, the FM signal in (2.11) becomes:

$$y(t) = A_c \cos\left(2\pi f_c t + \frac{K\Gamma_f}{f_m} \sin(2\pi f_m t) + \phi_0\right) \tag{2.12}$$

FM consists of a carrier frequency and an infinite number of sideband components spaced at intervals equal to integer multiples of the modulating frequency. Therefore, theoretically, the bandwidth of an FM signal is infinite. However, 98% of the signal's energy exists in a limited band of the signal bandwidth, which is dictated by the

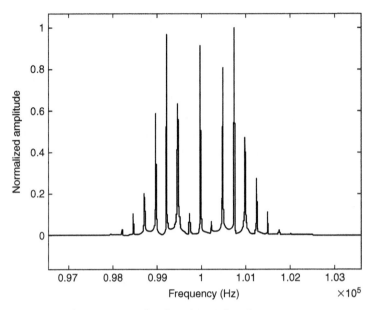

Figure 2.4 A simulated FM signal spectrum

magnitude of the modulating signal. Figure 2.4 depicts an example of an FM spectrum modulated on a 100-kHz carrier.

Define the frequency modulation index for single tone modulation only as:

$$\mu_{FM} = \frac{K\Gamma_f}{B_{max}} = \frac{f_{dev}}{B_{max}} \tag{2.13}$$

where B_{max} is the highest frequency component of the modulating signal $m(t)$, and $f_{dev} = K\Gamma_f$ is the frequency deviation. The relation in (2.12) can be further expressed in terms of Bessel functions as:

$$y(t) = A_c \left\{ \begin{array}{l} J_0(\mu_{FM})\cos(2\pi f_c t) + \displaystyle\sum_{n=1}^{\infty} (-1)^n J_n(\mu_{FM}) \\ \left[\cos(2\pi f_c t - 2\pi n f_m t) + (-1)^n \cos(2\pi f_c t + 2\pi n f_m t) \right] \end{array} \right\} \tag{2.14}$$

For a small μ_{FM}, as will be stated in Carson's rule concerning the narrowband case, the Bessel functions can be approximated as

$$J_0(\mu_{FM}) \approx 1 - \left(\frac{\mu_{FM}}{2}\right)^2 \qquad n = 0$$

$$J_n(\mu_{FM}) \approx \frac{1}{n!}\left(\frac{\mu_{FM}}{2}\right)^n \qquad n \neq 0 \qquad (2.15)$$

In this case, the relation in (2.14) becomes

$$y(t) \approx A_c\left\{\cos(2\pi f_c t) - \frac{\beta}{2}\cos(2\pi f_c t - 2\pi n f_m t) + \frac{\beta}{2}\cos(2\pi f_c t + 2\pi n f_m t)\right\}$$

$$y(t) \approx A_c\{\cos(2\pi f_c t) - \beta\sin(2\pi n f_m t)\sin(2\pi f_c t)\} \qquad (2.16)$$

Carson's rule gives an approximation of the FM signal bandwidth:

$$B_{FM} \approx \begin{cases} 2(\mu_{FM} + 1)f_m & B_{max} \gg \dfrac{1}{2\pi}K\Gamma_f \\[3mm] 2\mu_{FM}B_{max} & B_{max} \ll \dfrac{1}{2\pi}K\Gamma_f \end{cases} \text{Hz} \qquad (2.17)$$

Note that in the event where the modulating signal is sinusoidal, $B_{max} = f_m$.

The bandwidth for a PM signal is given as [2]:

$$B_{PM} \approx 2\left(\frac{\Gamma_p}{2\pi}\max\left(\frac{d}{dt}m(t)\right) + \alpha B\right) \quad 1 < \alpha < 2\,\text{Hz} \qquad (2.18)$$

The value of α depends on the modulation.

Example 2-1: Bandwidth of the Cellular AMPS FM Signal

Given a frequency modulation index of $\mu_{FM} = 3$ and a modulating frequency $f_m = 4000\,Hz$, using Carson's rule estimate the bandwidth of the advanced mobile phone system (AMPS) signal modulated on a carrier $f_c = 869\,MHz$.

In this case, neither the upper or lower bound condition on the bandwidth as stated in (2.17) is satisfied. Therefore, we compute both the upper and lower bandwidths as

$$
B_{FM} \approx \begin{cases} 2(\mu_{FM} + 1)f_m = 2(3 + 1)4000 = 32{,}000\,\text{Hz} \\ 2 \times \mu_{FM} \times B_{max} = 2 \times 3 \times 4000 = 24{,}000\,\text{Hz} \end{cases} \tag{2.19}
$$

The AMPS standard, however, specifies the signal bandwidth in terms of a spectral mask in which the signal must be -26 dBc at ± 20 kHz frequency offset and -45 dBc at ± 45 kHz frequency offset and does not rely on the required bandwidth stated in (2.19). The reason for the specification being written as it is has mainly to do with spectral management and efficiency in order to maximize capacity and minimize interference.

The predetection SNR (or CNR in this case) is dependent on the receive carrier power as well as the bandwidth of the IF selectivity filter, that is

$$
SNR_{\text{predetection}} = \frac{\frac{1}{2}A_c^2}{2N_0(\mu_{FM} + 1)B_{IF}} \tag{2.20}
$$

Note that the bandwidth used in (2.20) is the IF selectivity filter bandwidth, which tends to be different from the signaling bandwidth. The postdetection SNR is related to the predetection SNR by *Frutiger's relation*, namely [3]:

$$
SNR_{\text{postdetection}} = \frac{\left(\frac{3}{2}\right)\left(\frac{f_{dev}}{B_{max}}\right)^2 \left(\frac{B_{IF}}{B_{max}}\right) SNR_{\text{predetection}}}{1 + 0.9 \left(\frac{B_{IF}}{B_{max}}\right)^2 SNR_{\text{predetection}} \dfrac{e^{-SNR_{\text{predetection}}}}{\left(1 - e^{-SNR_{\text{predetection}}}\right)^2}} \tag{2.21}
$$

Example 2-2: Post Detection SNR for the Cellular AMPS FM System

Given a frequency modulation index of $\mu_{FM} = 3$ and the highest frequency component of the modulating signal to be $B_{max} = 4000$ Hz, what is the post detection SNR for an FM system where the predetection SNR ranges between 0 and 20 dB? Assume that the IF selectivity bandwidth is 30 kHz. Repeat the computation for an IF bandwidth of 100 kHz.

The results are depicted in Figure 2.5.

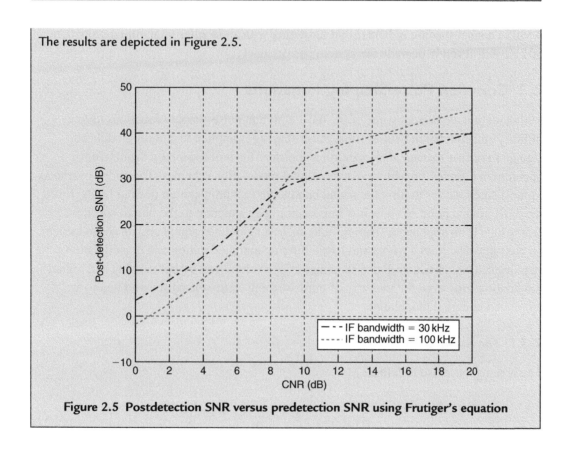

Figure 2.5 Postdetection SNR versus predetection SNR using Frutiger's equation

Note that, from the example above, choosing the bandwidth of the IF filter greatly depends on the CNR of the received signal. For received CNR $\leq$ 9 dB, choosing an IF filter bandwidth of 30 kHz results in better postdetection SNR than choosing an IF filter bandwidth of 100 kHz. For CNR $\geq$ 9 dB choosing the wider filter bandwidth of 100 kHz results in better detection SNR. Therefore, in order to design the optimum filter, one must know the condition in which the received signal will most likely operate. In reality, the AMPS system is designed with an IF bandwidth of 30 kHz. The required CNR for AMPS is 18 dB resulting in approximately 40 dB of postdetection SNR. Furthermore, it is important to note that the higher post detection SNR advantage for the 100 kHz IF bandwidth is only true in the absence of any interference due to adjacent channels (that is, in a purely thermal additive white Gaussian noise (AWGN) channel). In reality, the

AMPS channel spacing is 30 kHz and designing with an IF filter of anything more than 30 kHz will severely degrade the system performance.

2.3 Common Pulse-Shaping Functions

In this section we present some of the most common pulse shaping functions used in military and commercial waveforms. Pulse shaping is applied to certain modulated signals to further manage their spectral response. This will especially become more apparent in Chapter 3 when we discuss digital modulation schemes. The added overhead, in most cases, serves to limit the signal bandwidth to within certain desired limits. From the SDR system point of view, it is important to pay attention to the output spectrum that results from applying a shaping function to a modulated signal and its consequences on the radio design as a whole and to the filtering and modulation and demodulation requirements in particular. Pulse shaping has been employed by both analog and digital modulation. However, the majority of pulse-shaping techniques discussed herein are employed in digital modulation.

2.3.1 The Rectangular Pulse

The rectangular pulse is defined as

$$\Pi_{LT}(t) = \begin{cases} A & -\dfrac{LT}{2} \leq t \leq \dfrac{LT}{2} \\ 0 & \text{otherwise} \end{cases} \tag{2.22}$$

where A is the amplitude of the pulse and L is an integer. The frequency domain representation of the rectangular pulse is

$$F\left\{\Pi_{LT}(t)\right\} = A\int_{\frac{LT}{2}}^{\frac{LT}{2}} e^{-j2\pi ft}dt = ALT\frac{\sin(\pi LTf)}{\pi LTf} \tag{2.23}$$

The Fourier transform of the rectangular pulse is real and its spectrum, a sinc function, is unbounded. This is equivalent to an upsampled pulse-train of upsampling factor L. In real systems, rectangular pulses are spectrally bounded via filtering before transmission which results in pulses with finite rise and decay time.

2.3.2 The Sinc Pulse

Consider the brickwall filter:

$$\Pi_{\frac{1}{2T}}(f) = \begin{cases} T & -\frac{1}{2T} \leq f \leq \frac{1}{2T} \\ 0 & \text{otherwise} \end{cases} \qquad (2.24)$$

where T is the symbol duration. The equivalent time-domain pulse of (2.24) is

$$F^{-1}\left\{\Pi_{\frac{1}{2T}}(f)\right\} = T\int_{-1/2T}^{1/2T} e^{j2\pi ft}df = \text{sinc}\left(\frac{\pi t}{T}\right) \qquad (2.25)$$

Note that for $t = nT$

$$\text{sinc}\left(\frac{\pi t}{T}\right)\bigg|_{t=nT} = \begin{cases} 1 & n = 0 \\ 0 & n = \pm 1, \pm 2, \pm 3, \ldots \end{cases} \qquad (2.26)$$

The relation stated in (2.26), depicted by way of example in Figure 2.6, implies that the sinc pulse is a Nyquist pulse—that is, it is a pulse that attains its maximum value at $n = 0$ and is zero at all subsequent integer multiples of T. That is, a pulse is said to be a Nyquist pulse if, when sampled properly, it does not cause intersymbol interference (ISI) to all its neighboring symbols. The sinc function of (2.26) is not practical for several reasons: (1) The sinc pulse is infinite in duration, and any realistic implementation of it would automatically imply a time-domain truncation of the pulse, and (2) it is noncausal, and in order to enforce causality, a significant delay must be introduced into the pulse.

2.3.3 The Raised Cosine Pulse

Nyquist stated in his landmark paper [4] that the ISI-free property can be preserved if the sharp cutoff frequency of the brick-wall filter is modified by an amplitude characteristic having odd symmetry about the Nyquist cutoff frequency. One such filter is the raised cosine filter represented in the time domain as

$$g(t) = \frac{\cos(\alpha\pi t/T_s)}{1 - (2\alpha t/T_s)^2}\text{sinc}\,(t/T_s) = \frac{\pi}{4}\big\{\text{sinc}\,(\alpha t/T_s + 1/2) +$$

$$\text{sinc}\,(\alpha t/T_s - 1/2)\big\}\text{sinc}\,(t/T_s) \quad 0 \leq \alpha \leq 1 \qquad (2.27)$$

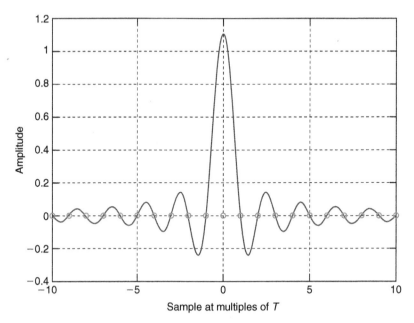

Figure 2.6 The ideal sinc function (solid line) and the respective samples at
nT, $n = 0,1,...$ **(circles)**

and shown in both time and frequency domains for two rolloff factors $\alpha = 1$ and $\alpha = 0.25$ in Figure 2.7 and Figure 2.8, respectively. The frequency domain representation of (2.27) is

$$G(f) = \begin{cases} T_s & |f| \leq (1-\alpha)\dfrac{F_s}{2} \\[2ex] T_s \dfrac{1 - \sin\left(\pi(f - F_s/2)/\alpha F_s\right)}{2} & (1-\alpha)\dfrac{F_s}{2} \leq |f| \leq (1+\alpha)\dfrac{F_s}{2} \\[2ex] 0 & \text{otherwise} \end{cases} \quad (2.28)$$

The symbol rate is related to the IF bandwidth B_{IF} as

$$R_{T_s} = \frac{B_{IF}}{1+\alpha} \tag{2.29}$$

Note that for both $\alpha = 1$ and $\alpha = 0.25$, the shaping functions are unity at $t = 0$ and zero at the integer multiples of the sampling period. Furthermore, note that for smaller

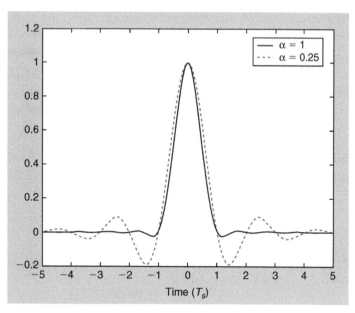

Figure 2.7 Normalized time-domain raised cosine pulse for $\alpha = 1$ and $\alpha = 0.25$

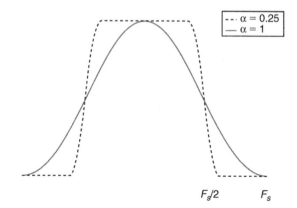

Figure 2.8 Frequency domain representation of the raised cosine pulse for $\alpha = 1$ and $\alpha = 0.25$

α the shaping function tends to have more contained spectrum at the expense of more oscillations in its response thus requiring a more robust timing recovery algorithm. The raised cosine pulse oscillations in general, however, decay at the rate of $1/t^3$ for $t > T_s$ as compared with the normal sinc function of the brick-wall filter. This rapid decay in magnitude response allows the designer to truncate the pulse response without any significant degradation to the performance. Another important observation from Figure 2.8 is excess bandwidth with respect to the Nyquist frequency. For $\alpha = 1$ the excess bandwidth is 100% or twice the Nyquist frequency, and for $\alpha = 0.25$ the excess bandwidth is 25%. Hence, α dictates the bandwidth of the filter. The null itself is placed at $(1 + \alpha) F_s/2$ as suggested in (2.28) and is dictated mainly by the amount of adjacent channel interference that can be tolerated by the system.

In many practical systems, a $\sin(x)/x$ correction[3] is applied to equalize for the sinc-like behavior of the square pulse, thus modifying the relation presented in (2.28) to:

$$G_m(f) = \begin{cases} \dfrac{\pi f T_s}{\sin(\pi f T_s)} & |f| \le (1-\alpha)\dfrac{F_s}{2} \\[2ex] \dfrac{\pi f T_s}{\sin(\pi f T_s)}\cos^2\left[\dfrac{\pi T_s}{2\alpha}\left(f - \dfrac{1-\alpha}{2T_s}\right)\right] & (1-\alpha)\dfrac{F_s}{2} \le |f| \le (1+\alpha)\dfrac{F_s}{2} \\[2ex] 0 & \text{otherwise} \end{cases} \qquad (2.30)$$

Furthermore, the raised cosine pulse may be realized as a combination of transmit root raised cosine filter in the transmitter and an identical root raised cosine filter in the receiver where the cosine squared term in (2.30) is replaced with its square root to realize the root raised cosine pulse.

2.3.4 The Cosine Pulse

The cosine pulse is as its name implies:

$$g(t) = \begin{cases} A\cos\left(\dfrac{\pi t}{T_s}\right) & -\dfrac{T_s}{2} \le t \le -\dfrac{T_s}{2} \\[2ex] 0 & \text{otherwise} \end{cases} \qquad (2.31)$$

[3]For certain modulation schemes, such as OFDM, it may be necessary to apply a correction to the $\sin(x)/x$ rolloff in order to flatten the spectrum and avoid unnecessary degradation to the tones at the edge of the band.

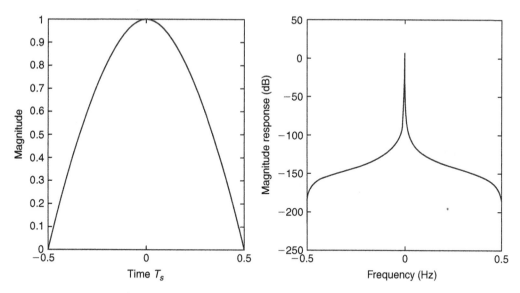

Figure 2.9 Cosine pulse and its respective spectrum

The spectrum of the cosine pulse is

$$G(f) = \frac{2AT_s}{\pi} \frac{\pi \cos(\pi f T_s)}{1 - 4f^2 T_s^2} \tag{2.32}$$

The time domain cosine pulse and its respective spectrum are shown in Figure 2.9.

2.3.5 The Triangular Pulse

The triangular pulse is the convolution of two square pulses, which implies that the triangular pulse spectrum is the product of two sinc functions:

$$G(f) = \frac{AT_s}{2} \left[\frac{\sin(\pi f T_s/2)}{\pi f T_s/2} \right]^2 \tag{2.33}$$

The equivalent time domain pulse of (2.33) is

$$g(t) = \begin{cases} A\left(1 - \dfrac{2t}{T_s}\right) & 0 \le t \le \dfrac{T_s}{2} \\ \\ 0 & t > \dfrac{T_s}{2} \end{cases} \tag{2.34}$$

2.3.6 The Gaussian Pulse

The Gaussian pulse is defined in the time domain as

$$g(t) = \frac{\sqrt{\pi}}{\alpha} e^{-\left(\frac{\pi}{\alpha}t\right)^2} \tag{2.35}$$

The rolloff factor α is related to the 3-dB bandwidth as

$$\alpha = \frac{\sqrt{\ln(4)}}{B_{3dB}} \tag{2.36}$$

In the frequency domain, the Gaussian pulse is given as

$$G(f) = e^{-(\alpha f)^2} \tag{2.37}$$

References

[1] Couch LW. Digital and Analog Communication Systems. 4th ed. New York, NY.: Macmillan; 1993.

[2] Carson AB. Communication Systems. 3rd ed. New York, NY: McGraw-Hill; 1986.

[3] Frutiger P. Noise in FM receivers with negative frequency feedback. Proc. IEEE, vol. 54, pp. 1506–1516, November 1966; Correction vol. 55, pp 1674, October 1967.

[4] Nyquist H. Certain topics in telegraph transmission theory. Trans. AIEE April 1928;47:617–44.

CHAPTER 3

Common Digital Modulation Methods

When designing a wireless communication system, there is always a tradeoff between bandwidth efficiency and power efficiency when it comes to selecting the appropriate modulation scheme. GSM, for example, uses GMSK which, as will be discussed later on, is a constant envelope modulation scheme that can operate in conjunction with highly efficient power amplifiers. This is true despite the variations introduced to the modulated signal due to filtering. GSM occupies 200 kHz of bandwidth and offers a raw data rate of 22.8 Kbps. *Enhanced Data-rate for GSM Evolution* (EDGE), on the other hand, occupies the same bandwidth as GSM but offers a raw data rate of 300 Kbps. The reason for the increased data rate is the use of the more bandwidth-efficient 8-PSK modulation. This spectral efficiency, however, comes at the cost of increased power due to the linear power amplifier required for differential 8-PSK signaling.

The factors presented above offer by themselves a challenge for the designer who aims to design a modem that uses a single modulation scheme, such as a plain GSM phone, for example. The design becomes much more complicated, and meeting the various requirements much more difficult when designing an SDR modem that supports multiple waveforms that often operate simultaneously. Therefore, a deep understanding of the nuances of the various modulation schemes is essential.

The advent of VLSI and DSP has enabled the communication engineer to apply more cost-effective and robust digital modulation techniques to support modern data and multimedia applications such as voice and video. Unlike analog modulation, digital modulation empowers the designer to use many digital signal processing techniques such as forward error correction, thus enabling communication under harsh channel conditions. In this chapter we discuss some but not all of the more common digital modulation schemes.

The aim of this chapter is to present an overview of common digital modulation methods and to highlight the parameters which affect the SDR design. In Section 3.1, the channel capacity theorem is discussed. Sections 3.2 through 3.9 discuss common digital modulation schemes ranging from the simple, such as binary phase shift keying (BPSK) and quadrature phase shift keying (QPSK), to the more complicated such as spread spectrum and orthogonal frequency-division multiplexing (OFDM). Section 3.10 contains the Appendix.

3.1 Channel Capacity Interpreted

Shannon-Hartley's channel capacity theorem is often applied at the beginning of any waveform and link budget analysis to provide the communication analyst with an upper bound on the data rate given a certain bandwidth and SNR. The achievable data rate, however, greatly depends on many parameters, as will be seen later on in the chapter. It is the influence of these parameters that dictates the deviation of the system data rate from the upper bound.

In this section, we present a simple interpretation of the Shannon-Hartley capacity theorem. A detailed treatment of this subject is beyond the scope of this book; however the interested reader should consult the work done in references [1]–[3]. The intent here is to show the set of assumptions that are normally made when estimating the channel capacity in the presence of AWGN.

Consider a redundancy-free zero-mean baseband signal $x(t)$ with random-like characteristics. In the absence of any coding, the no-redundancy assumption implies that any two samples $x(t_1)$ and $x(t_2)$ will appear to be random if they are sampled at the Nyquist rate or above [4]:

$$|t_1 - t_2| \geq \frac{1}{2B} \tag{3.1}$$

where B is the bandwidth of the signal. This type of signal is said to be communicating with maximum efficiency over the Gaussian channel. Furthermore, consider the received signal $y(t)$ of the signal $x(t)$ corrupted by noise:

$$y(t) = x(t) + n(t) \tag{3.2}$$

where $n(t)$ is AWGN with zero mean. The signal and noise are uncorrelated, that is

$$E\{x(t)\ n(t)\} = E\{x(t)\}E\{n(t)\} \tag{3.3}$$

where $E\{.\}$ is the expected value operator as defined in [5]. Consequently, the power of the signal $y(t)$ is the sum of the power of signal $x(t)$ and noise $n(t)$:

$$E\{(x(t) + n(t))^2\} = E\{x^2(t)\} + E\{n^2(t)\}$$
$$P_y = P_x + P_n \tag{3.4}$$

Note that any signal voltage content that is within the noise voltage range may not be differentiated from the noise itself. Therefore, we can assume that the smallest signal that can be represented is statistically at noise voltage level, and can be represented by

$$2^b = \frac{y(t)}{n(t)} \tag{3.5}$$

where b is the number of bits representing a binary number. The relation in (3.5) can be *loosely* re-expressed as

$$2^b = \frac{y(t)}{n(t)} = \sqrt{\frac{P_y}{P_n}} = \sqrt{\frac{E\{x^2(t)\} + E\{n^2(t)\}}{E\{n^2(t)\}}} = \sqrt{1 + \frac{E\{x^2(t)\}}{E\{n^2(t)\}}} \tag{3.6}$$

Define the SNR as

$$SNR = \frac{E\{x^2(t)\}}{E\{n^2(t)\}} \tag{3.7}$$

Then the relation in (3.6) becomes

$$2^b = \sqrt{1 + SNR} \Rightarrow b = \log_2\left(\sqrt{1 + SNR}\right) \tag{3.8}$$

In a given time $T = 1/2B$, the maximum number of bits per second that the channel can accommodate is the channel capacity C

$$C = \frac{b}{T} = 2B \log_2\left(\sqrt{1 + SNR}\right) = B \log_2(1 + SNR) \text{ bits per second} \qquad (3.9)$$

At this point, let's re-examine the assumptions made thus far:

- The noise is AWGN and the channel itself is memoryless.[1]

- The signal is random.

- The signal contains no redundancy—that is, the signal is not encoded and the samples are not related.

However, in reality, the received signal waveform is encoded and shaped and the samples are not completely independent from one another. Therefore, the Shannon capacity equation serves to offer an upper bound on the data rate that can be achieved. Given the channel environment and the application, it is up to the waveform designer to decide on the data rate, encoding scheme, and waveform shaping to be used to fulfill the user's needs. Therefore, when designing a wireless link, it is most important to know how much bandwidth is available, and the required SNR needed to achieve a certain BER.

Example 3-1: Channel Capacity

The voice band of a wired-telephone channel has a bandwidth of 3 kHz and an SNR varying between 25 dB and 30 dB. What is the maximum capacity of this channel at this SNR range?

The capacity can be found using the relation in (3.9). For example, for the case where the SNR = 25 dB and 30 dB the capacity is

$$C = B \log_2(1 + SNR) = 3000 \log_2(1 + 10^{25/10}) \approx 24.28 \text{ kbps}$$
$$C = B \log_2(1 + SNR) = 3000 \log_2(1 + 10^{30/10}) \approx 29.9 \text{ kbps} \qquad (3.10)$$

---

[1]Memoryless channel is defined in Chapter 5.

Note that at SNR = 25 dB with 3 kHz of channel bandwidth, a data rate of more than 20 Kbps cannot be attained. For the entire SNR range, the capacity is depicted in Figure 3.1.

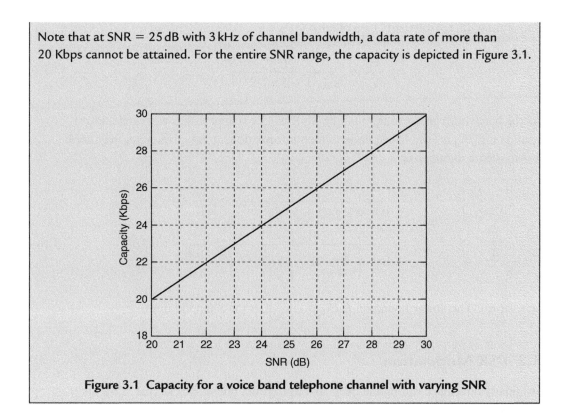

Figure 3.1 Capacity for a voice band telephone channel with varying SNR

The relation in (3.9) can be further manipulated as

$$C = SNR \times B \log_2 (1 + SNR)^{1/SNR}$$

$$SNR = \frac{C}{B} \frac{1}{\log_2 (1 + SNR)^{1/SNR}} \tag{3.11}$$

The SNR presented in (3.11) can be expressed in terms of energy per bit divided by the noise normalized by the bandwidth as [6]:

$$\frac{E_b}{N_0} = \frac{B}{C} SNR \tag{3.12}$$

Substituting (3.12) into (3.11) we obtain

$$\frac{E_b}{N_0} = \frac{1}{\log_2(1 + SNR)^{1/SNR}} \tag{3.13}$$

As the bandwidth grows to infinity, the SNR in (3.13) approaches zero and a lower bound on E_b/N_0 is established under which information, at any given rate, may not be transmitted without error:

$$\left.\frac{E_b}{N_0}\right|_{\text{lower bound}} = \lim_{SNR \to 0} \frac{1}{\log_2(1 + SNR)^{1/SNR}}$$

$$= \frac{1}{\log_2\left\{\lim_{SNR \to 0}(1 + SNR)^{1/SNR}\right\}} = \frac{1}{\log_2 e} \approx 0.69 \tag{3.14}$$

or $-1.6\,\text{dB}$. This lower bound on E_b/N_0 is known as the Shannon limit.

3.2 PSK Modulation

Originally developed during the deep space program, phase shift keying (PSK) is a popular class of modulation schemes used in a variety of military and civilian applications. A PSK waveform can be classified as either coherent or noncoherent (differential). A coherent PSK waveform can be expressed in the time domain as

$$x(t) = \sqrt{\frac{2E_s}{T_s}} \sin\left(\omega_c t + \frac{2\pi(m-1)}{M}\right), \quad m = 1, 2, \ldots, M, \quad -\frac{T_s}{2} \le t \le \frac{T_s}{2} \tag{3.15}$$

where E_s is the average power over the signaling interval, T_s is the symbol period, $M = 2^N$ is the number of allowable phase states, and N is the number of bits needed to quantize M. The symbol duration of M-PSK is related to the bit duration T_b as

$$T_s = \log_2(M)T_b \tag{3.16}$$

The signal constellations for BPSK, QPSK, and 8-PSK are shown in Figure 3.2.

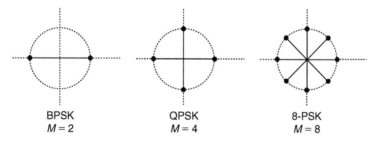

BPSK
$M = 2$

QPSK
$M = 4$

8-PSK
$M = 8$

Figure 3.2 Signal constellations for BPSK, QPSK, and 8-PSK

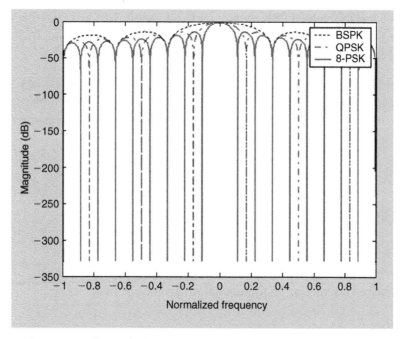

Figure 3.3 Theoretical spectra of BPSK, QPSK, and 8-PSK signal

The power spectral density of an *M*-PSK signal can be derived as

$$S_x(f) = 2E \left[\frac{\sin[\pi T_s(f - f_c)]}{\pi T_s(f - f_c)} \right]^2 \tag{3.17}$$

The normalized spectra of BPSK, QPSK, and 8-PSK are computed at the same given sampling rate, as shown in Figure 3.3. Each modulation scheme exhibits nulls at multiples of the symbol rate. Minimum BER occurs in PSK when the carrier frequency is an integer multiple of the symbol rate, that is $f_c = n/T_s$, and perfect synchronization exists between carrier and symbol timing. In fact, coherent PSK results in the smallest BER compared to other digital modulation schemes for a given E_b/N_o, provided that the received waveform symbol timing and carrier are perfectly synchronized. In BPSK, for example, the condition stated above implies that the initial phase shift of 0 or π depends on whether the transmitted bit is a 1 or a 0. The phase shifts become different from 0 or π if f_c is not an integer multiple of $1/T$. Note, however, that in the event where f_c is much larger than $1/T$ the BER degradation becomes minimal. Finally, it is important to note that coherent PSK and differential PSK require an E_b/N_o that is 3 dB less than that required for coherent FSK and differential FSK, respectively, to produce the same BER. FSK and differential FSK will be discussed later on.

The spectra of M-PSK signals are distinguished with a main lobe bounded by spectral nulls. This portion of the bandwidth is known as the null-to-null bandwidth. For PSK, the null-to-null bandwidth contains most of the signal power as will be stated for each modulation scheme below. The null-to-null bandwidth is given as

$$B_{null\text{-}to\text{-}null} = \frac{2}{T_s} = \frac{2}{\log_2(M)T_b} \tag{3.18}$$

The bandwidth efficiency of M-PSK signals is

$$\varepsilon = \frac{\log_2(M)}{2} \tag{3.19}$$

and is summarized for various PSK modulation schemes in Table 3.1.

3.2.1 BPSK

Like FSK, a BPSK waveform has a constant envelope and is represented by two antipodal signals [7] with a correlation coefficient of -1. Assuming that the received signal is

Table 3.1 Bandwidth efficiency of various M-PSK modulation schemes

M	2	4	8	16	32
ε (bits/s/Hz)	0.5	1	1.5	2	2.5

synchronized in phase and frequency, then the probability of error for coherent BPSK in AWGN is given as

$$P_{e,BPSK} = Q\left(\sqrt{\frac{2E_b}{N_0}}\right) = \frac{1}{2}\,\mathrm{erfc}\left(\sqrt{\frac{E_b}{N_0}}\right) \tag{3.20}$$

where $Q(.)$ is the Marcum Q function defined as[2]

$$Q(x) = \frac{1}{\sqrt{2\pi}} \int_x^\infty e^{-\alpha^2/2}\,d\alpha \tag{3.21}$$

In a slowly varying Rayleigh fading channel, the probability of error for BPSK is

$$P_{e,BPSK} = \frac{1}{2} - \frac{1}{2}\sqrt{\frac{E_b/N_0}{1 + E_b/N_0}} \tag{3.22}$$

In comparison with (3.46), coherent BFSK is 3 dB inferior to coherent BPSK. The bandwidths containing null-to-null, 90% of the power, and 99% of the power for BPSK are shown in Table 3.2.

The bandwidth relations expressed in Table 3.2 imply that 99% of the energy of a BPSK signal requires a filter bandwidth nearly ten times wider than a filter that captures 90% of the energy. Furthermore, the filter that passes 99% of the energy of the BPSK signal typically allows more degradation due to thermal noise, RF anomalies, and interference than the filter that passes 90% of the energy of the signal. The filter with wider bandwidth could provide an output signal with lower SNR, and therefore higher BER, than the filter with smaller bandwidth. The important observation is that the designer must take into

---

[2]Note that the function $erfc(x) = 1 - erf(x) = 2Q(\sqrt{2}x) \Rightarrow Q(x) = 0.5erfc(x/\sqrt{2})$

and

$$erf(x) = \frac{2}{\sqrt{\pi}} \int_0^x e^{-\alpha^2}\,d\alpha \approx 1 - \frac{e^{-x^2}}{\sqrt{\pi}x} \quad \text{for } x \gg 1 \text{ resulting in}$$

$$Q(x) = \frac{1}{\sqrt{2\pi}} \int_x^\infty e^{-\alpha^2/2}\,d\alpha \approx 1 - \frac{e^{-x^2}}{\sqrt{\pi}x} \quad \text{for } x \gg 1$$

**Table 3.2 Bandwidths between null-to-null, 90%
of the energy and 99% of the energy for BSPK**

Parameter	Bandwidth
Null-to-null bandwidth	$B_{null\text{-}to\text{-}null} = \dfrac{2}{T_s}$
90% bandwidth	$B_{90\%} \approx \dfrac{1.7}{T_s}$
99% bandwidth	$B_{99\%} \approx \dfrac{20}{T_s}$

account multiple factors when designing a selectivity filter to improve the desired signal SNR for a given waveform in SDR.

Unlike coherent BPSK, differential BPSK or DBPSK does not require a coherent signal in order to be demodulated. DBPSK uses the previous symbol as a reference to demodulate the current symbol. In other words, demodulation of the current symbol is done based on the phase information inferred from the current and previous symbols [8]. Frequency coherency is still necessary in differential PSK and is maintained typically by employing an AFC algorithm, as will be discussed in a later chapter. The demodulator can be either optimal resulting in the probability of error

$$P_{e,DBPSK} = \frac{1}{2} e^{-E_b/N_0} \quad \text{Optimum demodulator} \tag{3.23}$$

or suboptimum as shown in [9] where the frequency coherency requirement is somewhat relaxed and can be related to the IF selectivity filter bandwidth B_{IF} as

$$P_{e,DBPSK} = \frac{1}{2} e^{-0.76 E_b/N_0} \quad B_{IF} \approx 0.5/T_s$$

$$P_{e,DBPSK} = \frac{1}{2} e^{-0.8 E_b/N_0} \quad B_{IF} \approx 0.57/T_s \quad \text{Suboptimum demodulator} \tag{3.24}$$

$$P_{e,DBPSK} = Q\left(\sqrt{\frac{E_b}{N_0}}\right) \quad B_{IF} > 1/T_s$$

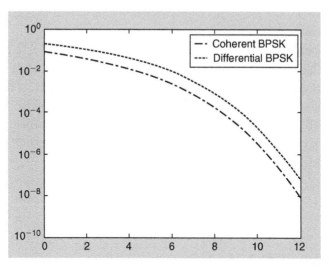

Figure 3.4 Performance comparison between a perfectly synchronized coherent BPSK and DBPSK with optimum demodulator in AWGN

Typically, for a large B_{IF}, the difference in performance between a suboptimal versus optimal demodulator is 2 dB in favor of the latter. The probability of error for DBPSK in a slow Rayleigh distributed channel is

$$P_{e,DBPSK} = \frac{0.5}{\frac{E_b}{N_0} + 1} \qquad (3.25)$$

Finally, note that the power spectral density of DBPSK is the same as that of coherent BPSK. Figure 3.4 shows the expected BER in AWGN between a perfectly synchronized BPSK system and a DBPSK with optimum demodulator in the presence of AWGN. DBPSK performs 0.5 to 1 dB worse than coherent BPSK, whereas suboptimum DBPSK performs anywhere between 1.5 and 2 dB worse than coherent BPSK.

In practical systems, the BPSK signal is filtered or shaped at baseband in order to narrow the spectrum. Shaping filters, however, if not properly designed could cause degradation in the form of ISI and/or spectral regrowth.

3.2.2 QPSK

QPSK is a PSK modulation scheme where the data bits are divided into groups of two, known as dibits. The mapping of dibits into phase can be arbitrary. Coherent QPSK is another popular variant of coherent PSK. Unlike higher order PSK schemes where increasing the bandwidth efficiency results in degraded BER performance, QPSK does not suffer from BER degradation despite the increased bandwidth efficiency (twice the bandwidth efficiency of BPSK). In other words, the BER of coherent QPSK is the same as that of coherent BPSK in AWGN:

$$P_{e,QPSK} = Q\left(\sqrt{\frac{2E_b}{N_0}}\right) \tag{3.26}$$

In a slowly varying Rayleigh fading channel, the probability of error for a Gray-coded QPSK is identical to that of BPSK:

$$P_{e,QPSK} = \frac{1}{2} - \frac{1}{2}\sqrt{\frac{E_b/N_0}{1 + E_b/N_0}} \tag{3.27}$$

Similar to BPSK, QPSK is a constant envelope waveform with discontinuous phases at the symbol boundaries. The bandwidths containing the null-to-null, 90% of the power, and 99% of the power for QPSK are shown in Table 3.3.

Table 3.3 Bandwidths between null-to-null, 90% of the energy, and 99% of the energy for QPSK

Parameter	Bandwidth
Null-to-null bandwidth	$B_{null-to-null} = \dfrac{1}{T_s}$
90% bandwidth	$B_{90\%} \approx 0.75 T_s$
99% bandwidth	$B_{99\%} \approx \dfrac{8}{T_s}$

A comparison between Table 3.2 and Table 3.3 reveals that capturing 90% of the energy of a BPSK signal requires the IF filter bandwidth to be almost twice as wide as the IF filter bandwidth needed to filter a QPSK signal. Furthermore, note that capturing 99% of the energy of a BPSK or QPSK signal requires the respective filter bandwidth to be approximately ten times wider than the filter that captures 90% of the energy. A wider band filter, as discussed earlier, allows for more thermal noise to degrade the desired signal as well as more interference and RF anomalies such as spurs to be present in the band. This tradeoff in increased signal energy versus degradation may lean in favor of filters that only accommodate 90% of the signal energy.

Similarly, on the transmit side, filtering or wave-shaping the QPSK signal transforms it from a constant envelope signal into a time-varying envelope signal. Filtering the transmitted signal is necessary in order to contain the signal spectrum as well as any spectral regrowth due to the transmitter nonlinearity from spilling into adjacent bands. For instance, a 180° phase shift, which in a nonfiltered ideal QPSK signal with no I/Q imbalance[3] occurs momentarily without affecting the constant envelope property of the signal, causes a filtered QPSK signal envelope to vary considerably by crossing the zero constellation point. When using a C-class fully saturated amplifier or hard-limited amplifier[4], the zero-crossing phenomenon leads to a considerable spectral regrowth and a poor spectral efficiency often resulting in degraded BER performance. Therefore, linear amplifiers are typically used with QPSK waveforms in order to prevent spectral regrowth and signal degradation. If power consumption is a critical parameter, modified QPSK modulation schemes, such as offset-QPSK (OQPSK) and π/4-DQPSK, could be used to limit spectral regrowth. OQPSK limits phase transitions to a maximum $\pm 90°$, whereas π/4-DQPSK limits phase transitions to a maximum of $\pm 135°$.

Unlike QPSK where the in-phase and quadrature bit-transitions are aligned in time, in OQPSK, the in-phase and quadrature bit-streams are offset from each other by half a symbol period (or one bit) as depicted in Figure 3.5. The OQPSK modulator differs from

---

[3]I/Q imbalance, as we shall see later on, causes fluctuation in the signal magnitude known as residual AM, which further degrades the constant envelope property of a PSK signal. In SDR, this phenomenon is minimized since the quadrature mixing is done in the digital domain, and any imbalance is caused by the numerical accuracy of the sinusoidal generation mechanism.

[4]Class C and D amplifiers are highly desirable in portable devices for their high efficiency and low power consumption.

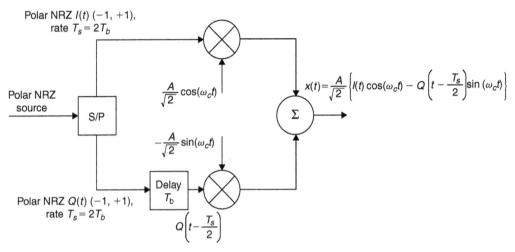

Figure 3.5 Conceptual OQPSK modulator

the QPSK modulator by the extra delay of half a symbol in the quadrature data path. The modulated signal output can be expressed as

$$x(t) = \frac{A}{\sqrt{2}}\left\{I(t)\cos(\omega_c t) - Q\left(t - \frac{T_s}{2}\right)\sin(\omega_c t)\right\}$$ (3.28)

In other words, either the in-phase signal or the quadrature signal can transition at any given bit boundary but not both simultaneously. Consequently, in OQPSK a maximum phase transition of $\pm 90°$ can occur once every half a symbol period. This in essence eliminates the $180°$ maximum phase shift that can occur in QPSK, thus preventing the signal from zero-crossing. Shaped OQPSK causes less envelope variation than shaped QPSK. Consequently nonlinear amplifiers can be used with OQPSK and become less prone to producing spectral regrowth. As a result, the regeneration of high-frequency sidelobes is also greatly reduced. This modulation scheme enables the RF designer the use of more efficient RF power amplifiers as compared to linear amplifiers needed for QPSK. Finally, the spectrum of OQPSK is identical to that of QPSK despite the delay in the quadrature path, and its BER performance is also identical to that of QPSK.

Differential encoded QPSK (DEQPSK) or noncoherent QPSK is a special case of QPSK where the need for a coherent reference at the receiver is not needed to properly decode

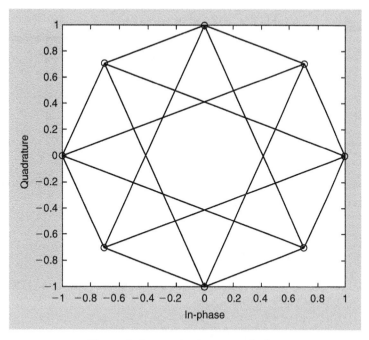

Figure 3.6 π/4-QPSK constellation

the received signal. A transition from a set of dibits to another is represented by a phase difference. π/4-QPSK, a form of DEQPSK, is of particular interest and has been adopted in cellular phone standards in the USA and Japan. π/4-QPSK, depicted in Figure 3.6, is considered a middle ground between QPSK and OQPSK. The maximum allowable phase transition in π/4-QPSK is $\pm 3\pi/4$ radians. This allows it to hold its constant envelope property in the presence of filtering better than QPSK but it is still more prone to envelope variation than OQPSK. π/4-QPSK can be coherently or noncoherently (differentially) detected. Differentially detected π/4-QPSK suffers a 3-dB degradation in performance compared to coherent π/4-QPSK detection in AWGN [10].

π/4-QPSK is generated by splitting the incoming bit stream into in-phase and quadrature symbols $d_{I,k}$ and $d_{Q,k.}$ at half the bit rate. The in-phase I_k and quadrature Q_k pulses are generated over the time period $kT_s \leq t \leq (k + 1)T_s$ based on previous values I_{k-1} and Q_{k-1}. The time variable is dropped for notational convenience since no pulse shaping or

Table 3.4 π/4-QPSK bit to phase mapping

$d_{I,k}$	$d_{Q,k}$	ϕ_k
1	1	$\pi/4$
0	1	$3\pi/4$
0	0	$-3\pi/4$
1	0	$-\pi/4$

filtering is yet assumed. Consider the mapping in Table 3.4, the in-phase and quadrature binary symbols can be generated via the recursive relation:

$$
\begin{aligned}
I_k &= I_{k-1}\cos\phi_k - Q_{k-1}\sin\phi_k = \cos\delta_k \\
Q_k &= I_{k-1}\sin\phi_k + Q_{k-1}\cos\phi_k = \sin\delta_k
\end{aligned}
\tag{3.29}
$$

where the initial condition $\delta_0 = 0$ applies. After pulse shaping, the π/4 QPSK signal at RF becomes

$$
x(t) = I(t)\cos(\omega_c t) - Q(t)\sin(\omega_c t)
\tag{3.30}
$$

where

$$
\begin{aligned}
I(t) &= \sum_{k=0}^{K-1} I_k h\left(t - kT_s - \frac{T_s}{2}\right) \\
Q(t) &= \sum_{k=0}^{K-1} Q_k h\left(t - kT_s - \frac{T_s}{2}\right)
\end{aligned}
\tag{3.31}
$$

and $h(.)$ is the pulse-shaping function. The topic of pulse shaping will be addressed in the next section.

There are four different methods by which π/4 QPSK can be demodulated:

- Coherent demodulation
- IF differential detector
- FM discriminator, and
- Baseband differential detection

The coherent demodulation method typically outperforms the other three methods, which incidentally have similar performance, by 2 to 3 dB [11]. The probability of error performance of the coherent π/4-QPSK demodulator is identical to that of the coherent QPSK demodulator in AWGN. The probability of error for noncoherent π/4-QPSK is similar to that of DQPSK:

$$P_{e,\frac{\pi}{4}DQPSK} = \left\{ \sum_{k=0}^{\infty} \left(\frac{\beta}{\alpha} \right)^k \Psi_k(\alpha\beta) - \frac{1}{2} \Psi_0(\alpha\beta) \right\} e^{\frac{\alpha^2 + \beta^2}{2}} \qquad (3.32)$$

where

$$\alpha = \sqrt{2 \left(1 - \sqrt{\frac{1}{2}} \right) \frac{2E_b}{N_0}}$$

$$\beta = \sqrt{2 \left(1 + \sqrt{\frac{1}{2}} \right) \frac{2E_b}{N_0}} \qquad (3.33)$$

and $\psi_k(.)$ is the kth order modified Bessel function:

$$\Psi_k(x) \equiv \sum_{n=0}^{\infty} \frac{(x/2)^{k+2n}}{n! \Gamma(k + n + 1)} \qquad (3.34)$$

The gamma function is defined as

$$\Gamma(m) = \int_0^{\infty} p^{m-1} e^{-p} \, dp \quad \text{for } m > 0 \qquad (3.35)$$

In the event where the channel is Rayleigh distributed rather than Gaussian distributed, the probability of error is stated as [12]:

$$P_{e,\frac{\pi}{4}DQPSK} = \frac{1}{2} \left\{ 1 - \frac{E_b/N_0}{\sqrt{(E_b/N_0)^2 + 2(E_b/N_0) + 1/2}} \right\} \qquad (3.36)$$

Table 3.5 Gray-code mapping
of binary triplet bits into octal
symbols

$d_{3k}\, d_{3k+1}\, d_{3k+2}$	c_k
000	3
001	4
010	2
011	1
100	6
101	5
110	7
111	0

3.2.3 M-PSK

For higher order modulation, the probability of error of coherent *M-PSK* demodulation is

$$P_{e,M-PSK} = 2Q\left(\sqrt{\frac{2E_b \log_2 M}{N_0}}\, \sin\frac{\pi}{M}\right) \text{ for } M > 4 \tag{3.37}$$

Similarly, the probability of error for noncoherent or *M-DPSK* demodulation is

$$P_{e,M-DPSK} = 2Q\left(\sqrt{\frac{2E_b \log_2 M}{N_0}}\, \sin\frac{\pi}{\sqrt{2}M}\right) \text{ for } M > 4 \tag{3.38}$$

An interesting practical example of higher order PSK modulation is EDGE. EDGE uses $3\pi/8$ shifted 8-PSK modulation [13]. Incoming binary bits are grouped in three and Gray-encoded into octal-valued symbols as shown in Table 3.5.

The output of the 8-PSK modulator, depicted in Figure 3.7, is further shifted by the symbol index n multiplied by $3\pi/8$:

$$x_n = s_n e^{j\frac{3\pi}{8}n} = e^{j\pi\left(\frac{3}{8}n + \frac{2}{8}c_n\right)} \tag{3.39}$$

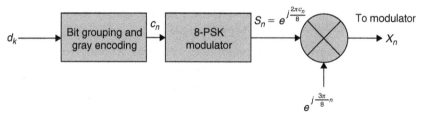

Figure 3.7 Conceptual 8-PSK modulator

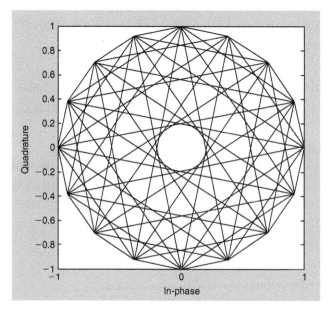

Figure 3.8 Constellation of shifted 8-PSK EDGE modulation

The additional phase shift serves to limit the envelope phase variations after waveform shaping using 0.3 GMSK and analog filtering. A typical nonshaped EDGE constellation is shown in Figure 3.8.

3.3 FSK Modulation

FSK is a digital FM modulation scheme that is most common in the HF band [14]. Modems using FSK modulation are considered the first modems using digital signaling. In FSK, the data bits dictate the frequency at which transmission will occur.

Traditionally, noncoherent asynchronous FSK was used in analog modems where timing information was not required. This advantage carries on to digital modem implementations with considerable oversampling over the data rate. Noncoherent FSK is particularly attractive for frequency hopping spread spectrum systems where the determination of the current information bit does not have to rely on previous information bits that may have occurred at different frequencies.

3.3.1 Binary FSK

In general, a binary FSK modulated signal uses two tones with distinct frequencies f_0 and f_1 to express a binary 0 or 1:

$$
\begin{aligned}
x_0(t) &= A\cos(2\pi f_0 t + \theta_0) \quad kT_s \leq t < (k+1)T_s \text{ for binary 0} \\
x_1(t) &= A\cos(2\pi f_1 t + \theta_1) \quad kT_s \leq t < (k+1)T_s \text{ for binary 1}
\end{aligned}
\tag{3.40}
$$

where T_s is the symbol or bit period, and θ_0 and θ_1 are arbitrary phases. The signal power is $P_{fsk} = A^2/2$. In the event where θ_0 and θ_1 are not equal phases, then the FSK signal is said to be discontinuous or noncoherent. This type of FSK modulation is not common. A more common type is coherent FSK modulation in which the phases θ_0 and θ_1 are equal ($\theta_0 = \theta_1 = \theta$). Furthermore, in order to enable coherent FSK demodulation, the frequencies f_0 and f_1 are chosen such that

$$
r(t) = \int_{kT_s}^{(k+1)T_s} x_0(t)\,x_1(t)\,dt \cong A^2 \int_{kT_s}^{(k+1)T_s} \cos(2\pi f_0 t + \theta)\cos(2\pi f_1 t + \theta)\,dt = 0
\tag{3.41}
$$

This type of FSK signaling is known as orthogonal FSK. In this case, the two tones f_0 and f_1 are said to be orthogonal. From the appendix in Section 3.10, f_0 and f_1 can be expressed as

$$
\begin{aligned}
f_0 &= \frac{2m+n}{4T_s} \\
f_1 &= \frac{2m-n}{4T_s}
\end{aligned}
\tag{3.42}
$$

The difference between f_0 and f_1 is

$$f_0 - f_1 = \frac{n}{2T_s} \tag{3.43}$$

The relations in (3.42) and (3.43) imply that in order to ensure orthogonality between $x_0(t)$ and $x_1(t)$, the frequencies f_0 and f_1 must be chosen such that they are integer multiples of $1/4T_s$ and their difference must be an integer multiple of $1/2T_s$. To ensure phase continuity, the frequency separation between f_0 and f_1 must be an integer multiple of $1/T_s$. The special case where the frequency separation is exactly $1/T_s$ is known as *Sunde's* FSK and can be expressed as

$$x(t) = A \cos\left(2\pi f_c t + \frac{d_k}{2T_s} t\right) \tag{3.44}$$

where $d_k = \pm 1$ represents the binary data stream. The baseband equivalent PSD of (3.44) is

$$S_x(f) = \frac{4A^2 T_s}{\pi^2} \left|\frac{\cos(\pi T_s f)}{1 - 4T_s^2 f^2}\right|^2 + \frac{A^2}{4}\left\{\delta\left(f - \frac{1}{2T_s}\right) + \delta\left(f + \frac{1}{2T_s}\right)\right\} \tag{3.45}$$

The null-to-null bandwidth, the bandwidth containing 90% of the waveform energy, and the bandwidth containing 99% of the bandwidth energy are shown in Table 3.6.

A typical coherent FSK signal waveform with discontinuous and continuous phase and their relative spectra are depicted in Figure 3.9, Figure 3.10, and Figure 3.11, respectively. An important difference between discontinuous phase and continuous phase FSK is the widening of the spectrum of the former over the latter. The spectral containment property of continuous phase FSK makes it a more desirable signaling scheme. The error probabilities for orthogonal coherently demodulated FSK (Sunde's FSK) and orthogonal non-coherently demodulated FSK in AWGN are

$$P_e = Q\left(\sqrt{\frac{E_b}{N_0}}\right) \quad \text{for coherently demodulated FSK}$$

$$P_e = \frac{1}{2}e^{-\frac{1}{2}\frac{E_b}{N_0}} \quad \text{for non-coherently demodulated FSK} \tag{3.46}$$

Table 3.6 Bandwidths between null-to-null, 90% of
the energy, and 99% of the energy for FSK

Parameter	Bandwidth
Null-to-null bandwidth	$B_{null-to-null} = \dfrac{3}{T_s}$
90% bandwidth	$B_{90\%} \approx \dfrac{1.23}{T_s}$
99% bandwidth	$B_{99\%} \approx \dfrac{2.12}{T_s}$

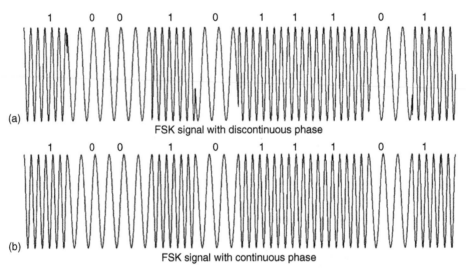

(a)
FSK signal with discontinuous phase

(b)
FSK signal with continuous phase

Figure 3.9 FSK signal waveform (a) with discontinuous phase,
(b) with continuous phase

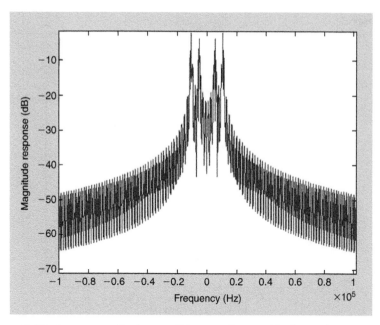

Figure 3.10 Spectrum of coherent FSK waveform with discontinuous phase

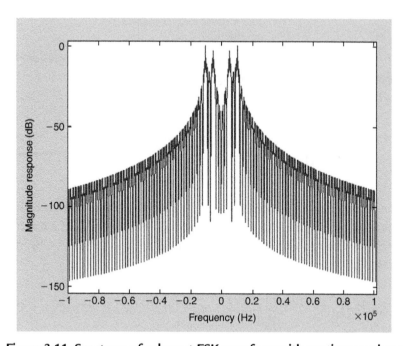

Figure 3.11 Spectrum of coherent FSK waveform with continuous phase

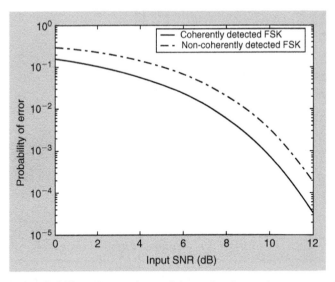

Figure 3.12 Probability of error for orthogonal coherently versus orthogonal noncoherently demodulated FSK

The probability of errors for orthogonal and coherent BFSK and noncoherent FSK is depicted in Figure 3.12. In Rayleigh fading, the probability of error for noncoherently demodulated FSK is

$$P_e = \frac{1}{2\left(1 + \dfrac{1}{2}\dfrac{E_b}{N_0}\right)} \quad \text{Rayleigh fading condition} \tag{3.47}$$

To ensure phase continuity and prevent spectral widening, a third form of FSK, known as continuous phase FSK (CPFSK) was developed. CPFSK belongs to a wider class of modulation schemes known as continuous phase modulation (CPM). A CPFSK waveform can be expressed as

$$x(t) = A\cos\left(2\pi f_c t + \frac{\pi\mu d_k(t - kT_b)}{T_b} + \pi\mu\sum_{n=0}^{k-1} d_n\right) \quad \text{for } kT_b \le t \le (k+1)T_b \tag{3.48}$$

where μ is the modulation index, and $d_k = \pm 1$ represents the binary data stream. The frequency deviation is the absolute value of the derivative of the second term in the parenthesis of (3.69):

$$f_{dev} = \frac{1}{2\pi} \left| \frac{d}{dt} \left\{ \frac{\pi \mu d_k (t - kT_b)}{T_b} \right\} \right| = \left| \frac{\mu d_k}{2T_b} \right| = \frac{\mu}{2T_b} \tag{3.49}$$

which implies that the modulation index

$$\mu = 2T_b f_{dev} \tag{3.50}$$

is the ratio of the frequency deviation over the bit rate. The two tone frequencies are then given as

$$
\begin{aligned}
f_0 &= f_c - \frac{\mu}{2T_b} \\
f_1 &= f_c + \frac{\mu}{2T_b}
\end{aligned}
\tag{3.51}
$$

Note that Sunde's FSK refers to the case where $\mu = 1$.

3.3.2 M-ary FSK

An *M-ary* FSK signal has the general form

$$x_m(t) = A \cos(2\pi f_m t + \theta_m) \quad kT_s \le t \le (k+1)T_s \quad m = 1, 2, \dots, M \tag{3.52}$$

where T_s is the symbol period defined in terms of the bit period as $T_s = NT_b$. The data stream is divided into N-tuples where $N = \log_2(M)$. Similar to binary FSK, in the event where all the phases are equal, the *M-ary* FSK modulation scheme is said to be coherent and the demodulation scheme can be either coherent or non-coherent. To ensure orthogonality between any pair of frequencies of the various signal messages, the frequency separation must be $k/2T_s$ for the coherent case and k/T_s for the non-coherent case. Assuming the frequency separation is uniform between the various messages and M is even, then the relation in (3.52) can be rewritten as

$$x_m(t) = A\cos\left(2\pi f_c t + \mu \frac{\pi}{T_s}(2m - (M+1))(t - KT_s) + \theta_m\right)$$
$$kT_s \leq t \leq (k+1)T_s \quad m = 1, 2, \ldots, M \tag{3.53}$$

where again μ is the modulation index. The power spectral density of (3.53), given the messages are equiprobable, is

$$S_x(f) = \frac{A^2 T_s}{M} \sum_{m=1}^{M} \left\{ \frac{1}{2} \frac{\sin^2(\delta_m)}{\delta_m^2} + \frac{1}{M} \sum_{l=1}^{M} \beta_{l,m}(f) \frac{\sin(\delta_m)\sin(\delta_l)}{\delta_m \delta_l} \right\} \tag{3.54}$$

where

$$\delta_m = \pi \left\{ fT_s - \frac{\mu}{2}(2m - (M+1)) \right\} \tag{3.55}$$

and

$$\beta_{l,m}(f) = \frac{\cos(\delta_m + \delta_l) - \rho\cos(2\pi fT_s - \delta_m - \delta_l)}{1 + \rho^2 - 2\rho\cos(2\pi fT_s)}$$
$$\rho = \frac{2}{M} \sum_{n=1}^{M/2} \cos(\mu\pi(2l-1)) \tag{3.56}$$

Note that for a small modulation index, the MFSK spectra tend to be narrow and gracefully diminish in magnitude. As μ increases in value, the spectrum widens and becomes centered around $fT_s = 0.5$ and its odd multiples. Finally, for M-ary ($M > 2$) noncoherently demodulated (detected) FSK, the probability of error is

$$P_e = \frac{1}{M} \sum_{m=2}^{M} (-1)^m \frac{M!}{m!(M-m)!} e^{\left(\frac{(m-1)E_s}{mN_0}\right)} \tag{3.57}$$

where

$$E_s = E_b \times \log_2(M) \tag{3.58}$$

3.3.3 Sinusoidal FSK (S-FSK)

S-FSK is another constant envelope modulation scheme worth mentioning with a shaping function defined as

$$g(t) = \begin{cases} \cos\left\{\dfrac{\pi t}{2T_s} - \dfrac{1}{4}\sin\left(\dfrac{2\pi t}{T_s}\right)\right\} & -T_s \leq t \leq T_s \\ 0 & \text{otherwise} \end{cases} \tag{3.59}$$

The modulated signal waveform can be expressed as

$$x(t) = A\cos\left(2\pi f_c t - I_k Q_k\left(\dfrac{\pi}{2T_s}t - \beta\sin\left(\dfrac{2\pi t}{T_s}\right)\right) + \dfrac{\pi}{2}(1 - I_k)\right) \tag{3.60}$$

where $I_k = \pm 1$ and $Q_k = \pm 1$ are the in-phase and quadrature symbols and β is the shaping factor. A comparison between the MSK pulse-shaping function, discussed in the next section, and the pulse-shaping function presented in (3.59) is shown in Figure 3.13. The probability of error for S-FSK is the same as that of MSK.

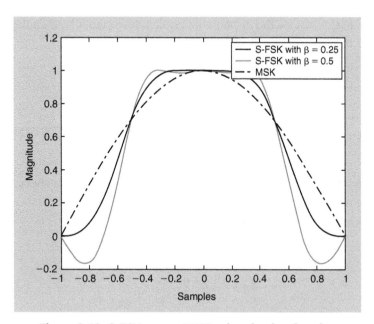

Figure 3.13 S-FSK versus MSK pulse-shaping function

3.3.4 Minimum Shift Keying (MSK)

MSK is a continuous phase modulation scheme that can also be interpreted as a form of FSK [15]. The signal is generated in a manner similar to OQPSK with the added pulse shaping given as a half-cycle sinusoid in the in-phase and quadrature data paths. There are two types of MSK: type 1 uses a weighting function that alternates between positive and negative half sinusoids, or simply a full sinusoidal waveform; type 2 on the other hand employs a weighting function with positive only half-cycle sinusoids. The power spectral density and probability of error are identical for both types. The signaling waveform for a type 1 MSK modulation scheme is[5]

$$
x(t) = A \left\{ I(t) \cos\left(\frac{\pi t}{2T_s}\right) \cos(2\pi f_c t) + Q(t - T_s) \sin\left(\frac{\pi t}{2T_s}\right) \sin(2\pi f_c t) \right\}
\qquad (3.61)
$$

Note that, similar to OQPSK, the in-phase and quadrature symbol durations occupy two symbol durations T_s each. Figure 3.14 depicts the baseband in-phase and quadrature of an MSK waveform. The normalized spectrum of this MSK waveform is depicted in Figure 3.15.

Let $I_k = \pm 1$ and $Q_k = \pm 1$ be the in-phase and quadrature symbols; then the MSK signal waveform in (3.61) can be expressed as

$$
x(t) = \pm A \cos\left(\frac{\pi t}{2T_s}\right) \cos(2\pi f_c t) + \pm A \sin\left(\frac{\pi t}{2T_s}\right) \sin(2\pi f_c t)
$$

$$
x(t) = A \cos\left[2\pi \left(f_c - I_k Q_k \frac{1}{4T_s} \right) t + \frac{\pi}{2}(1 - I_k) \right] \quad kT_s \le t \le (k+1)T_s
$$

$$(3.62)$$

From the relation in (3.62), it is apparent that an MSK waveform is a special case of an FSK waveform with two frequencies:

$$
f_{low} = f_c - \frac{1}{4T_s}
$$

$$
f_{high} = f_c + \frac{1}{4T_s}
$$

$$(3.63)$$

---

[5] MSK can also be viewed as a special case of continuous phase FSK with modulation index $\mu = 0.5$.

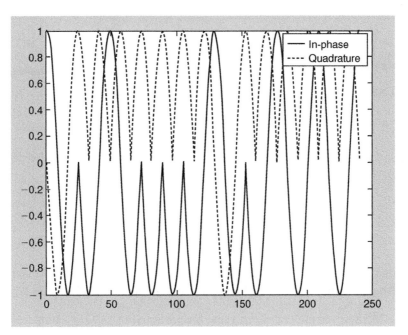

Figure 3.14 MSK simulated in-phase and quadrature baseband waveforms

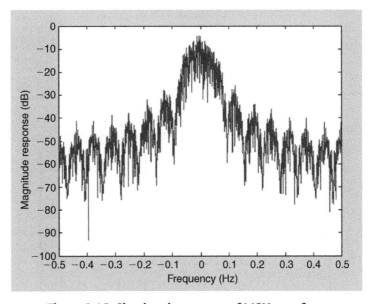

Figure 3.15 Simulated spectrum of MSK waveform

The minimum in MSK is implied from the frequency separation between f_{low} and f_{high} which is $1/2T_s$. This frequency separation is the minimum required frequency separation for the two FSK waveforms to be orthogonal. Unlike FSK, which could exhibit either continuous or discontinuous phase at the bit boundaries, MSK is always continuous at the bit boundaries.

The power spectral density of MSK, depicted along with power spectral densities of BPSK and QPSK in Figure 3.16, can be expressed as

$$S_x(f) = \frac{16A^2T_s}{\pi^2} \left| \frac{\cos(2\pi T_s f)}{1 - 16T_s^2 f^2} \right|^2 \tag{3.64}$$

The null-to-null bandwidth, the bandwidth containing 90% of the waveform energy, and the bandwidth containing 99% of the bandwidth energy are shown in Table 3.7.

Note that the MSK spectrum decays at a rate proportional to $1/(fT_s)^4$ compared to BPSK, QPSK, and OQPSK which fall at a rate of $1/(fT_s)^2$. The null-to-null bandwidth of the MSK is at $\frac{3}{2}fT_s$ compared to that of BPSK which is at $2fT_s$ and QPSK which is at fT_s. The MSK spectrum's main lobe is wider than that of QPSK and narrower than that of BPSK. The probability of error for MSK in AWGN is

$$P_{e,MSK} = Q\left(\sqrt{\frac{2E_b}{N_0}} \right) \tag{3.65}$$

Table 3.7 Bandwidths between null-to-null, 90% of the energy, and 99% of the energy for MSK

Parameter	Bandwidth
Null-to-null bandwidth	$B_{null-to-null} = \dfrac{1}{T_s}$
90% bandwidth	$B_{90\%} \approx \dfrac{0.76}{T_s}$
99% bandwidth	$B_{99\%} \approx \dfrac{1.2}{T_s}$

which is similar to that of BPSK, QPSK, and OQPSK. Note, however, that if MSK is coherently detected as an FSK signal with decisions made over symbol duration T_s, then its performance is 3 dB worse than that of QPSK or BPSK.

Furthermore, a comparison between Table 3.2, Table 3.3, and Table 3.7 reveals that selectivity at the analog IF filter (since in most common architectures selectivity filtering occurs at IF in SDR) can be designed to accommodate 99% of the bandwidth of MSK ($B_{99\%} \approx \frac{1.2}{T_s}$), so MSK will provide better performance than BPSK, QPSK, or OQPSK which require much higher bandwidth. This is true from the energy per bit in AWGN perspective only. If, on the other hand, the filter can be designed only to accommodate a bandwidth around $\frac{3}{4T_s}$ plus the maximum anticipated frequency offset due to oscillator mismatch or Doppler, then QPSK and OQPSK will outperform MSK in AWGN. Note that the IF filter selectivity is dictated to a great extent by the adjacent channel interference, co-channel interference, and other interference from other wireless devices operating on nearby frequencies.

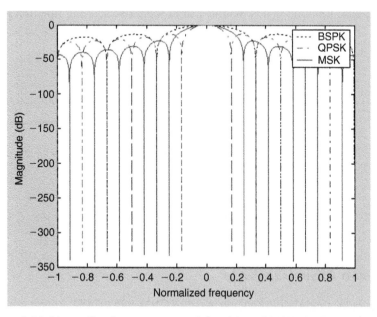

Figure 3.16 Normalized power spectral densities of MSK, QPSK, and BPSK

3.4 Continuous Phase Modulation (CPM)

CPM methods have been used in a variety of military and civilian applications ranging from satellite systems to cell phones. CPM is a bandwidth and power efficient class of modulation schemes particularly attractive for use with nonlinear power amplifiers. A CPM signal is defined in the general sense as

$$x(t) = A \cos \left(2\pi f_c t + 2\pi \underbrace{\sum_{n=-\infty}^{\infty 1} \mu_n d_n \rho(t - nT_s)}_{\text{Excess phase}} \right) \quad \begin{array}{l} -\infty \le t \le \infty \\ \text{for } d_n \text{ is an } M\text{-}ary \\ \text{valued data bit} \end{array} \quad (3.66)$$

The excess phase expression under the bracket in (3.66) implies that a CPM signal may be influenced by more than one modulation symbol. The modulation index μ_n can vary cyclically from one symbol to the next, as in multi-h phase (or multi-μ as notation adopted in this book) phase modulation, or stay a constant number over the entire time axis. When μ is chosen as a rational number, it limits the number of phase states to a finite number thus allowing for maximum likelihood detection schemes such as the Viterbi algorithm to be used. The phase function $\rho(.)$ is the integral of the pulse shaping function $g(.)$

$$\rho(t) = \int_{-\infty}^{t} g(\varsigma)d\varsigma \quad (3.67)$$

The pulse-shaping function is finite in time and in frequency. As mentioned in the previous section, for a rectangular pulse-shaping function:

$$g(t) = \begin{cases} \dfrac{1}{2T_s} & 0 \le t \le T_s \\ 0 & \text{otherwise} \end{cases} \quad (3.68)$$

and $M > 2$, the modulation scheme is CPFSK and for $M = 2$ and for $\mu = 0.5$ the modulation scheme is MSK.

3.5 Gaussian MSK (GMSK)

GMSK is a popular modulation scheme, used in GSM and CDPD among other applications, which can be demodulated using coherent or noncoherent methods.

The main advantages of GMSK are its spectral efficiency, its constant phase property, which allows it to be used with nonlinear power-efficient amplifiers, as well as its robust performance. The pulse-shaping Gaussian filter of GMSK is defined as

$$g(t) = B\sqrt{\frac{2\pi}{\ln 2}} e^{-\frac{2\pi^2 B^2}{\ln 2} t^2} \tag{3.69}$$

B is the 3-dB bandwidth of the Gaussian shaped filter, BT_b is the normalized bandwidth. The ideal or infinite order Gaussian lowpass filter has a constant group delay and exponentially decaying amplitude response:

$$G(f) = e^{-\frac{\ln 2}{2}\left(\frac{fT_b}{BT_b}\right)^2} = e^{-\frac{\ln 2}{2}\left(\frac{f}{B}\right)^2} \tag{3.70}$$

The input to the Gaussian filter is a balanced NRZ data signal defined as

$$d(t) = \sum_{n=-\infty}^{\infty} d_n \Pi\left(\frac{t - nT_b}{T_b}\right) \quad d_n \in \{-1, +1\} \tag{3.71}$$

where

$$\Pi\left(\frac{t}{T_b}\right) = \begin{cases} 1 & 0 \le t \le T_b \\ 0 & \text{otherwise} \end{cases} \tag{3.72}$$

is a rectangular pulse. The output of the Gaussian pulse-shaping filter in response to a rectangular pulse is

$$p(t) = \Pi\left(\frac{t}{T_b}\right) * g(t) = \frac{1}{2T_b}\left\{Q\left(2\pi B \frac{t - \frac{T_b}{2}}{\sqrt{\ln 2}}\right) - Q\left(2\pi B \frac{t + \frac{T_b}{2}}{\sqrt{\ln 2}}\right)\right\} \tag{3.73}$$

where $0 \le BT_b \le 1$ and $Q(.)$ is the Marcum Q function previously defined as

$$Q(x) = \frac{1}{\sqrt{2\pi}} \int_t^\infty e^{-\alpha^2/2} d\alpha \tag{3.74}$$

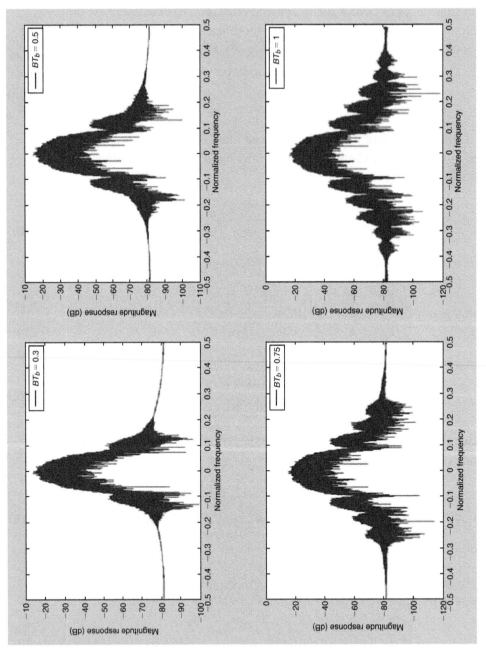

Figure 3.17 Power spectral density of GMSK with BT_b = 0.3, 0.5, 0.75, and 1

Table 3.8 Percentage of power in occupied RF bandwidth for GMSK and MSK

Modulation	BT_b	90% of Power	99% of Power	99.99% of Power
GMSK	0.2	$0.52/T_b$	$0.79/T_b$	$1.22/T_b$
GMSK	0.25	$0.57/T_b$	$0.86/T_b$	$1.37/T_b$
GMSK	0.5	$0.69/T_b$	$1.04/T_b$	$2.08/T_b$
MSK	N/A	$0.78/T_b$	$1.2/T_b$	$6/T_b$

In practice, the pulse-shaping Gaussian filter is implemented as an FIR filter with known latency.

The power spectral density of GMSK for various BT_b factors is shown in Figure 3.17. The extreme case where $BT_b = \infty$ corresponds to MSK. GSM uses GMSK with $BT_b = 0.3$ whereas Bluetooth uses a GFSK modulation scheme with a Gaussian filter with $BT_b = 0.5$. As the BT_b factor becomes smaller, the spectrum becomes more compact and the sidelobe levels decrease. For $BT_b = 0.5$, the first sidelobe is approximately 30 dB below the main lobe as compared to MSK whose first sidelobe is only 20 dB below the main lobe. However, as the spectrum becomes more compact, the ISI degradation worsens. Therefore, the choice of BT_b constitutes a compromise between error rate performance due to ISI and spectral efficiency. The occupied RF bandwidth for GMSK as a function of the percentage of power is shown in Table 3.8.

The probability of error for GMSK in AWGN can be approximated as

$$P_{e,GMSK} \approx Q\left(\sqrt{\frac{2\alpha E_b}{N_0}}\right) \tag{3.75}$$

where α is a factor that depends on BT_b. For example, for $BT_b = 0.25$, α is equal to 0.68. Figure 3.18 draws a comparison between GMSK and MSK. Other pulse shaping functions, such as raised cosine and spectrally raised cosine filters, will be discussed in a later section.

3.6 On-Off Keying (OOK)

An *M-ary* ASK signal can be expressed as

$$x(t) = \begin{cases} A_i \cos(2\pi f_c t) & 0 \le t \le T \\ 0 & \text{otherwise} \end{cases} \tag{3.76}$$

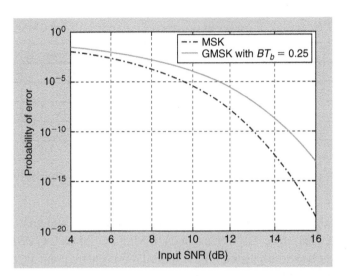

Figure 3.18 Probability of error for GMSK $BT_b = 0.25$ and MSK

where $A_i = A(2i - (M - 1))$, $i = 0, 1,..., M - 1$, and $M \geq 4$. On-off keying (OOK) is a special case of amplitude shift keying (ASK). The OOK signal can be represented as

$$x(t) = \begin{cases} A \cos(2\pi f_c t) & \text{for binary 1} \quad 0 \leq t \leq T \\ 0 & \text{for binary 0} \quad 0 \leq t \leq T \end{cases} \tag{3.77}$$

For uncorrelated binary data, the power spectral density of OOK is

$$S_x(f) = \frac{A^2 T_s}{4} \left| \frac{\sin(\pi f T_s)}{\pi f T_s} \right|^2 + \frac{A^2}{4} \delta(f) \tag{3.78}$$

Note that the OOK spectrum is similar to that of BPSK except for the line spectrum at the carrier. Assuming coherent demodulation at the receiver, the probability of OOK in AWGN is

$$P_{e,OOK} = Q\left(\sqrt{\frac{E_b}{N_0}} \right) \tag{3.79}$$

The advantage of OOK is its simplicity compared to other close modulation schemes such as FSK. In OOK, the transmitter is idle when transmitting a zero, thus allowing

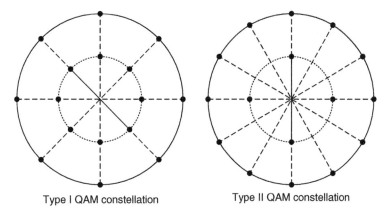

Type I QAM constellation   Type II QAM constellation

Figure 3.19 Type I and Type II QAM constellations

the radio to conserve power. It is more spectrally efficient than FSK which requires roughly 1½ times the bandwidth. OOK's main weakness is its vulnerability to cochannel interference.

3.7 Quadrature Amplitude Modulation (QAM)

QAM is a bandwidth efficient signaling scheme that, unlike CPM, does not possess a constant envelope property.[6] Unlike CPM waveforms, QAM needs to operate in the linear region of a power amplifier to avoid any signal compression and hence degradation. For this reason, among others, when choosing QAM as a signaling scheme, the waveform designer is foregoing power efficiency for bandwidth efficiency. QAM is a widely used modulation scheme in applications ranging from short-range wireless communications to telephone systems.

QAM can be realized in various constellation types: Type 1 and Type II, for instance, have circular constellations, as shown in Figure 3.19, whereas Type III, the most popular of the three, is a square constellation [16]–[18]. There are many other QAM constellations; however, in this chapter we are only concerned with Type III QAM or square QAM as shown in Figure 3.20.

[6]QAM is more bandwidth efficient than *M-ary* PSK for the same average power.

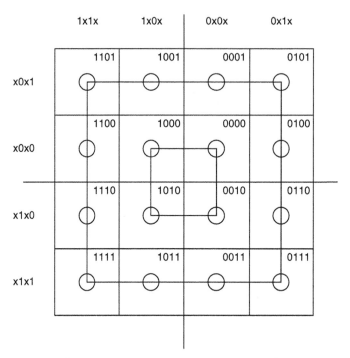

Figure 3.20 16-square QAM constellation

In the most general sense, a QAM signal can be represented by

$$x_n(t) = A_n g(t) \cos(2\pi f_c t + \theta_n) \quad \text{for } n = 1, \dots, M \tag{3.80}$$

where $g(t)$ is a pulse-shaping function. The relation can be further expanded as a linear combination of two orthonormal functions:

$$x_n(t) = A_n \cos(\theta_n) g(t) \cos(2\pi f_c t) - A_n \sin(\theta_n) g(t) \sin(2\pi f_c t), \ 0 \leq t \leq T_s \tag{3.81}$$

The power spectral density of a square QAM signal without any pulse shaping is similar to that of MPSK except for the average power:

$$S_x(f) = 2E_{\min}(M-1)\left|\frac{\sin(\pi f T_s)}{\pi f T_s}\right|^2 = 2E_{\min}(M-1)\left|\frac{\sin(\pi f N T_b)}{\pi f N T_b}\right|^2 \tag{3.82}$$

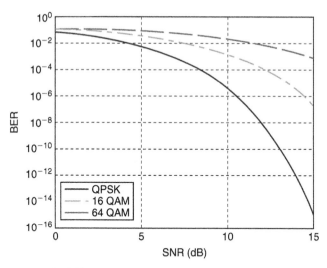

Figure 3.21 Probability of symbol error for QAM for M = 4, 16, and 64

where $N = \log_2 M$, $T_b = T_s/N$, and $E_{\min}$ is the energy of the signal with smallest magnitude in the constellation. The symbol error probability for a square Gray-coded QAM in AWGN is

$$P_{e,symbol,QAM} = 1 - \left\{1 - \frac{2\left(\sqrt{M}-1\right)}{\sqrt{M}} Q\left(\sqrt{\frac{3\log_2 M}{(M-1)}\frac{\hat{E}_s}{N_0}}\right)\right\} \quad M = 2^L \ L \text{ is even} \quad (3.83)$$

where $\hat{E}_s/N_0$ is the average symbol SNR. The bit error probability for QAM, depicted for $M = 4$, 16, and 64 in Figure 3.21, is

$$P_{e,bit,QAM} = \frac{P_{e,symbol,QAM}}{\log_2 M} \quad (3.84)$$

QAM is particularly vulnerable to outside interference or imperfections in the transceiver such as:

- Phase noise

- I/Q imbalance, which in SDR is due to the digital quadrature modulation before digital to analog conversion

- CW interference

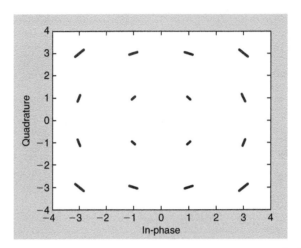

Figure 3.22 Effect of phase noise on the QAM constellation

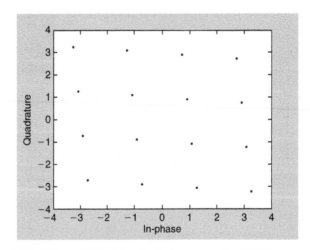

Figure 3.23 Effect of IQ phase imbalance on the QAM constellation

- PA nonlinearity

- DC offset

The above degradations will be studied in detail in the next few chapters. The effects of phase noise, IQ phase imbalance, and nonlinear compression on a 16-QAM signal are shown by way of an example in Figure 3.22, Figure 3.23, and Figure 3.24.

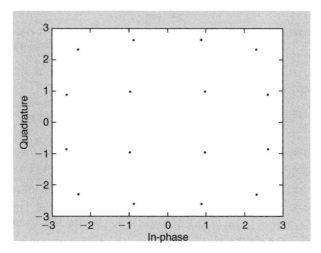

Figure 3.24 Effect of nonlinear compression in the form of AM/AM and AM/PM on the QAM constellation

3.8 Orthogonal Frequency Division Multiplexing (OFDM)

3.8.1 Basic OFDM

OFDM was first proposed by Chang in the 60s [19]. Weinstein and Ebert [23] first proposed the use of the fast Fourier transform (FFT) to modulate the data symbols on orthogonal equally spaced carriers as it is commonly used today. In its early days, OFDM was limited in its use to waveforms in military applications. Today, OFDM has gained wide acceptance and become the waveform of choice for many applications ranging from digital audio and video broadcasting (DAB and DVB) to wireless short range high data rate applications as well as next-generation high data rate wide area standards such as WiMAX and LTE.

The basic premise of OFDM is the use of N-orthogonal subcarriers in the frequency domain to transmit a block of serial data symbols in parallel. This is done by mapping (assigning) a carrier to each data symbol before transforming the data into the time domain via the IFFT. To do so, N-source symbols with period T_s each are buffered into a block of N-parallel modulated symbols with period $T = NT_s$. The block length is chosen such that NT_s is much greater than the RMS delay spread of the channel [20]. Nonetheless, despite the choice of a much larger modulated symbol period T compared to the RMS delay spread, the dispersive nature of a multipath channel will cause adjacent

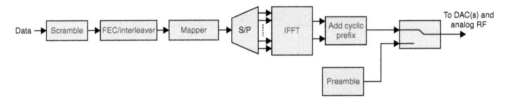

Figure 3.25 Conceptual OFDM digital transmitter

blocks to overlap, thus degrading the performance of the system in the form of ISI. The ISI degradation can be equalized, however, by inserting guard-time intervals between blocks either in the form of a cyclic prefix or zero-padding. A conceptual OFDM transmitter is shown in Figure 3.25. The S/P block in Figure 3.25 is a serial to parallel data block which also serves to place the mapped data into the proper IFFT bins. This placement of symbols into IFFT bins can be viewed as modulating the symbols with various orthogonal tones.

Assume the data mapped onto the N-subcarrier IFFT input is of the form:

$$X\left(-\frac{N}{2}\right), X\left(-\frac{N}{2}+1\right), \cdots, X(k), \cdots, X\left(\frac{N}{2}-1\right) \tag{3.85}$$

The modulated data itself can be PSK, differential PSK, or QAM. The discrete-time domain signal is

$$x(n) = \frac{1}{\sqrt{N}} \sum_{k=-N/2}^{N/2-1} X(k)e^{j2\pi\frac{k}{N}n} \quad n \in \left[-\frac{N}{2}, \cdots, \frac{N}{2}-1\right] \tag{3.86}$$

At the output of the PA, after quadrature modulation, the signal can be expressed as

$$x(t) = \mathrm{Re}\left\{\frac{1}{\sqrt{N}} \sum_{k=-N/2}^{N/2-1} X(k)e^{j2\pi\left(f_c\frac{k+0.5}{T}\right)t}\right\} \quad n \in \left[-\frac{N}{2}, \cdots, \frac{N}{2}-1\right] \tag{3.87}$$

At the receiver, the information modulated on the subcarriers is obtained by performing the FFT on the received block of data:

$$X(k) = \frac{1}{\sqrt{N}} \sum_{k=-N/2}^{N/2-1} x(n)\, e^{-j2\pi\frac{k}{N}n} \quad k \in \left[-\frac{N}{2}, \cdots, \frac{N}{2}-1\right] \tag{3.88}$$

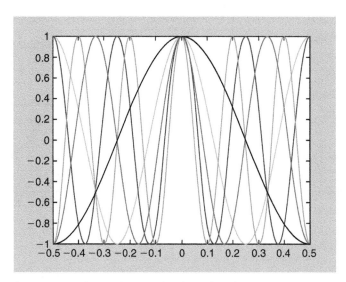

Figure 3.26 Five orthogonal subcarriers in one OFDM symbol

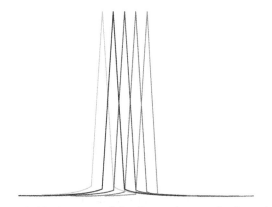

Figure 3.27 Spectra of the five orthogonal subcarriers

The modulated orthogonal subcarriers are separated in frequency by $1/T$. Theoretical time and frequency domain representations of five orthogonal sinusoidal subcarriers are depicted in Figure 3.26 and Figure 3.27, respectively.

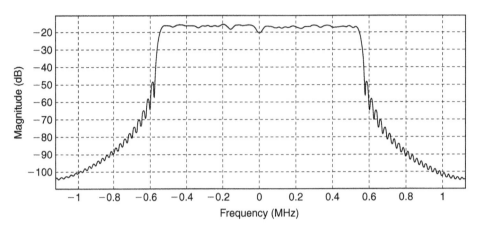

Figure 3.28 Spectrum of an OFDM waveform

In most practical OFDM systems, the modulated signal is DC-free—that is, the DC bin input to the IFFT is null. However, the depth of the null or even its existence depends greatly on the performance of the transceiver. For instance, local oscillator (LO) leakage on the transmitter or DC offset on the receiver may decrease the depth of the null or even replace it with a significant DC signal. The spectrum of a generic OFDM waveform with a shallow null is depicted in Figure 3.28.

At the input to the receiver, in an indoor channel, for example, the received signal is plagued by multipath Rayleigh fading as shown in Figure 3.29. The QPSK signal constellation after gain and phase (zero-forcing) equalization is shown in Figure 3.30.

3.8.2 Cyclic Prefix

The purpose of the cyclic prefix is to combat ISI and intra-symbol interference. The cyclic prefix is a guard time made up of a replica of the time-domain OFDM waveform. The basic premise is to replicate part of the back of the OFDM signal to the front to create the guard period, as shown in Figure 3.31.

The duration of the cyclic prefix is chosen such that it is longer than the maximum delay spread τ_{max} caused by the multipath channel. The starting point for sampling of the OFDM symbol on the receiver side must be somewhere in the interval (τ_{max}, T_{cp})

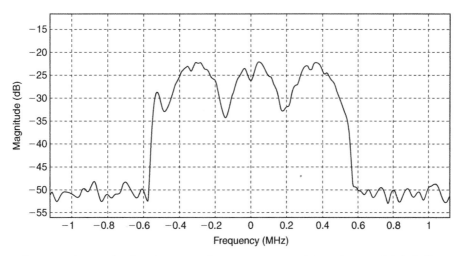

Figure 3.29 Received OFDM waveform spectrum subjected to Rayleigh fading

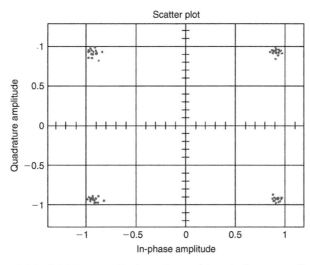

Figure 3.30 QPSK constellation after gain and phase equalization

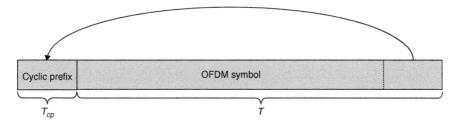

Figure 3.31 Cyclic-prefix and OFDM symbol

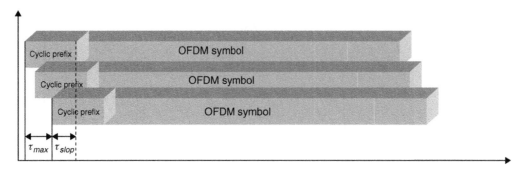

Figure 3.32 Overlapped OFDM symbols due to multipath in the channel

as shown in Figure 3.32. This ensures that the received signal contains all the channel multipath effects, and the channel estimate used to equalize the output of the FFT is *theoretically* sufficient to restore the spectrum to its prechannel condition. Channel frequency estimation and frequency offset compensation are further discussed in Section 3.10.2. In addition to protecting the modulated symbol from ISI, a cyclic prefix gives the receiver added robustness against timing error.

3.8.3 OFDM Performance

AWGN in the time domain is statistically equivalent to AWGN with the same average power in the frequency domain. Therefore, modulation in the frequency domain on OFDM subcarriers is equivalent to modulation in conventional temporal systems. As mentioned earlier, the typical modulation scheme of the subcarriers is PSK, differential PSK, or QAM with either coherent or noncoherent demodulation. Consequently, the probability of error versus SNR is dictated by the frequency domain modulation scheme employed.

OFDM as a modulation scheme is robust in the face of narrowband fading and narrowband interference. This is true only if a small number of subcarriers are affected. OFDM is also ideal for dispersive low delay spread indoor channels. Furthermore, another attractive feature of OFDM is its resilience to sample-timing errors afforded it by the length of the cyclic prefix, which is presumed to be larger than the maximum delay spread.

OFDM, on the other hand, is very sensitive to phase noise, especially when using higher order modulations such as QAM or *M-ary* PSK. It is also sensitive to frequency offset errors. However, a frequency offset error can be algorithmically corrected for in the modem. OFDM is also sensitive to nonlinearities in the form of signal compression and clipping. In most OFDM applications, a linear amplifier is utilized to ensure that the peak to average power ratio of the system is accommodated and that no unnecessary distortions are caused to the waveform.[7]

3.9 Spread Spectrum Modulation

Spread spectrum (SS) was mainly developed in the 50s for military and space applications. Since then, spread spectrum is used heavily in both military and civilian applications. Spread spectrum in all of its flavors offers some unique benefits namely [21]:

- Robust performance in the presence of narrowband interference and jamming whether it is intentional or otherwise

- Excellent performance in the presence of multipath frequency selective fading

- Low probability of intercept

- Allows for multi-user random access communications

- High resolution ranging and accurate universal timing

In spread spectrum, the desired signal's bandwidth is spread or stretched by certain means, such as with the aid of a pseudo-noise (PN) sequence, beyond the minimum bandwidth necessary to send information. This has been traditionally done either via direct sequence (DS) spreading, frequency hopping (FH), time hopping (TH), or any combination of the three techniques. An example of DSSS spreading and despreading based on CDMA2000 is discussed in Section 3.10.3.

3.9.1 PN Codes and Maximum-Length Sequences

A PN code is a sequence of binary numbers with certain autocorrelation properties. These sequences are typically periodic. A maximum-length sequence is a periodic PN sequence with the longest possible period for a given length M of the shift register. The period of such a sequence is $N = 2^M - 1$. The three properties of maximum-length sequences are:

- *The balance property*: The number of logic 1 s is always one more than the number of logic 0 s

---

[7] Predistortion techniques have also been successful in lowering the linearity requirements of the PA for OFDM.

- *The run property*: A run is simply defined as a subsequence of identical symbols. One half the runs in a maximum-length sequence is of length 1, one fourth of the runs is of length 2, one eighth of the runs is of length 3, and so on. The total number of runs for a maximum length sequence generated using a shift register of length M is $(M + 1)/2$.

- *The correlation property*: The autocorrelation sequence of a maximum-length sequence is periodic. The autocorrelation of the maximum-length sequence $\{p_n\}$ is as described in [22]:

$$r(\tau) = -\frac{1}{N} + \frac{N + 1}{N} \rho_{T_c}(\tau)* \sum_{n=-\infty}^{\infty} \delta(\tau + NnT_c) \tag{3.89}$$

where N is the length of the sequence and

$$\rho_{T_c}(\tau) = \begin{cases} 1 - \dfrac{|\tau|}{T_c} & |\tau| \leq T_c \\ 0 & \text{otherwise} \end{cases} \tag{3.90}$$

The power spectral density of (3.90), depicted in Figure 3.33, is

$$R(f) = \frac{1}{N^2}\delta(f) + \frac{N + 1}{N^2} \sum_{\substack{n=-\infty \\ n \neq 0}}^{\infty} \sin c^2\left(\frac{n\pi}{N}\right)\delta\left(f + \frac{n}{NT_c}\right) \tag{3.91}$$

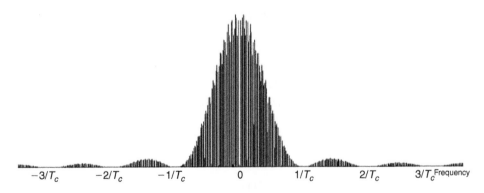

Figure 3.33 Power spectral density of a PN waveform

3.9.2 Direct Sequence Spread Spectrum (DSSS)

The processing gain PG of a spread spectrum signal can be defined as the ratio of the spreading bandwidth B_{ss} to the minimum required signal bandwidth B_d:

$$PG \cong \frac{B_{ss}}{B_d} \tag{3.92}$$

In DSSS, a narrowband jammer is spread by the PN code at the receiver, thus spreading its power across the band. On the other hand, due to the PN code's excellent autocorrelation properties, the desired signal is restored back to its original spectrum after despreading as illustrated in Section 3.10.3. Define the spreading factor as $N = T_b/T_c$ where T_b is the bit duration and T_c is the chip duration. A chip represents a single PN code symbol. The average signal power at the receiver input is E_b/T_b corrupted by the jammer noise J, thus resulting in an input SNR of

$$SNR_i = \frac{E_b/T_b}{J} \tag{3.93}$$

The output SNR after despreading can be expressed as

$$SNR_o = \frac{2E_b}{JT_c} \tag{3.94}$$

The gain factor of 2 in (3.94) is due to coherent detection and is independent of despreading. Define the processing gain then as the ratio

$$PG = \frac{1}{2}\frac{SNR_o}{SNR_i} = \frac{1}{2} \times 2\frac{\left(\dfrac{E_b}{JT_c}\right)}{\left(\dfrac{E_b}{JT_b}\right)} = \frac{T_b}{T_c} = \frac{B_{ss}}{B_d} \tag{3.95}$$

For a BPSK signal spread using direct sequence and employing a coherent detector, the probability of error is

$$P_{DS,BPSK} = Q\left(\sqrt{\frac{2E_b}{JT_c}}\right) \tag{3.96}$$

Note that after despreading, the jammer may be treated as wideband noise, and hence the equivalence

$$\frac{N_0}{2} = \frac{JT_c}{2} \qquad (3.97)$$

Therefore we can express the bit energy to noise density ratio as

$$\frac{E_b}{N_0} = PG \times \frac{P}{J} \qquad (3.98)$$

where $P = E_b/T_b$ is the average signal power. The jamming margin can then be defined as

$$\Gamma = 10 \log_{10}\left(\frac{PG}{E_b/N_0}\right) \qquad (3.99)$$

Example 3-2: Processing Gain and Required Chipping Rate

Assume that in order to obtain a decent BER, the required E_b/N_o is 12 dB. Furthermore, assume that the duration of the information bit T_b is 0.5 ms. Given that the required jamming margin is approximately 21.11 dB, what is the required processing gain? What is the resulting chip-duration? What is the required length of the shift register that produces the corresponding maximum length PN sequence?

From (3.99) the processing gain is the jamming margin plus E_b/N_o and hence $PG|_{dB} = \Gamma + E_b/N_0 = 21.11 + 12$ dB $= 33.11$ dB. In the linear domain, this corresponds to a processing gain of $10^{33.11/10} \approx 2047$. From the relation in (3.95), the chip-duration is the information bit duration divided by the linear processing gain

$$PG = \frac{T_b}{T_c} \Rightarrow T_c = \frac{T_b}{PG} = \frac{0.5 \text{ ms}}{2047} = 0.2443 \text{ } \mu s$$

The length of the shift register is given as $M = \log_2(N + 1) = \log_2(2047 + 1) = 11$.

3.9.3 Frequency Hopping Spread Spectrum (FHSS)

In DSSS, the PN sequence spreads the spectrum of the signal by the chipping rate, resulting in the instantaneous widening of the spectrum. On the other hand, in FHSS, the PN sequence is used to drive the synthesizer and pseudo-randomly hop the signal

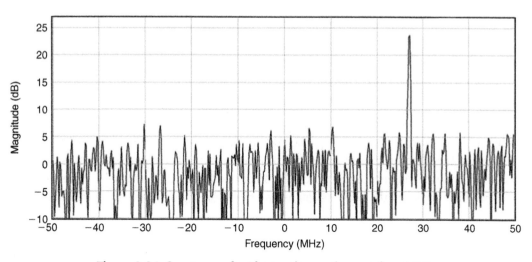

Figure 3.34 Spectrum of a Bluetooth waveform at $f_c - 4\,\text{MHz}$

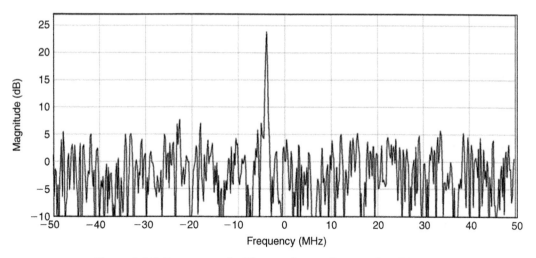

Figure 3.35 Spectrum of a Bluetooth waveform at $f_c + 27\,\text{MHz}$

bandwidth across a much wider band. In this case, the synthesizer is being sequentially programmed with different frequencies by the PN sequence to modulate the desired signal over a much wider frequency band. One such example is Bluetooth where frequency hopping is utilized to spread the spectrum as shown in Figure 3.34 and Figure 3.35. Unlike

DS systems, FH systems can cover a spectrum of several GHz, which is challenging for DS systems still in today's technology. However, the implication of such large bandwidths is phase incoherency. Frequency synthesizers are unable to maintain frequency coherency over such wide bandwidths. This presents a particular challenge for fast FH systems. This in turn forces the designer to resort to noncoherent modulation schemes.

There are two types of FHSS systems, namely slow frequency hopping and fast frequency hopping. In slow frequency hopping, the symbol rate of the data signal is an integer multiple of the hopping rate. In this scenario, one or more data symbols are transmitted each hop. In fast frequency hopping, on the other hand, each data symbol is transmitted over multiple hops and the hopping rate is an integer multiple of the data symbol rate.

3.10 Appendix

3.10.1 Coherent FSK Orthogonality Principles

Consider the relation

$$\int_{kT}^{(k+1)T} \cos(2\pi f_0 t + \theta)\cos(2\pi f_1 t + \theta)\,dt = 0 \tag{3.100}$$

Using the trigonometric identity

$$\cos a \cos b = \frac{1}{2}\{\cos(a+b) + \cos(a-b)\} \tag{3.101}$$

The relation in (3.100) can be expanded

$$\int_{kT}^{(k+1)T} \cos(2\pi f_0 t + \theta)\cos(2\pi f_1 t + \theta)\,dt =$$

$$\frac{1}{2}\left\{\int_{kT}^{(k+1)T} \cos(2\pi(f_0 + f_1)t + 2\theta)\,dt + \cos(2\pi(f_0 - f_1)t)\,dt\right\} =$$

$$\frac{1}{2}\left\{\int_{kT}^{(k+1)T} \cos(2\theta)\cos(2\pi(f_0 + f_1)t)\,dt -\right.$$

$$\left. \sin(2\theta)\sin(2\pi(f_0 + f_1)t)\,dt + \cos(2\pi(f_0 - f_1)t)\,dt\right\} 0 \tag{3.102}$$

After integration, (3.102) becomes

$$\frac{1}{2}\frac{\cos(2\theta)}{2\pi(f_0 + f_1)}\sin(2\pi(f_0 + f_1)t)\Big|_{kT}^{(k+1)T} + \frac{1}{2}\frac{\sin(2\theta)}{2\pi(f_0 + f_1)}\cos(2\pi(f_0 + f_1)t)\Big|_{kT}^{(k+1)T} +$$

$$\frac{1}{2}\frac{1}{2\pi(f_0 - f_1)}\sin(2\pi(f_0 - f_1)t)\Big|_{kT}^{(k+1)T} = 0 \tag{3.103}$$

The relation in (3.103) is true provided that

$$2\pi(f_0 + f_1)T = 2m\pi$$
$$2\pi(f_0 - f_1)T = n\pi \tag{3.104}$$

where m and n are integers. Solving for f_0 and f_1, we obtain

$$f_0 = \frac{2m + n}{4T}$$

$$f_1 = \frac{2m - n}{4T} \tag{3.105}$$

The difference between f_0 and f_1 is

$$f_0 - f_1 = \frac{n}{2T} \tag{3.106}$$

The relations between f_0 and f_1 set the condition for orthogonality between the two tones.

3.10.2 OFDM Channel Estimation and Frequency Offset Estimation Algorithms

In this section, the basic frequency and channel estimation algorithms used in OFDM are discussed.

3.10.2.1 Frequency Offset Estimation

The autocorrelation function $r(\tau)$ is related harmonically to the power spectral density via the Wienner-Khinchine theorem as

$$r(\tau) = \int_{-\infty}^{\infty} R(f)e^{j2\pi f\tau}df \tag{3.107}$$

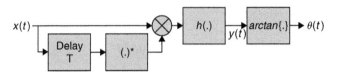

Figure 3.36 Conceptual frequency estimator

The frequency can then be obtained as

$$f_{estimate} = \frac{1}{2\pi\tau} \arg\{r(\tau)\} \tag{3.108}$$

The noiseless OFDM preamble can be expressed as a sum of equal-frequency shift harmonics:

$$x(t) = \sum_{n=0}^{N-1} \alpha_n e^{j2\pi n \Delta F t} + \eta(t) \tag{3.109}$$

where in (3.109), the DC ($n = 0$) case the coefficient is assumed to be zero and ΔF is the intercarrier frequency offset and $\eta(t)$ is additive white Gaussian noise.

Next, consider the conceptual frequency estimator depicted in Figure 3.36; the filter $h(.)$ is a moving average filter. Assume that the received signal frequency is offset by f_o; then the output of the moving average filter over M symbols can be expressed as

$$y(t) = \frac{1}{MT} \int_{t_{initial}}^{t_{initial}+MT} x(t)\, x^*(t-T)\, dt \tag{3.110}$$

$$y(t) = \frac{1}{MT} \int_{t_{initial}}^{t_{initial}+MT} \sum_{n=0}^{N-1} \alpha_n e^{j2\pi(n\Delta F+f_0)t} \sum_{l=0}^{N-1} \alpha_l^* e^{-j2\pi(l\Delta F+f_0)(t-T)}\, dt$$

$$+ \frac{1}{MT} \int_{t_{initial}}^{t_{initial}+MT} \sum_{n=0}^{N-1} \alpha_n e^{j2\pi(n\Delta F+f_0)t} \eta^*(t-T)\, dt$$

$$+ \frac{1}{MT} \int_{t_{initial}}^{t_{initial}+MT} \eta(t) \sum_{n=0}^{N-1} \alpha_l^* e^{-j2\pi(l\Delta F+f_0)(t-T)}\, dt$$

$$+ \frac{1}{MT} \int_{t_{initial}}^{t_{initial}+MT} \eta(t)\eta^*(t-T)\, dt$$

where $T = 1/\Delta F$. Furthermore, we know that

$$\int_{t_{initial}}^{t_{initial}+MT} \alpha_n \alpha_l^* e^{j2\pi(n-l)(\Delta F + f_0)t} dt = \begin{cases} |\alpha_n|^2 & n = l \\ 0 & \text{otherwise} \end{cases} \tag{3.111}$$

Then the relation in (3.110) becomes

$$y(t) = \frac{1}{MT} e^{j2\pi f_0 T} \sum_{n=0}^{N-1} |\alpha_n|^2 + \sigma_\eta^2 \tag{3.112}$$

where σ_η^2 is the noise variance. From (3.112), we observe that the energy in the frequency estimator depends on the channel profile.

The *arctan* function itself can be implemented via a look-up table or approximated as

$$\Psi(t) = \tan^{-1}\{f(t)\} = \tan^{-1}\left[\frac{Q(t)}{I(t)}\right] = \frac{I(t)Q(t)}{I^2(t) + 0.28125Q^2(t)} \tag{3.113}$$

for angles between 0 and 90 degrees. Rewrite the relation in (3.112) as

$$y(t) = \frac{1}{MT} \sum_{n=0}^{N-1} |\alpha_n|^2 \{\cos(2\pi f_0 T) + j\sin(2\pi f_0 T)\} \tag{3.114}$$

Frequency offset estimation is part of the OFDM synchronization process, in which frame boundary detection and sampling error correction take place before channel estimation and data demodulation.

3.10.2.2 Channel Estimation

A training sequence is typically used at the beginning of a data frame to predict the channel behavior and estimate the coefficients for equalization. After frame acquisition and frequency offset adjustment, the time domain signal is transformed into the frequency domain via the FFT as shown in Figure 3.37. For each subcarrier in the signal we have

$$Y(k) = C(k)X(k) + Z(k) \tag{3.115}$$

where $C(.)$ is the channel transfer function at a given subcarrier index k, and $Z(.)$ is the frequency domain representation of an additive noise component at the *kth* subcarrier,

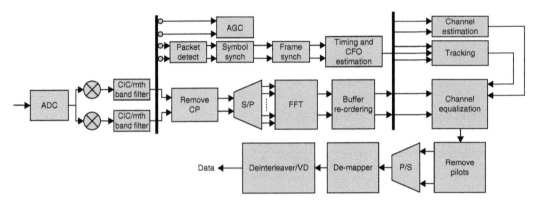

Figure 3.37 A conceptual OFDM receiver

and $X(.)$ is the known transmitted data at the receiver. The channel estimate for the *kth* subcarrier is simply obtained as

$$\hat{C}(k) = \frac{Y(k)}{X(k)} \tag{3.116}$$

where $\hat{C}(.)$ is the noisy estimate of $C(.)$. In this case, it is assumed that the channel is slowly varying such that its behavior is static over an entire frame.

3.10.3 CDMA-2000 Waveform Model

3.10.3.1 Received Signal Model and Ideal Symbol Recovery

At the output of the base station, the combined transmitted signal can be modeled as follows

$$\sum_{k=0}^{K-1} \sqrt{\frac{E_c^k}{2}} \sum_{n=-\infty}^{\infty} x^k(t - nT_c)w^k(t - nT_c) \times$$
$$\left[a^I(t - nT_c)\cos(\omega_c t + \theta) + a^Q(t - nT_c)\cos(\omega_c t + \theta)]*g(t) \right] \tag{3.117}$$

where K is the total number of available channels, w^k is the *kth* Walsh code, and $g(t)$ models the cumulative effect of the transmit filters. Figure 3.38 presents a depiction of the base-station modulator for a single channel. Multiple channels are typically modulated

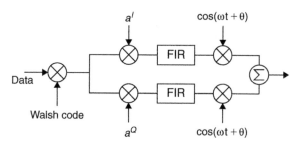

Figure 3.38 Conceptual modulation of single-channel forward-link signal

by the various Walsh codes, summed, and then further modulated by the in-phase and quadrature PN codes. The complex data signal $x^k(t)$ can be expressed as

$$x^k(t - nT_c) = x_r^k(t - nT_c) + jx_i^k(t - nT_c) \tag{3.118}$$

where $x^0(t - nT_c) = 1|_{\forall n \in Z}$ is the pilot signal. On the user side, the received signal can be expressed as

$$r(t) = \sum_{k=0}^{K-1} \sqrt{\frac{E_c^k}{2}} \sum_{n=-\infty}^{\infty} x^k(t - nT_c)w^k(t - nT_c) \times$$

$$[a^I(t - nT_c)\cos(\omega_c t + \theta) + a^Q(t - nT_c)\sin(\omega_c t + \theta)]*g(t)*h(t) + \hat{I}_{oc} + \eta(t) \tag{3.119}$$

where $h(t)$ is the channel response, $\hat{I}_{oc}$ is outer-cell interference, and $\eta(t)$ is additive noise. At the receiver, given an ideal sampling and frequency estimates, the in-phase and quadrature components are first mixed down to baseband:

$$r(t)\cos(\omega_c t) = \frac{1}{2}\sum_{k=0}^{K-1} \sqrt{\frac{E_c^k}{2}} \sum_{n=-\infty}^{\infty} x^k(t - nT_c)w^k(t - nT_c)$$

$$\times [a^I(t - nT_c)\cos(\theta) + a^Q(t - nT_c)\sin(\theta)]*g_I(t)*h_I(t)*f_I(t)$$

$$r(t)\sin(\omega_c t) = \frac{1}{2}\sum_{k=0}^{K-1} \sqrt{\frac{E_c^k}{2}} \sum_{n=-\infty}^{\infty} x^k(t - nT_c)w^k(t - nT_c)$$

$$\times [-a^I(t - nT_c)\sin(\theta) + a^Q(t - nT_c)\cos(\theta)]*g_Q(t)*h_Q(t)*f_Q(t)$$

$$\tag{3.120}$$

where $f(t)$ is the impulse response of the receive filters at baseband, and the interference and noise terms have been omitted to simplify the analysis and will be used in later sections.

Despreading the relation in (3.120) by the PN codes, we obtain the following four relations:

$$
r(t)\cos(\omega_c t)a^I(t - nT_c) = \frac{1}{2}\sum_{k=0}^{K-1}\sqrt{\frac{E_c^k}{2}}\sum_{n=-\infty}^{\infty} x^k(t - nT_c)w^k(t - nT_c)
$$
$$
\times [\cos(\theta) + a^I(t - nT_c)a^Q(t - nT_c)\sin(\theta)]
$$
$$
*g_I(t)*h_I(t)*f_I(t)
$$

(3.121)

$$
r(t)\cos(\omega_c t)a^Q(t - nT_c) = \frac{1}{2}\sum_{k=0}^{K-1}\sqrt{\frac{E_c^k}{2}}\sum_{n=-\infty}^{\infty} x^k(t - nT_c)w^k(t - nT_c)
$$
$$
\times [a^I(t - nT_c)a^Q(t - nT_c)\cos(\theta) + \sin(\theta)]
$$
$$
*g_I(t)*h_I(t)*f_I(t)
$$

(3.122)

$$
r(t)\sin(\omega_c t)a^I(t - nT_c) = \frac{1}{2}\sum_{k=0}^{K-1}\sqrt{\frac{E_c^k}{2}}\sum_{n=-\infty}^{\infty} x^k(t - nT_c)w^k(t - nT_c)
$$
$$
\times [-\sin(\theta) + a^I(t - nT_c)a^Q(t - nT_c)\cos(\theta)]
$$
$$
*g_Q(t)*h_Q(t)*f_Q(t)
$$

(3.123)

$$
r(t)\sin(\omega_c t)a^Q(t - nT_c) = \frac{1}{2}\sum_{k=0}^{K-1}\sqrt{\frac{E_c^k}{2}}\sum_{n=-\infty}^{\infty} x^k(t - nT_c)w^k(t - nT_c)
$$
$$
\times [-a^I(t - nT_c)a^Q(t - nT_c)\sin(\theta) + \cos(\theta)]
$$
$$
*g_Q(t)*h_Q(t)*f_Q(t)
$$

(3.124)

3.10.4 Ideal Symbol Recovery

Assuming that the data being sent is BPSK modulated on QPSK and furthermore ignoring the filtering and channel effects, using the relations in (3.121) through (3.124) and despreading according to Figure 3.39, we obtain

$$
I_d(t) = \sum_{k=0}^{K-1}\sqrt{\frac{E_c^k}{2}}\sum_{n=-\infty}^{\infty} x^k(t - nT_c)w^k(t - nT_c)w(t - nT_c)\cos(\theta)
$$

(3.125)

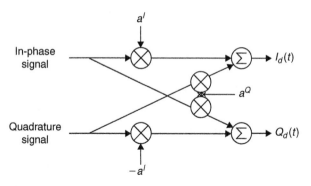

Figure 3.39 Basic CDMA-2000 short PN despreader

and

$$Q_d(t) = \sum_{k=0}^{K-1} \sqrt{\frac{E_c^k}{2}} \sum_{n=-\infty}^{\infty} x^k(t - nT_c)w^k(t - nT_c)w(t - nT_c)\sin(\theta) \tag{3.126}$$

where $w(.)$ is the desired Walsh code. Demodulating over one symbol of N-chips, we obtain

$$I_s(t) = \sqrt{\frac{E_c^k}{2}}d_s \cos(\theta) \tag{3.127}$$

$$Q_s(t) = \sqrt{\frac{E_c^k}{2}}d_s \sin(\theta) \tag{3.128}$$

where d_s is the data symbol integrated over N-chips and I_s and Q_s are the in-phase and quadrature symbols. The sine and cosine terms are further removed during the coherent demodulation process with the pilot signal.

References

[1] Shannon CE. A Mathematical Theory of Communication. Bell Syst. Tech J. October 1948;27:623–56.
[2] Blahut RE. Principles and Practice of Information Theory. Reading, Mass: Addison-Wesley; 1987.

[3] Verdu S. Fifty years of Shannon theory. IEEE Transaction on Information Theory October 1998;44(6):2057–78.

[4] Proakis J, Manolakis D. Digital Signal Processing. 2nd Edition. New York, NY: Macmillan; 1992.

[5] Papoulis A. Probability, Random Variables, and Stochastic Processes. New York, NY: McGraw-Hill; 1965.

[6] Feher K. Wireless Digital Communications. Upper Saddle River, NJ: Prentice Hall; 1995.

[7] Proakis J. Digital Communications. 2nd Edition. New York, NY: McGraw-Hill; 1989.

[8] Steber M. PSK Demodulation (Part 1). Watkins-Johnson Technical Note March/ April 1984;11(2).

[9] Park JH. On binary DPSK detection. IEEE Trans. Commun. April 1978;26(4): 484–6.

[10] Liu, C. and Feher, K., "Noncoherent detection of $\pi/4$-shifted systems in a CCI-AWGN combined interference environment," *Proceedings of the IEEE 40th Vehicular Technology Conference,* San Francisco, 1989.

[11] Anvari K, Woo D. Susceptibility of the $\pi/4$ DQPSK TDMA channel to receiver impairments. RF Design Feb. 1991:49–55.

[12] Miller L, Lee J. BER expressions for differentially detected $\pi/4$ DQPSK modulation. IEEE Transaction on Communications January 1998;46(1).

[13] Schell, S., "Implementation effects on GSM's EDGE modulation," White paper, Tropian Inc, Cupertino, CA.

[14] Watson, B., "FSK: Signals and demodulation," *Watkins-Johnson Technical Note,* San Jose, CA.

[15] Pasupathy S. Minimum shift keying: a spectrally efficient modulation. IEEE Communication Magazine July 1979.

[16] Cahn C. Performance of digital phase modulation communication systems. IRE Transaction on Communications Systems September 1960;CS-8:150–5.

[17] Hancock J, Lucky R. Performance of combined amplitude and phase modulated communications system. IRE Transaction on Communications Systems December 1960;CS-8:232–7.

[18] Campopiano C, Glazer B. A coherent digital amplitude and phase modulation scheme. IRE Transaction on Communications Systems 1962;CS-10:90–5.

[19] Chang R. Synthesis of band-limited orthogonal signals for multichannel data transmission. BSTJ December 1966;46:1775–96.

[20] Van Nee R, Prasad R. OFDM for Wireless Multimedia Communications. Boston: Artech House Publishers; 2000.

[21] Pickholtz R, Schilling D, Milstein L. Theory of spread-spectrum communications— a tutorial. IEEE Transactions on Communications May 1982;30(5).

[22] Holmes J. Coherent Spread Spectrum Systems. Malabar, FL: Krieger Publishing Company; 1982.

[23] Weinstein S, Ebert P. Data transmission by frequency division multiplexing using the discrete Fourier transform. IEEE Trans. Commun. Oct. 1971;19(10):628–34.

CHAPTER 4

High-Level Requirements and Link Budget Analysis

An ideal SDR offers a highly flexible and versatile platform that permits the instantiation of a variety of different waveforms[1] with widely varying specifications. More precisely, the radio must be capable of transmitting and receiving over a wide frequency range, at various power levels and bandwidths, using different modulation and multiple access schemes. The radio must also be capable of supporting various applications ranging, for example, from simple landing aid waveforms and secure voice communication to advanced data messaging or multimedia application over ad-hoc networks. This *versatility* imposes on the SDR multiple sets of requirements that must be met all in one platform. The aim of this chapter, then, is to discuss how high-level and link budget requirements can be derived and how they can be translated into low-level requirements and eventually into realizable system-design parameters.

This chapter is divided into three sections. In Section 4.1, high-level requirements and how they interrelate to one another is discussed in detail. The section starts off with a brief description of the Software Communications Architecture (SCA) and its role in future military and commercial software radios. This is followed by a discussion on usage scenarios and integrated platforms. The interdependence of high-level requirements on the environment, platform, and targeted application is highlighted.

In Section 4.2, the link budget analysis is discussed in detail. Path loss and the various degradations that impair the desired signal ranging from man-made to environmental are

---

[1]A waveform is sometimes used synonymously with a standard's air-interface or physical layer.

discussed. Finally, the sensitivity equation that determines the required noise figure of the receiver for a given CNR is explained in detail.

Section 4.3 discusses cascaded noise figure analysis—that is, the impact of noise figure and gain of the individual receiver blocks on the overall system noise figure.

4.1 High-Level Requirements

High-level requirements are usually driven by the applications that the SDR needs to support either simultaneously or separately in time. Military platforms, for instance, may require the radio to support multiple waveforms such as IFF (interrogate friend or foe) in the L-band, UFO (UHF follow-on) SATCOM in the UHF band, and WNW (wideband networking waveform). These waveforms support various applications encompassing voice, data, and multimedia. For instance, IFF, employed by both civilian (limited use) and military aircraft, is a pulsed-type waveform used to identify friendly aircraft over several miles. UFO SATCOM on the other hand operates in the UHF band and is intended for situations where over the horizon data (DAMA) and voice communication is required. This type of communication enables two terminals to communicate via a network of bent-pipe satellites situated several thousand miles away in earth orbit. Unlike IFF's pulse-based signaling, UFO SATCOM utilizes a shaped BPSK modulation scheme intended for spectral containment. WNW, on the other hand, is comprised of a new set of military waveforms that utilizes sophisticated ad-hoc networking schemes that enable data sharing for future *digital battlefield* applications. WNW OFDM in particular is a waveform with multiple bandwidths and transmit-power requirements intended for applications that use high data rates also over several miles between fixed or fast-moving terminals. Given these widely diverse waveforms, it is obvious that the SDR must support multiple requirements on the signal processing level as well as on the data processing level. More specifically, the SDR must accommodate different high-level requirements concerning bandwidths, dynamic ranges, sensitivities, linearity requirements, selectivity, frequency bands, to name a few parameters, all in one platform.

The situation presented above, which requires a single radio to support multiple standards and requirements, is not unique to the military. In the commercial world, a single mobile device could cater to several wireless applications. For example, suppose that a mobile device supports Digital Enhanced Cordless Telecommunications (DECT) for cordless applications in the home, GSM and UMTS for voice and data coverage in a cellular network, and 802.11(b) for high data rate indoor wireless LAN. The idea is to allow the handset user to roam between GSM/UMTS voice and, say, 802.11a, g, or h voice over IP in an indoor environment in a seamless manner. In order to accommodate these standards, a widely varying set of requirements must be met by a single SDR platform. From this discussion, it is obvious that an SDR must accommodate all the requirements that individual application-specific radios must meet, while at the same time providing the added advantage of cost reduction, ease of use, and reduced form factor. Clearly, this implies that designing an SDR increases the complexity of the design several times over what one normally expects from an application-specific radio, such as a cellular phone or a WiFi device integrated into a laptop.

4.1.1 Software Communications Architecture

The Software Communications Architecture (SCA) was developed through the Joint Tactical Radio System (JTRS) Joint Program Executive Office (JPEO). The SCA is an open standard architecture pertinent to a variety of military SDR systems. It defines the SDR's primary operating environment. In principle, SCA maximizes the independence of software from hardware by mandating application and device portability, maximizing code reuse, and allowing for seamless technology insertion over time. Furthermore, SCA depicts the hardware architecture as a platform that can be described in object-oriented terms that are consistent with the software architecture description.

The SCA was defined by the JTRS Joint Program Office (JPO) in order to enable the various radios to meet the following goals [1],[2]:

- Interoperable and affordable family of radios for various tiers of users

- Programmable and reprogrammable including over the air configuration and software downloads

- Software portability and reuse for the various supported waveforms, which enables lower development costs and extends the life cycle of the radio

- Hardware-independent software, thus permitting software reuse with various hardware platforms

- Enable rapid technology insertion for new software and hardware technologies that become available over time

- Scalable architecture that enables the development of low-capability handheld devices to high-capability platform-specific radio equipment

SCA has numerous key features that can be organized into four major categories:

- Software architecture

- Hardware architecture

- Security architecture

- Application program interfaces (APIs)

The software architecture is based on an embedded distributed computing environment mainly defined by an operating environment (OE), applications, and logical devices. The OE itself is comprised of a real-time operating system (RTOS), a real-time object request broker (ORB), core framework (CF) defined by the SCA, and services. The RTOS utilizes a POSIX[2]-based processing infrastructure in order to enable full JTRS functionality. The interfaces defined by the SCA for the ORB are based on a subset of CORBA[3] (common object request broker architecture) in order to seamlessly integrate and enable the interoperability of various software and hardware components from the same or various vendors.

The hardware architecture, depicted in Figure 4.1, is an SCA-defined platform designed to support all the software, security, and radio functionality needed for JTRS. The hardware platform is divided into three main sections: the black section where the data is encrypted between the various interfaces, the crypto section that encrypts and de-encrypts the data going from the red section to the black section and vice versa, and the red section where the data flows in de-encrypted format between various devices.

The security architecture is comprised of various security elements of the SCA that are relevant to both commercial and military applications. Security requirements pertinent to

---

[2]POSIX stands for *portable operating system interface*. The IX denotes the UNIX heritage from which these standards evolved. POSIX is developed by the IEEE as a family of system standards based on UNIX.
[3]The object management group (www.omg.org) defines CORBA middleware in the following terms: "Using the standard protocol IIOP, a CORBA-based program from any vendor, on almost any computer, operating system, programming language, and network, can interoperate with a CORBA-based program from the same or another vendor, on almost any other computer, operating system, programming language, and network."

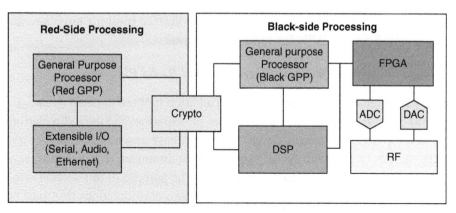

Figure 4.1 Conceptual JTRS SCA compliant hardware

military radios are discussed in detail in the SCA security supplement [3],[4]. The various security requirements can be classified into key categories as listed in [5]:

- Encryption and decryption services

- Information integrity

- Authentication and nonrepudiation

- Access control

- Auditing and alarms

- Key and certification management

- Security policy enforcement and management

- Configuration management

- Memory management

Note, however, that JTRS security is not only concerned with the software aspects of the radio, since that alone is not sufficient. Some security measures must be enforced by the hardware.

Waveform APIs are defined such that maximum portability can be attained on any given JTRS platform. Separate APIs are designated for each waveform or class of waveforms.

SCA defines the basic building blocks in such a way that the resulting template can be used in developing specific APIs for various waveforms.

4.1.2 Usage Scenarios, Modes of Operation, and Radio Platform

A set of key requirements and specifications in addition to those dictated by the waveform(s) ensue from usage scenarios, modes of operation, and the radio platform itself. To develop a usage scenario, the designer must have answers to some key questions concerning various factors such as the propagation environment, simultaneous waveform operations, and coexistence, just to mention a few. The purpose of this section is to be thorough but not exhaustive in discussing the various considerations that must be addressed when designing the radio platform and developing the usage scenarios.

4.1.2.1 Propagation Environment and Interference

The first key component of this factor is the propagation environment. The question that comes to mind concerns the location and environment where the radio will be operating. For a commercial application, one might ask if the radio will be used mostly indoors or outdoors. Will it operate in a cluttered office environment surrounded by metal cabinets, or in a typical living room? Is it going to be used in a highly mobile environment such as an airplane, or in a relatively slow-moving environment at pedestrian speed? In a military application, is the radio going to be integrated on an aircraft, or on the backpack of a soldier? Will it be mounted on a fixed platform on a ship, or is it required to be operational on a rotary wing aircraft?

The answers to these questions will allow the wireless designer to develop, with the aid of engineering analysis and simulations, a set of key, high-level parameters. A simple example would be: given a range (distance) and a certain fade margin, a link budget analysis can be performed and the noise figure of the radio as well as the antenna gains can be derived.

The second component of the operating environment has to do with interference. Interference in a military application is either intended or unintended. For instance, a radio operating in an adjacent channel causes unintended interference to the desired signal. On the other hand, intentional interference is varied and can be caused by enemy jammers to confuse and disable the desired receiver from conducting proper radio communication. In a commercial application, broadband noise from one application operating in a nearby band at high power could desensitize the radio and the designer must account for sufficient band-selection filtering in order to prevent the receiver from being desensitized.

4.1.2.2 Physical Platform

The physical platform on which the radio will be mounted is a key element in defining the radio's performance. For example, consider a radio mounted on a stealthy military aircraft communicating via a UHF waveform in a line-of-sight (LOS) environment. The UHF antenna is integrated into the skin of the plane, thus resulting in negative antenna gain. Compare this to a radio mounted on a nonstealthy aircraft communicating with the same waveform in a similar LOS environment using an external antenna with sufficient gain. In the former case, the SDR must account for the negative antenna gain with better noise figure than the latter to obtain the same receiver performance. Similarly, if a certain commercial waveform is used in a cellular phone where the range of communication may suffer due to the device's proximity to the body, this may require the radio to have lower noise figure and higher PA output power compared to a radio using the same waveform integrated into a laptop computer in a similar channel environment. The latter, however, may suffer from coexistence interference due to other waveforms running simultaneously on the same laptop with antennas close by, thus requiring the radio to have better interference rejection and selectivity than the former.

Another important aspect of the physical platform is the size, weight, and power consumption (SWaP). For example, a handheld device must conform to a certain form factor in its size and weight. It must be light and small enough to be carried in a purse or a pocket, or be held by hand during operation. Its power consumption must be low enough that it can operate without charging the battery for a reasonable amount of time. The SWaP parameters tend to be the most difficult to meet in any design, be it commercial or military.

Other fundamental parameters that are byproducts of the device's usage scenarios include rain, temperature, durability, and altitude, among others. A military SDR mounted on a jet aircraft must operate under extreme conditions at high or low altitudes, at high or low speeds, in the desert or high mountains, in rain or sandstorms. All of these situations resulting from the usage scenarios must be accommodated when designing the radio.

4.1.2.3 Simultaneity

A key feature of SDR is simultaneity. Simultaneity is the ability of the radio to operate several applications using multiple waveforms at the same time, thus eliminating the need for multiple radios and reducing cost and power. For example, a cellular phone user could be exchanging data over a wireless USB connection using a UWB OFDM waveform with another wireless device while at the same time speaking to another person using a GSM waveform. In this case, the SDR is running two separate physical layer waveforms using the same hardware at the same time. A commercial multistandard SDR radio may have to support several different protocols for different applications. These applications could be categorized as [6]:

- Short-range connectivity such as Bluetooth, Zigbee (802.15.4), MBOA-UWB

- Wireless broadband such as WLAN (802.11a,b,g,n), WiMax (802.16e)

- Cellular such as GSM, WCDMA, and CDMA2000

- Broadcast such as AM, FM, DAB-T, DVB-H, DMB-T

- Positioning such as GPS

One example of such a multistandard SDR could support 802.11g, WiMax, GSM, EDGE, WCDMA, and MBOA-UWB (802.15.3a). The frequency bands this radio must cover are shown in Table 4.1.

Simultaneity is common in military communications, where an SDR could be operating UHF voice along with Link-16, IFF, and a landing aid waveform at the same time. In such usage scenarios, the system designer must be mindful of cosite interference issues, where a transmitter might impair a receiver due to lack of signal isolation internal to the radio or external between the antennas. Design measures must then be taken to provide enough signal isolation, internal and external, to prevent desensitization of the victim waveform in the receiver.

4.1.2.4 Crossbanding

Crossbanding is an enabling technology that solves compatibility issues between various waveforms operating at different frequencies and bandwidths, data rates, coding and modulation schemes, and possibly security levels. In an SDR-based handheld radio, an image could be received via a WCDMA network and transmitted using a MBOA-UWB waveform to a nearby printer for printing. In the military, crossbanding is necessary to abridge communications between the various waveforms in a digital battlefield scenario. Today, SCA compliant military waveforms can achieve a certain amount of crossbanding between applications.

Table 4.1 Partial requirements for a multistandard radio

Parameter	802.11a,b,g	802.15.3a	LTE	GSM/EDGE	WCDMA
Access Technique	CSMA-CA	CSMA-CA	FDD	TDMA	Full Duplex CDMA
Frequency Band	802.11b,g: 2.4–2.483 GHz ISM 802.11a: 5.15–5.35 GHZ UNII	3.1–10.6 GHz	Depends on region[4]	GSM900: 880–915 MHz	Tx: 1920 –1980 MHz Rx: 2110–2170 MHz
Bandwidth	Possible bandwidths: 10, 20, and 30 MHz	528 MHz	1.4, 3, 5, 10, 15, and 20 MHz	200 kHz	3.8 MHz
Modulation	OFDM/DSS[5]	OFDM	OFDM on the downlink and single carrier OFDM on the uplink	GMSK/ Differential Gaussian 8-PSK	QPSK
Sensitivity	a: −82 dBm[6] b: −93 dBm g: −95 dBm	−80 dBm for lowest data rate	Varies per band. For band 1 and 1.4 MHz −102.5 for 64 QAM −84.7	−102 dBm for both	−106.7 dBm
Required CNR	b: 11 dB a,g: varies with data rate	−3.16 for lowest data rate	Depends on modulation rate: 2.2 dB for QPSK and 22 dB for 64 QAM (minimum)	9 dB for both	0.9 dB for a BER < 0.1% to properly demodulate 12.2 Kbps
Peak Tx Power	b,g: 20 dBm	−10 dBm TFI −14 dBm FFI	Approx 21 dBm	Mobile: 33 dBm	Mobile: 24 dBm

[4]For example, Region 1: Tx 1920-1980 MHz, Rx: 2110-2170 MHz and Region 2: Tx 1850-1910 MHz, Rx: 1930-1990 MHz.

[5]For 802.11b only.

[6]This sensitivity level for 802.11a and g is for 6 Mbps without ACI.

4.1.2.5 Transceiver, Antenna, and Channel Impairments

In this section, the source of impairments in the transceiver, antenna, and the channel is briefly discussed. These impairments will be discussed in detail in the coming sections and chapters where tradeoffs between various design methods and approaches as well as detailed analysis for various architectures will be conducted.

The source of impairments and degradation in the transmitter exist in both the digital and analog portion. The sources of degradation are due, for example, to numerical fixed point implementations, the DAC, filters (analog and digital), mixers, gain stages, LO and frequency generation, PA nonlinearity and broadband noise. A summary of these possible degradations is shown in Figure 4.2.

Signal degradations in the receiver are due to the anomalies in the analog front end, which are similar in nature to those found in the transmitter chain, in the LNA, and in the band definition filters, duplexers, and diplexers. Farther down the chain, the mixer, LO, frequency generation, gain stages, the ADC, and the digital-receive portion of the modem are also sources of degradation as summarized in Figure 4.3. The degradations in the modem on the receiver side, however, are far more critical than on the transmitter side.

The function of the antenna is to convert an electric energy signal into an electromagnetic energy signal and vice versa. The antenna is also impacted with imperfections that lead to performance loss in the system, as depicted in Figure 4.4.

The channel characteristics depend mainly on the environment and frequency of operation as summarized in Figure 4.5. An indoor channel may be a flat-fading channel or a frequency selective fading channel. Over cable, for instance, the losses are purely thermal. A satellite link may suffer from galactic noise and atmospheric loss in addition to frequency selective fading and shadowing. The channel characteristics under various conditions are well described in the literature and will not be discussed further in this book. However, when discussing link budgets, we will keep in mind a certain fade margin when computing receiver sensitivities and ranges. Adjacent and cochannel interference will be discussed in a later section.

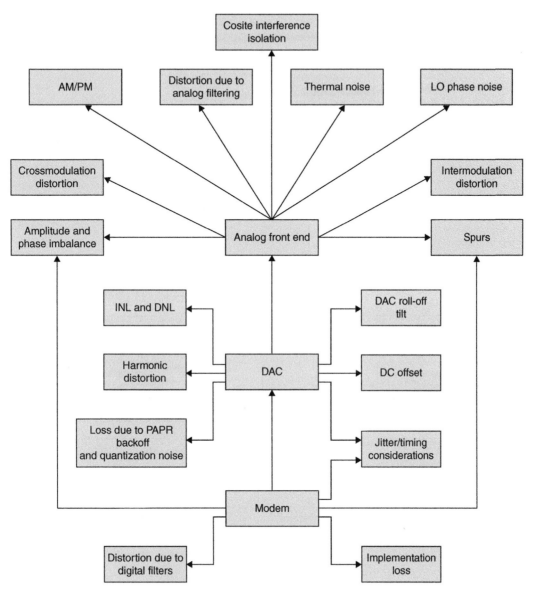

Figure 4.2 Pictorial representation of signal degradation in the transmitter

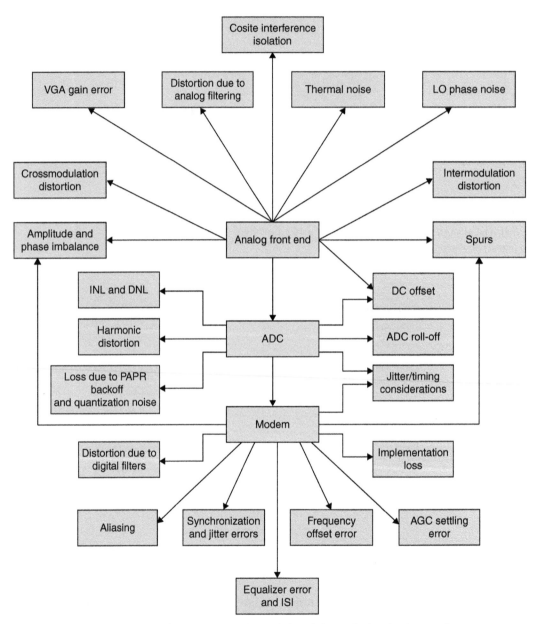

Figure 4.3 Pictorial representation of signal degradation in the receiver

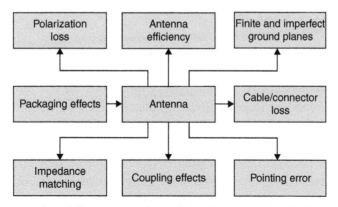

Figure 4.4 Pictorial representation of signal degradation in the antenna

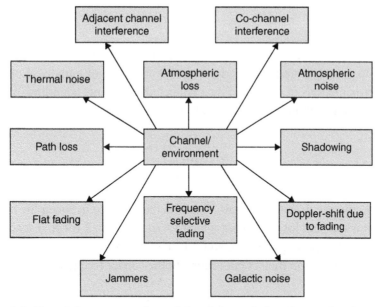

Figure 4.5 Pictorial representation of signal degradation due to the channel and environment

4.2 Link Budget Analysis

Link budget analysis is concerned with several important channel and transceiver parameters such as transmit power, system noise figure, carrier frequency and channel bandwidth, transmit and receive antenna gains, temperature, the environment (indoor, outdoor, LOS, etc.), and the required SNR. Ultimately, given the environment and some of the key parameters mentioned above, one might predict the range[7] and performance of the radio. Or similarly, given the range and transmit power, one might predict the system noise figure needed to meet a certain SNR requirement. Link budget analysis is performed by the radio designer early on in the design process to determine some of these essential design parameters. Other parameters, such as allowable transmit power, channel bandwidth, and carrier frequency, are obtained from the waveform specification through a combination of waveform analysis, standardization bodies, and government regulations as designated by the FCC.

4.2.1 Path Loss

Path loss is intimately related to the environment where the transmitter and receiver are located. Path loss models are developed using a combination of numerical methods and empirical approximations of measured data collected in channel sounding experiments. In general, propagation path loss increases with frequency as well as distance:

$$P_l = 10 \log_{10} \left(\frac{16\pi^2 d^n}{\lambda^2} \right) \tag{4.1}$$

where P_l is the average propagation path loss, d is the distance between the transmitter and receiver, n is the path loss exponent which varies between 2 for free space to 6 for obstructed in building propagation [7], and λ is the free space wavelength defined as the ratio of the speed of light in meters per second to the carrier frequency in Hz

$$\lambda = \frac{c}{f_c} = \frac{2.9979.10^8 \text{ meters/second}}{f_c \text{ Hz}} \tag{4.2}$$

---

[7]The maximum range between two transceivers is the distance where the two radios can communicate with an acceptable BER.

Example 4-1: Path Loss

What is the path loss of a UWB MBOA signal transmitted at the carrier frequencies of 3432, 3960, and 4488 MHz as the range varies between 1 and 10 meters? Assume the path loss exponent for in-building line of sight to be 2.7.

First, let's compute the path loss at 4 meters for the carrier frequency of 3960 MHz using the relations in (4.1) and (4.2):

$$P_l = 10 \log_{10}\left(\frac{16\pi^2 d^n}{\lambda^2}\right) = 10 \log_{10}\left(\frac{16\pi^2 4^{2.7}}{0.0757^2}\right) = 60.66\,\text{dB} \tag{4.3}$$

For the complete range of 1 to 10 meters, the path loss is shown in Figure 4.6.

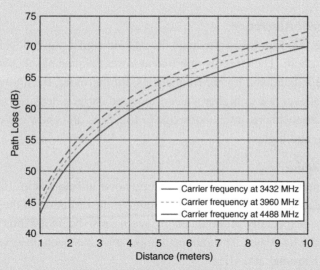

Figure 4.6 Path loss for a UWB MBOA signal with path loss exponent of 1.7 at the various carrier frequencies

4.2.2 Man-Made Noise, Atmospheric Effects, and Excess Path Loss

In addition to path loss, the desired signal incurs other types of losses lumped in the form of excess path loss and atmospheric losses. These losses must be factored in when preparing a link budget. Examples of such losses are listed below.

4.2.2.1 Man-Made Noise

Man-made noise is produced by power lines, electrical equipment, ignition systems, neon lights, and communication devices, just to name a few. These interferers and their byproducts of spurs, harmonics, etc. can propagate via ground waves, power lines, and over the air to the receiver. Man-made noise decreases as frequency increases and can be computed with respect to the man-made temperature at 100 MHz as

$$T_m = T_{m\_100} \left(\frac{100}{f_{MHz}} \right)^{2.5} \tag{4.4}$$

where T_{m_100} is the man-made temperature at 100 MHz. Man-made noise is typically determined via measurement in a given environment.

4.2.2.2 Atmospheric Absorption

Atmospheric absorption serves to attenuate the desired signal power level as it propagates through the atmosphere. The effect is minimal at lower frequencies; however, it is more pronounced at the X-band and higher frequencies. Furthermore, the attenuation effect is more severe at sea level due to higher water-vapor and oxygen densities compared to high altitudes. Further atmospheric attenuation due to fog, rain, and snow are more prominent at 1 GHz (L-band) and higher.

4.2.2.3 Galactic Noise

Galactic noise is more dominant than atmospheric noise at frequencies between 30 MHz and 400 MHz. Galactic noise reaches its maximum in the direction of the galactic center. The effect is especially pronounced given a narrow-beam antenna pointing in the direction of the galactic center. The antenna temperature has been observed to be higher than 18,000 kelvin around 100 MHz.

4.2.2.4 Other Types of Losses and Noise

Some other types of losses and noise are:

- Multipath loss, which depends on the environment, frequency, and bandwidth. Multipath loss is either in the form of frequency selective or flat fading across the whole bandwidth.

- Diffraction loss refers to the process in which an electromagnetic wave is bent (diffracted) by an obstacle (e.g., a tall building). The propagation loss due to knife-edge diffraction depends on the height of the obstacle and the wavelength.

- Refraction loss is due to the varying density of the atmosphere versus elevation. As altitude varies so does the propagation velocity of the electromagnetic wave.[8]

- Scatter propagation loss is categorized as ionospheric scatter, tropospheric scatter, and meteor burst scatter. These scatter phenomena enable over the horizon VHF, UHF, and SHF communication.

- Ionospheric propagation loss happens at HF frequencies which varies with latitude, time of day, season, and the 11-year sunspot cycle.

- Ground wave loss occurs at frequencies below HF. Horizontally polarized electric field waves that are close to the earth surface are attenuated.

4.2.3 Receiver Sensitivity

When evaluating a wireless link, the four most important questions that must be answered in order to evaluate the performance are:

- What is the maximum allowable transmit power?

- What are the signal and channel bandwidths?

- What is the desired range and environment?

- What is the CNR (or corresponding SNR) required to close link?

In order to answer the questions presented above, we must estimate the sensitivity and link budget of the system.

4.2.3.1 The Basic Sensitivity Equation

The system sensitivity is defined as the minimum desired received signal power needed to successfully demodulate and decode the received waveform at a certain minimum required BER. Theoretically, the sensitivity of the system will depend on the system noise

---

[8]Note that electromagnetic waves propagate in a curved path, and consequently over-the-horizon communication is possible.

figure, the required CNR, the channel bandwidth, and the system temperature. The CNR can be expressed in terms of E_b/N_o (or SNR) as

$$CNR_{dB} = 10 \log_{10} \left(\frac{E_b}{N_0} \frac{R_b}{B} \right) = SNR_{dB} + 10 \log_{10} \left(\frac{R_b}{B} \right) \qquad (4.5)$$

where R_b is the data rate and B is the noise equivalent bandwidth, to be distinguished from signal bandwidth or modulation bandwidth. In order to account for implementation losses, the relation in (4.5) is expressed as

$$CNR_{dB} = 10 \log_{10} \left(\frac{E_b}{N_0} \frac{R_b}{B} \right) + L_{i,dB} \qquad (4.6)$$

where L_i is the implementation loss of the system. The relations above can also be expressed in dB-Hz by eliminating the bandwidth factor.

Define the sensitivity level required at a room temperature of 290 kelvins to achieve a certain CNR as

$$\begin{aligned} P_{dBm} &= 10 \log_{10}(kT)_{dBm/Hz} + NF_{dB} + 10 \log_{10}(B) + CNR_{dB} \\ P_{dBm} &= -174_{dBm/Hz} + NF_{dB} + 10 \log_{10}(B) + CNR_{dB} \end{aligned} \qquad (4.7)$$

where CNR is the carrier to noise ratio in dB, $K = 1.381 \times 10^{-23}$ joules/kelvin is Boltzmann's constant, and $T = 290$ degrees is room temperature in kelvins.[9] Note that the amount of thermal noise present is closely tied to the channel bandwidth. The system noise figure NF is defined as

$$NF = 10 \log_{10}(F) = 10 \log_{10} \left(\frac{CNR_{input}}{CNR_{output}} \right) \qquad (4.8)$$

where F is the noise factor (or noise figure when expressed in dB), which is the ratio of the input SNR of a system or block to the output SNR of the system or block. In most cases, a gain (or loss) block may be conceptually thought of as a combination of a gain

[9]290 kelvins is equivalent to 62° Fahrenheit.

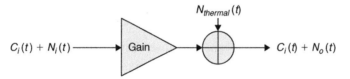

Figure 4.7 A conceptual RF gain block

stage that amplifies (or loss stage that attenuates) the input signal plus noise by a gain (or loss) g and a thermal noise (Johnson noise) stage which adds white Gaussian noise to the amplified noisy input signal as shown in Figure 4.7. Thermal noise is a phenomenon, also known as thermal agitation, which is due to the random motion of free electrons in a conducting material caused by heat. Thermal noise is directly proportional to temperature and is additive in nature and statistically follows a Gaussian distribution.

At this juncture, it is important to note that the designer must not solely rely on (4.7) in computing the sensitivity of the receiver. There are many other factors that contribute to degrading the receiver sensitivity. Phase noise, image noise due to multi-stage conversion, and PA broadband noise in full duplex systems such as WCDMA are some of the causes of degradation to the receiver's sensitivity. These factors must be taken into account when computing the true sensitivity of the receiver. At the beginning, the analyst may add a degradation factor to (4.7) to accommodate these anticipated anomalies.

4.2.3.2 Noise Power and Noise Floor

Define the noise power as the power delivered to a matched load resistor delivered by a source resistor at room temperature:

$$P_{noisepower} = 10\log_{10}(kTB) = -174\,\text{dBm/Hz} + 10\log_{10}(B) \qquad (4.9)$$

Consequently, the noise floor, in dBm, of the system is defined as

$$P_{noisefloor} = -174\,\text{dBm/Hz} + NF + 10\log_{10}(B) \qquad (4.10)$$

Note that the sensitivity relation expressed in (4.7) does not take into account the noise contribution due to image noise or LO broadband noise, for example, which affect the noise factor as

$$F_{total} = F_{thermal\_noise} + F_{image\_noise} + F_{broadband\_noise} \qquad (4.11)$$

Sensitivity, in the form presented in (4.7), is only meant to offer a lower bound on the minimal signal power level that achieves an acceptable BER. Another interesting point is the bandwidth. Early on in the design, the chosen bandwidth does not reflect the channel bandwidth, which can only be determined by computing the noise equivalent bandwidth at the output of the final selectivity filter. We will elaborate on the effect of selectivity on the final system sensitivity in a subsequent chapter. The received signal CNR is then defined as

$$CNR_{received} = P_{signal} - P_{noisefloor} \text{ dB} \tag{4.12}$$

4.2.3.3 Minimum Discernable Signal

Define the minimum discernable signal (MDS) at the output of the receiver as the signal level that causes an SNR of 0 dB. In other words, the signal power is equal to that of the noise power at the output of the receiver. The output MDS of certain analog blocks is related to the input MDS minus the gain:

$$MDS_{out} = MDS_{in} + G_{dB} \text{ dBm} \tag{4.13}$$

Example 4-2: Sensitivity

Compute the noise power, noise floor, and sensitivity of an SDR that supports both GSM and DECT with parameters listed in Table 4.2. Assume the temperature to be 290 kelvins.

Table 4.2 System parameters of an SDR supporting DECT and GSM

Standard	Required CNR (dB)	System Noise Figure (dB)	Channel Bandwidth (kHz)
GSM	9	9.8	200
DECT	10.3	19.0	1728

The noise power as defined in (4.9) for GSM is given as

$$P_{noisepower} = 10\log_{10}(kTB) = -174 \text{ dBm/Hz} + 10\log_{10}(200e3) = -120.964 \text{ dBm} \tag{4.14}$$

and, similarly, the noise power for DECT can also be found as

$$P_{noisepower} = 10\log_{10}(kTB) = -174 \text{ dBm/Hz} + 10\log_{10}(1.728e3) = -111.5986 \text{ dBm} \tag{4.15}$$

The noise floor can then be estimated using (4.10) as

$$P_{noisefloor} = P_{noisepower} + NF = -120.964 + 9.8 = -111.1638 \, dBm \text{ for GSM}$$
$$P_{noisefloor} = P_{noisepower} + NF = -111.5986 + 19.0 = -92.5986 \, dBm \text{ for DECT}$$

(4.16)

And finally, the sensitivity can be obtained using (4.7):

$$P_{noisefloor} = P_{noisefloor} + CNR = -120.964 + 9.0 = -102.1638 \, dBm \text{ for GSM}$$
$$P_{sensitivity} = P_{noisefloor} + CNR = -111.1638 + 19.0 = -82.2986 \, dBm \text{ for DECT}$$

(4.17)

Example 4-3: System Noise Figure

The occupied bandwidth of an 802.11b system is 22 MHz with a chip rate to data rate ratio of 11, thus implying a 2-Mbps data rate. In order to achieve a BER of 10^{-6}, an E_b/N_o of 11 dB is required. Furthermore, the desired receiver sensitivity of the system is -93 dBm. What is the required system noise figure at room temperature? Assume 0 dB implementation loss.

The first step is to compute the CNR. The CNR is given as

$$CNR_{dB} = 10 \log_{10}\left(\frac{E_b}{N_0}\frac{R_b}{B}\right) = 11 + 10 \log_{10}\left(\frac{1}{11}\right) = 0.5861 \, dB$$

(4.18)

The noise figure of the system can then be computed according to (4.7) as

$$NF_{dB} = P_{dBm} + 174_{dBm/Hz} - 10 \log_{10}(B)_{dB} - CNR_{dB}$$
$$NF_{dB} = -93 + 174 - 10 \log_{10}(22 \times 10^6) - 0.5861 = 6.9897 \, dB$$

(4.19)

Note that in certain applications, such as SATCOM, system noise figures are typically derived from equipment temperature and antenna temperature. The equipment temperature can be derived from the equipment noise figure as

$$T_{equipment} = T_o(10^{NF_{dB}/10} - 1)$$

(4.20)

where T_o is the ambient temperature in kelvins. The system temperature is the sum of the equipment temperature plus the antenna temperature:

$$T_{system} = T_{equipment} + T_{antenna} \text{ kelvins} \tag{4.21}$$

From the system temperature, the system noise figure can then be derived

$$NF_{system} = 10 \log_{10} \left(\frac{T_{system}}{290\,K} + 1 \right) dB \tag{4.22}$$

Example 4-4: System and Antenna Noise Figure

Given a UFO Satellite airborne terminal with equipment noise figure of 5.2 dB and antenna temperature of 320 kelvins, find the system NF.

The equipment temperature is given as

$$T_{equipment} = T_o(10^{NF_{dB}/10} - 1) = 290(10^{5.2/10} - 1) = 670.28\,K \tag{4.23}$$

The system temperature is the sum of the equipment temperature plus antenna temperature

$$T_{system} = T_{equipment} + T_{antenna} = 670.28 + 320 = 990.28\,K \tag{4.24}$$

Consequently, the system noise figure is

$$NF_{system} = 10 \log_{10} \left(\frac{T_{system}}{290\,K} + 1 \right) = 10 \log_{10} \left(\frac{990.28\,K}{290\,K} + 1 \right) = 6.45\,dB \tag{4.25}$$

Note that the antenna noise temperature is frequency and elevation dependent. Changing the frequency or antenna orientation may have a significant impact on antenna temperature. Furthermore, the antenna noise temperature is also dictated to a large extent by whether the antenna is exposed to the sun or night sky. The sun tends to be a large noise source for satellite antennas.

4.2.4 The Link Budget

Given certain system parameters such as required CNR to achieve a certain data rate, transmit signal power, bandwidth, transmit and receive antenna gains, path loss and channel conditions, one could determine the required system noise figure of the receiver. Similarly, given the noise figure of the system, one could determine the distance under certain conditions in which a signal could be transmitted at a certain power and received reliably by the receiver. The link budget and the ensuing sensitivity analysis are typically the first high-level signal processing requirements analyzed by the waveform designer.

4.2.4.1 Link Analysis

The link budget equation can be expressed as [8]:

$$CNR_{\text{required,dB}} = P_{tx} + G_{tx} + G_{rx} - P_l - 10\log_{10}(kT) - 10\log_{10}(B) - NF_{\text{system}} \text{ dB} \quad (4.26)$$

where

P_{tx} is the transmit power in dBm
G_{tx} is the transmitter antenna gain in dB
G_{rx} is the receiver antenna gain in dB
P_l is the average path loss as defined in (4.1)

The relation in (4.26) excludes the losses due to cables and VSWR mismatches. Furthermore, the link margin is defined as the received signal power minus the receiver sensitivity:

$$P_{rx} - P_{\text{sensitivity}} = P_{tx} + G_{tx} + G_{rx} - P_l - P_{\text{sensitivity}} \text{ dB} \quad (4.27)$$

We interpret the relation in (4.27) as follows. The transmitted signal is radiated at a given output power, which can be determined accurately via measurement. The transmit-signal power level will increase or decrease depending on the antenna gain or loss. Multiple or highly directive antennas can increase the effective radiated power (ERP). Once the signal is radiated out of the antenna, it begins to attenuate. The rate of attenuation depends on the environment of operation. Indoor signals, for example, are attenuated by objects such as metal cabinets, computers, people walking around, etc. In sky-wave propagation, ionospheric conditions, antenna pointing errors, precipitation, Faraday rotation, and scintillation are among the parameters that can attenuate a signal.

Once the desired signal reaches the receiver antenna, it may increase or decrease in power depending on the receiver antenna gain and directivity. The received signal is then further corrupted by thermal noise among other factors in the receiver by the time it reaches the ADC.

Example 4-5: Link Budget Analysis of an 802.11b System

The occupied bandwidth of an 802.11(b) spread spectrum system is 22 MHz. Each bit is spread over 11 chips. The system's noise figure is assumed to be 7 dB. The required E_b/N_0 to achieve a BER of 10^{-6} is 11 dB. The transmitter and receiver antenna gains including connector and VSWR losses are -0.5 dBi. At the ISM band of 2.4 GHz, let the distance between the transmitter and receiver be 30 meters. What is the transmit required power to close the link at this distance? Assume a path loss exponent of $n = 4.2$.

Recall that CNR is related to E_b/N_0 or SNR as

$$CNR_{dB} = 10\log_{10}\left(\frac{E_b}{N_0}\frac{1}{PG}\right) = 10\log_{10}\left(\frac{E_b}{N_0}\right) - 10\log_{10}(PG)$$
$$CNR_{dB} = 11 - 10\log_{10}(11) = 0.5861\,dB$$

(4.28)

where PG is the processing gain after despreading.

Assume that the experiment is done at room temperature of 290 K and that the channel bandwidth is equivalent to the modulation bandwidth. Then sensitivity of the system is given as

$$P_{dBm} = 10\log_{10}(kT)_{dBm/Hz} + NF_{dB} + 10\log_{10}(B)_{dB} + CNR_{dB}$$
$$P_{dBm} = -174_{dBm/Hz} + 7 + 10\log_{10}(22\,MHz) + 0.5861 = -93\,dBm$$

(4.29)

The required transmit power including fade margin is

$$P_{tx} = -G_{tx} - G_{rx} + 10\log_{10}\left(\frac{4\pi d^{n/2}}{\lambda}\right)^2 + P_{dBm}$$
$$P_{tx} = +0.5 + 0.5 + 102.091 - 93 = 10.091\,dBm$$

(4.30)

The required transmit power in watts is

$$P_{tx,Watts} = \frac{10^{P_{tx}/10}}{1000} = \frac{10^{10.091/10}}{1000} = 0.0102\,W$$

(4.31)

Example 4-6: Link Budget Analysis of a UWB MBOA System

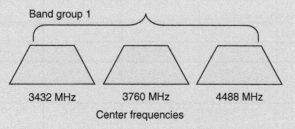

Band group 1

3432 MHz   3760 MHz   4488 MHz

Center frequencies

Figure 4.8 Band-group 1 of the UWB MBOA

UWB-MBOA's band group 1 is divided into three channels centered at 3432, 3760, and 4488 MHz, as shown in Figure 4.8. Based on the parameters listed in Table 4.3 compute the sensitivity, path loss, and link margin parameters. According to your analysis, at which data rates does this system close link, given the required ranges below?

Table 4.3 UWB MBOA link budget parameters

Parameter					
Information Rate (Mb/s)	53.3	80	106.7	160	200
Average Tx Power (dBm)	−14.23	−14.23	−14.23	−14.23	−14.23
Nominal Tx Antenna Gain (dBi)	−1	−1	−1	−1	−1
Nominal Rx Antenna Gain (dBi)	−1	−1	−1	−1	−1
Channel Bandwidth (MHz)	528	528	528	528	528
Center Frequency (MHz)	4488	4488	4488	4488	4488
Required CNR (dB)	−3.16	−1.00	−0.04	2.31	3.28
Range (meters)	9	7	5	5	5
Noise Figure (dB)	7.40	7.40	7.40	7.40	7.40

The analysis for this example follows the analysis presented in the previous example. The results are summarized in Table 4.4.

Table 4.4 Link budget analysis for a UWB MBOA system

Global Parameter					
Boltzman's constant (J/K)	1.381E − 23	1.381E − 23	1.381E − 23	1.381E − 23	1.381E − 23
Temperature (degrees)	20	20	20	20	20
Temperature (kelvins)	293.15	293.15	293.15	293.15	293.15
KT (dBm/Hz)	−173.9	−173.9	−173.9	−173.9	−173.9
Parameter					
Information Rate (Mb/s)	53.3	80	106.7	160	200
Average Tx Power (dBm)	−14.23	−14.23	−14.23	−14.23	−14.23
Nominal Tx Antenna Gain (dBi)	−1	−1	−1	−1	−1
Nominal Rx Antenna Gain (dBi)	−1	−1	−1	−1	−1
Channel Bandwidth (MHz)	528	528	528	528	528
Center Frequency (MHz)	4488	4488	4488	4488	4488
Speed of light (meter/second)	299,800,000	299,800,000	299,800,000	299,800,000	299,800,000
Lambda (meter)	0.067	0.067	0.067	0.067	0.067
Path loss exponent	2.3	2.3	2.3	2.3	2.3
Required CNR (dB)	−3.16	−1.00	−0.04	2.31	3.28
Range (meters)	9	7	5	5	5
Noise Figure (dB)	7.40	7.40	7.40	7.40	7.40
Sensitivity (dBm)	−82.46	−80.30	−79.35	−76.99	−76.02
Path Loss (dB)	67.44	64.93	61.56	61.56	61.56
Average Rx Power (dBm)	−83.668	−81.158	−77.797	−77.797	−77.797
Link Margin	−1.208	−0.862	1.548	−0.811	−1.780

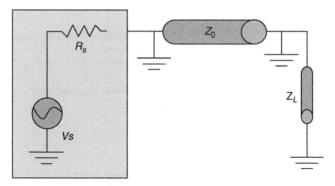

Figure 4.9 Conceptual model of a transmission line

4.2.4.2 Voltage Standing Wave Ratio (VSWR)

In order to ensure maximum power transfer between the antenna, the cable, and the transceiver, the transmitter or receiver load impedances must match that of the cable, and the cable load impedance must match that of the antenna. Any load mismatch will result in loss of power transfer. Assume that the characteristic impedance of the transmission line circuit depicted in Figure 4.9 is Z_0; then for a complex load, the reflection coefficient can be expressed as

$$\rho = \frac{Z_L - Z_0}{Z_L + Z_0} = \frac{V_{reflected}}{V_{incident}} \qquad |\rho| \leq 1 \qquad (4.32)$$

where $Z_L = R_L + jX_L$ is the complex load, with R_L being the load resistance and X_L the load reactance. The incident voltage $V_{incident}$ is the voltage magnitude sent through the transmission line to the load whereas $V_{reflected}$ is the magnitude of the voltage reflected off the load. A complex reflection coefficient implies that the incident wave suffers both magnitude and phase degradation.

Define VSWR as the ratio of the absolute maximum RMS voltage to the absolute minimum RMS voltage along a mismatched transmission line as

$$VSWR = \frac{1 + |\rho|}{1 - |\rho|} = \frac{V_{max}}{V_{min}} \qquad 1 \leq VSWR < \infty \qquad (4.33)$$

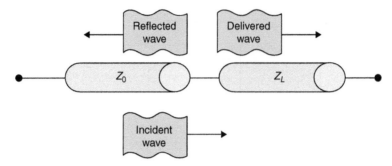

Figure 4.10 Two unmatched loads with wave representing incident, delivered, and reflected voltages

Likewise, the reflection coefficient can be expressed in terms of VSWR as

$$|\rho| = \left|\frac{1 - \text{VSWR}}{1 + \text{VSWR}}\right| \tag{4.34}$$

Then the return loss Γ can be defined in terms of the reflection coefficient as the ratio of power P_R reflected off the load to the power incident onto the transmission line P_I:

$$\Gamma = -10 \log_{10}(|\rho|^2) \tag{4.35}$$

Furthermore, define the mismatch loss as the ratio of power delivered P_D to the load to the power available P_A from the source:

$$\Psi = 10\log_{10}\left(\frac{P_D}{P_A}\right) = 10\log_{10}\left(\frac{P_A - P_R}{P_A}\right) = 10\log_{10}(1 - |\rho|^2) \tag{4.36}$$

where P_R is the power reflected off the load.

Next consider a transmission line whose characteristic impedance is Z_i feeding another transmission line that is matched to its own impedance Z_o as shown in Figure 4.10, then define the transmission coefficient κ as

$$\kappa = 1 + \rho = \frac{2Z_L}{Z_L + Z_0} \tag{4.37}$$

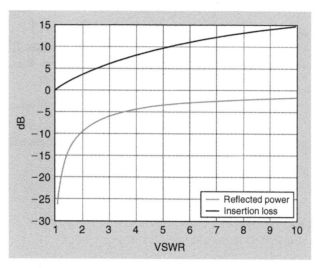

Figure 4.11 Reflected power and insertion loss as a function of VSWR

Then the insertion loss, in a lossless system,[10] can be expressed in dB as

$$\Lambda_{dB} = 10 \log_{10}\left(\frac{1}{|\kappa|^2}\right) \tag{4.38}$$

Figure 4.11 shows an example of the relationship between reflected power, insertion loss, and VSWR.

The overall VSWR at the input of a cascade of series elements, as depicted in Figure 4.12, is upper-bounded as [9]:

$$\text{VSWR}_{total} \leq \text{VSWR}_{\max} = \prod_{l=1}^{L} \text{VSWR}_L \tag{4.39}$$

---

[10] The relation for insertion loss expressed in (4.38) is only true for transmission lines. The insertion loss of a filter, on the other hand, is defined as a function of frequency as $IL_{Filter,dB} = 10 \log_{10}\left(\frac{1}{1 - |\rho(\omega)|^2}\right)$. Provided that the source and load are matched, this relation is the reciprocal of the square of the scattering parameter $|S_{12}|^2$.

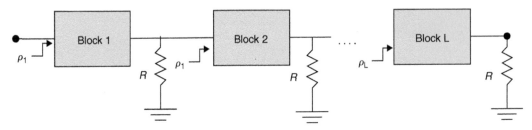

Figure 4.12 A cascade of lossless reciprocal series elements

and lower-bounded as

$$\text{VSWR}_{total} \geq \text{VSWR}_{min} = \max\left\{\frac{\text{VSWR}_1}{\prod_{l=2}^{L}\text{VSWR}_l}, 1\right\} \quad \begin{array}{l}\text{VSWR}_1 \geq \text{VSWR}_l \\ \forall l \in \{2,\ldots,L\}\end{array} \tag{4.40}$$

4.2.4.3 Cable Loss Estimation

Often, the cable manufacturer provides a cable loss specification at a certain frequency f_o for a given cable type and length. It is then the desire of the system analyst to provide the equivalent cable loss at a frequency f. The insertion loss of the cable assembly at frequencies lower than the upper limit is expressed by the relation:

$$L = L_0\sqrt{\frac{f}{f_0}} \tag{4.41}$$

where L is the insertion loss in dB at frequency f, L_0 is the insertion loss in dB at frequency f_o, f is the frequency at which the insertion loss will be estimated, and f_o is the frequency at which the insertion loss is specified. Note that this is only true for well-matched terminations. Poor matching, say between the antenna connector and the front of the radio for large cable lengths, for example, on airborne platforms, could create a situation where many more losses are created in the cable than are accounted for in (4.41).

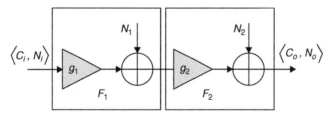

Figure 4.13 Cascaded analog gain blocks

4.3 Cascaded Noise Figure Analysis

Consider the definition of the noise factor and noise figure in (4.8). In this section, we will present a simple derivation of the Friis formula as it applies to system noise figure obtained via cascaded noise and gain analysis. Consider the simplified system made up of two analog gain stages connected in cascade as shown in Figure 4.13. Let the input to the system be carrier plus noise defined as the pair $\langle C_1, N_1 \rangle$ and the output of the system output carrier plus noise defined as the pair $\langle C_o, N_o \rangle$. Let N_1 and N_2 be additive thermal noise due to the two cascaded amplifiers whose gains are g_1 and g_2, respectively. According to (4.8), the noise factor of the first gain block can be expressed as

$$F_1 = \frac{CNR_i}{CNR_{output,1}} = \frac{\dfrac{C_i}{N_i}}{\dfrac{g_1 C_i}{g_1 N_i + N_1}} = \frac{\dfrac{1}{N_i}}{\dfrac{g_1}{g_1 N_i + N_1}}$$

$$= \frac{g_1 N_i + N_1}{g_1 N_i} = 1 + \frac{N_1}{g_1 N_i} \qquad (4.42)$$

Next, define the second noise factor also as the input signal CNR to the output signal CNR based on the second block, that is:

$$F_2 = \frac{CNR_i}{CNR_{output,2}} = \frac{\dfrac{C_i}{N_i}}{\dfrac{g_2 C_i}{g_2 N_i + N_2}} = 1 + \frac{N_2}{g_2 N_i} \qquad (4.43)$$

Next, let's relate the various block noise factors to the total noise factor, of the whole system. The signal at the output of the second block is amplified (or attenuated) by g_1 and g_2. Hence, the output signal itself is $S_o = g_1 g_2 S_i$. The output noise of the second block can be simply derived from Figure 4.13 as $N_o = (g_1 N_1 + N_1)g_2 + N_2$. The system output CNR then becomes $(S_o/N_o) = g_1 g_2 S_i/(g_1 N_i + N_1)g_2 + N_2$, and the noise factor of the two stage system can then be derived simply as

$$
F_T = \frac{CNR_i}{CNR_o} = \frac{\dfrac{S}{N}}{\dfrac{g_1 g_2 S}{(g_1 N_i + N_1)g_2 + N_2}}
$$

$$
= \frac{g_1 g_2 N_i + g_2 N_1 + N_2}{g_1 g_2 N_i} = 1 + \frac{N_1}{g_1 N} + \frac{N_2}{g_1 g_2 N_i} = F_1 + \frac{F_2 - 1}{g_1} \tag{4.44}
$$

In general, the total receiver noise factor may be defined as the Friis cascaded noise figure equation:

$$
F_T = F_1 + \frac{F_2 - 1}{g_1} + \frac{F_3 - 1}{g_1 g_2} + \ldots + \frac{F_L - 1}{g_1 g_2 \cdots g_L} = F_1 + \sum_{l=2}^{L} \left(\frac{F_l - 1}{\prod_{n=1}^{l-1} g_n} \right) \tag{4.45}
$$

and, hence, the total system receive noise figure in dB is given as

$$
NF = 10 \log_{10}(F_T) \tag{4.46}
$$

It is interesting to note from (4.45) that the noise factor of the first block F_1 is the most dominant, followed by the noise factor of the second block F_2, etc. This is especially true for higher gain values. Therefore, when designing a receiver for example, it is desirable that the front end components have the lowest noise figure as compared with the back end components close to the ADCs. At sensitivity, the receiver operates at maximum gain—that is, the various gains $\{g_i : i = 1 : L\}$ are set to maximum values, thus minimizing the noise impact of the later stages. This may be best illustrated via an example.

Example 4-7: Cascaded Noise Figure of LTE System

Consider the LTE direct conversion receiver line-up shown in Table 4.5. The individual gain stages are set to receive a 64-QAM signal in Band 4 at −78.5 dBm. The channel bandwidth in this case is 5 MHz and the required CNR is 20 dB. Assume that further degradations to the signal caused by phase noise, IQ-imbalance, etc. account for an additional 2-dB hit on the CNR. Can the current line-up meet the desired sensitivity requirement? If the sensitivity cannot be met with this line-up, what changes can be made to the LNA in terms of gain and noise figure in order to meet the requirement? Assume that the noise density is −174 dBm/Hz.

Table 4.5 Gain and noise-figure line-up of an LTE transceiver

Cumulative	Parameter	Band Switch	Duplexer	LNA 21	Mixer	Roofing/ Chebyshev Filter	VGA
9.481495027	NF (dB)	1.47	1.8	2.4	6.80	22.42844134	19.35341486
57.63	Gain (dB)	−1.47	−1.8	16	2.70	32.2	10

Using the Friis formula described in (4.45) and (4.46), the cascaded system noise figure of the system can be estimated as shown in Table 4.6 to be 9.48 dB.

Using the sensitivity equation shown in (4.7), we can derive P_{dBm} as

$$P_{dBm} = -174_{dBm/Hz} + NF_{dB} + 10\log_{10}(B) + CNR_{dB} + L_d$$
$$= -174_{dbm/Hz} + 9.48 + 10\log_{10}(5e6) + 20 + 2 = -75.53\,dBm \qquad (4.47)$$

where L_d is the loss due to degradation specified as 2 dB. In order to meet the sensitivity of −78.5 dBm, we require an additional 2.97 dB of improvement to the noise figure. However, lowering the LNA noise figure from 2.4 to 1.3 dB and increasing the gain from 16 to 21.1 dB results in the line-up shown in Table 4.7.

The resulting sensitivity due to this line-up meets the requirements at the expense of more current consumption, cost, and size of LNA.

Table 4.6 Cascaded noise figure and gain analysis of LTE receiver

Cumulative	Parameter	Band Switch	Duplexer	LNA 21	Mixer	Roofing/Chebyshev Filter	VGA
9.481495027	NF (dB)	1.47	1.8	2.4	6.80	22.42844134	19.35341486
57.63	Gain (dB)	−1.47	−1.8	16	2.70	32.2	10
	inband gain (lin)	1.47	3.27	5.67	5.901407634	9.480775673	9.481495027
		0.71285303	0.660693448	39.81071706	1.862087137	1659.586907	10
	Aggregate gain (lin)	1	0.71285303	0.470977326	18.74994508	34.91403155	57,942.86964
	cum NF (lin)	1.402813705	2.123244462	3.689775986	3.891712626	8.87314477	8.874614616
	NF (lin)	1.402813705	1.513561248	1.737800829	4.786300923	174.921879	86.16710176

Table 4.7 Improved line-up of LTE receiver as compared with the one in Table 4.6

Cumulative	Parameter	Band Switch	Duplexer	LNA 21	Mixer	Roofing/Chebyshev Filter	VGA
6.499621153	NF (dB)	1.47	1.8	1.3	6.80	22.42844134	19.35341486
62.73	Gain (dB)	−1.47	−1.8	21.1	2.70	32.2	10
		1.47	3.27	4.57	4.663607502	6.499179464	6.499621153
	inband gain (lin)	0.71285303	0.660693448	128.8249552	1.862087137	1659.586907	10
	Aggregate gain (lin)	1	0.71285303	0.470977326	60.67363296	112.9795915	187,499.4508
	cum NF(lin)	1.402813705	2.123244462	2.86417797	2.926582358	4.465992058	4.46446284
	NF (lin)	1.402813705	1.513561248	1.348962883	4.786300923	174.921879	86.16710176

References

[1] Eyermann P, Powel M. Maturing the software communications architecture for JTRS. IEEE Military Communications Conference (MILCOM) 2001;1:158–62.

[2] Joint Tactical Radio System (JTRS) Joint Program Office (JPO), Software communications architecture specification V3.0, JTRS-5000SCA V3.0, August 2004.

[3] Joint Tactical Radio System (JTRS) Joint Program Office (JPO), Security supplement to the software communications architectureV2.2.1, April 2004.

[4] Joint Tactical Radio System (JTRS) Joint Program Office (JPO), Security supplement to the software communications architectureV3.0, August 2004.

[5] Kurdziel M, Beane J, Fitton J. An SCA security supplement compliant radio architecture. IEEE Military Communication Conference (MILCOM) Oct. 2005;4:2244–50.

[6] Pärssinen A. System design for multi-standard radio. Tutorial on Wireless CMOS Transceivers, IEEE International Solid-State Circuits Conference (ISSCC), Feb. 2006.

[7] Rappaport T. Wireless Communications. Upper Saddle River, NJ: Prentice Hall; 1996.

[8] Chu T-S, Greenstein LJ. A quantification of link budget differences between cellular and PCS bands. IEEE Transactions on Vehicular Technology January 1999;48(1):60–5.

[9] Ragan G. Elementary line theory. Microwave Transmission Circuits, pp. 35, vol. 9, Radiation Laboratory Series, McGraw-Hill, New York, NY, 1948.

CHAPTER 5

Memoryless Nonlinearity and Distortion

Nonlinear circuits are considered to be either memoryless or with memory. In memoryless circuits, the output of the circuit at time t depends only on the instantaneous input values at time t and not on any of the past values of its input. The output of such a circuit can be expressed in the form of a power series:

$$y(t) = \sum_{n=0}^{\infty} \beta_n x^n(t) = \beta_0 + \beta_1 x(t) + \beta_2 x^2(t) + \beta_3 x^3(t) + \cdots \qquad (5.1)$$

The output of a nonlinear circuit with memory at time t, on the other hand, depends on the input signal at time t as well as inputs that occurred in previous instances before t. The largest time delay on which the output depends determines the memory of the circuit. In the event where the output signal has infinite memory, that is a system with infinite nonlinear impulse response, the output signal may be best represented using integrals and differentials of the input signal. Generally speaking, a nonlinear circuit with memory with reactive elements can be modeled using a Volterra series, which is a power series with memory. In this chapter, the system requirements addressed can be sufficiently modeled with memoryless polynomials.

This chapter is divided into nine sections. In Section 5.1, the 1-dB compression point in the context of memoryless circuits is introduced. Section 5.2 discusses signal degradation due to desensitization and blocking, whereas Section 5.3 introduces intermodulation distortion and analysis. Cross modulation distortion is presented in 5.5 and harmonics generated by nonlinear circuits are discussed in Section 5.6. Section 5.7 discusses the phase noise implication for the desired signal. Reciprocal mixing as it is associated with phase noise is also presented. Section 5.8 briefly mentions spurious signals and Section 5.9 contains the appendices.

5.1 1-dB Compression Point Due to Memoryless Nonlinearities

Let $x(t)$ be, as defined in 5.9.1, an input signal to a memoryless nonlinear device consisting of two tones at two distinct frequencies:

$$x(t) = \alpha_1 \cos(\omega_1 t) + \alpha_2 \cos(\omega_2 t) \tag{5.2}$$

and let $y(t)$ be the output of a nonlinear device, limited for simplicity to third order nonlinearity, in response to $x(t)$:

$$y(t) = \beta_1 x(t) + \beta_2 x^2(t) + \beta_3 x^3(t) \tag{5.3}$$

The polynomial in (5.3) is expanded as shown in Section 5.9.1. The resulting signal is comprised of components summarized in Table 5.1.

For small signal gain, the gain of a nonlinear memoryless device is mainly linear and depends on β_1 provided that the harmonics are negligible. However, as the input signal level increases, the compressive or saturating nature of the circuit does not allow the gain of the output signal to grow accordingly. This is true since β_3 is generally negative. In other words, the output power will reach a point at which the input-output power relationship is no longer linear. At this point, the amplifier can no longer supply the necessary output power to the node. The output power gently drops off from the linear output power curve. The point at which the linear output power and the output power of the amplifier or nonlinear device differ by 1 dB is known as the 1-dB compression point (1-dB-CP), depicted in Figure 5.1.

For a single tone input, that is, for $\alpha_2 = 0$ in (5.2), the signal plus distortion to signal ratio is:

$$\rho = 20 \log_{10} \left(\frac{\beta_1 \alpha_1 + \dfrac{3}{4} \beta_3 \alpha_1^3}{\beta_1 \alpha_1} \right) = 20 \log_{10} \left(1 + \frac{3}{4} \left| \frac{\beta_3}{\beta_1} \right| \alpha_1^2 \right) \tag{5.4}$$

Table 5.1 Harmonics, IMs, and their respective
amplitudes due to 2nd and 3rd order nonlinear system

Frequency	Amplitude
DC	$\frac{1}{2}\beta_2\left(\alpha_1^2 + \alpha_2^2\right)$
ω_1	$\beta_1\alpha_1 + \frac{3}{2}\beta_3\left(\frac{1}{2}\alpha_1^3 + \alpha_1\alpha_2^2\right)$
ω_2	$\beta_1\alpha_2 + \frac{3}{2}\beta_3\left(\frac{1}{2}\alpha_2^3 + \alpha_1^2\alpha_2\right)$
$2\omega_{i=1,2}$	$\frac{1}{2}\beta_2\alpha_i^2$ for $i = 1,2$
$3\omega_{i=1,2}$	$\frac{1}{4}\beta_3\alpha_i^3$ for $i = 1,2$
$\omega_1 \pm \omega_2$	$\beta_2\alpha_1\alpha_2$
$2\omega_1 \pm \omega_2$	$\frac{3}{4}\beta_3\alpha_1^2\alpha_2$
$2\omega_2 \pm \omega_1$	$\frac{3}{4}\beta_3\alpha_1\alpha_2^2$

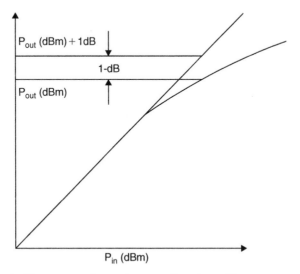

Figure 5.1 The 1-dB compression point as a function of input versus output power

For compressive nonlinearity, the relationship in (5.2) can be expressed as:

$$\rho = 20 \log_{10} \left(1 - \frac{3}{4} \left| \frac{\beta_3}{\beta_1} \right| \alpha_1^2 \right) \tag{5.5}$$

where in this case we assume that $\beta_1 = |\beta_1|$ and $\beta_3 = -|\beta_3|$. To compute the 1-dB-CP, the expression in (5.5) can be rewritten as:

$$\rho = 20 \log_1 \left(1 - \frac{3}{4} \left| \frac{\beta_2}{\beta_1} \right| \alpha_1^2 \right) = -1 \text{ dB} \Rightarrow$$

$$\alpha_{1dBCP} \cong \alpha_1 = \sqrt{\frac{4}{3} \frac{\beta_1}{\beta_3} (1 - 10^{-1/20})} = \sqrt{0.145 \left| \frac{\beta_1}{\beta_3} \right|} \tag{5.6}$$

Define the RMS input power with respect to a source resistor R_s as:

$$P_{ICP,dB} = 10 \log_{10} \left(\frac{1}{2} \frac{\alpha^2}{R_s} \right) \text{dBm} \tag{5.7}$$

Then the input power at which the 1-dB input compression point (ICP) occurs for a single tone input is given as:

$$P_{ICP,dB} = 10 \log_{10} \left(\frac{2}{3} \left| \frac{\beta_1}{\beta_3} \right| \frac{(1 - 10^{-1/20})}{R_s} \right) \text{dBm} \tag{5.8}$$

Extending the above analysis into the two-tone case, assume that the desired signal is at ω_1, then the in-band signal plus distortion to signal ratio is defined as:

$$\rho = 20 \log_{10} \left(\frac{\beta_1 \alpha_1 + \frac{3}{2} \beta_3 \left(\frac{1}{2} \alpha_1^3 + \alpha_1 \alpha_2^2 \right)}{\beta_1 \alpha_1} \right)$$

$$= 20 \log_{10} \left(1 - \frac{3}{2} \left| \frac{\beta_3}{\beta_1} \right| \left(\frac{1}{2} \alpha_1^2 + \alpha_2^2 \right) \right) \tag{5.9}$$

Note that some of the distortion due to third order nonlinearity occurs exactly in band of the desired signal or on top of the desired tone, as can be seen from this discussion.

For simplicity assume that $\alpha_1 = \alpha_2 = \alpha$, that is, the two tones have equal input power and $\rho = -1\,\text{dB}$; we obtain:

$$\alpha^2 = \frac{4}{9}\left|\frac{\beta_1}{\beta_3}(1 - 10^{-1/20})\right| \Rightarrow \alpha = \sqrt{\frac{4}{9}\left|\frac{\beta_1}{\beta_3}(1 - 10^{-1/20})\right|} \qquad (5.10)$$

where α in (5.10) is the 1-dB compression point. Again, let the RMS input power at the 1-dB-CP with respect to a source resistor R_s be as defined in (5.7):

$$P_{ICP,dB} = 10\log_{10}\left(\frac{1}{2}\frac{\alpha^2}{R_s}\right)\,\text{dBm} \qquad (5.11)$$

Then the 1-dB-ICP for a two tone test becomes:

$$P_{ICP,dB} = 10\log_{10}\left(\frac{2}{9}\left|\frac{\beta_1}{\beta_3}\right|\frac{(1 - 10^{-1/20})}{R_s}\right)\,\text{dBm} \qquad (5.12)$$

In the transmitter, the PA normally operates below the 1-dB compression point to avoid spectral regrowth and signal distortion. The average power level at which the PA operates on the desired signal without distortion is defined with reference to the 1-dB-ICP minus a certain back-off.

An interesting comparison between the 1-dB compression point of the single tone versus the two tone case will aid the reader in understanding the concept of desensitization and blocking discussed in the next section. Comparing the relations in (5.4) and (5.9) at the 1-dB compression point:

$$\rho = 20\log_{10}\left(1 - \frac{3}{4}\left|\frac{\beta_3}{\beta_1}\right|\alpha_{1dBCP}^2\right) = 20\log_{10}\left(1 - \frac{3}{2}\left|\frac{\beta_3}{\beta_1}\right|\left(\frac{1}{2}\alpha_1^2 + \alpha_2^2\right)\right) = -1\,\text{dB} \quad (5.13)$$

we obtain:

$$\frac{3}{4}\left|\frac{\beta_3}{\beta_1}\right|\alpha_{1dBCP}^2 = \frac{3}{2}\left|\frac{\beta_3}{\beta_1}\right|\left(\frac{1}{2}\alpha_1^2 + \alpha_2^2\right) \Rightarrow \alpha_1^2 = \alpha_{1dBCP}^2 - 2\alpha_2^2 \qquad (5.14)$$

As defined earlier, let $P_1 = \dfrac{1}{2}\dfrac{\alpha_1^2}{R_s}$, $P_2 = \dfrac{1}{2}\dfrac{\alpha_2^2}{R_s}$, and $P_{1dBCP} = \dfrac{1}{2}\dfrac{\alpha_{1dBCP}^2}{R_s}$, and then the relation in (5.14) implies the 1-dB compression point of the desired signal in the two-tone case occurs at lower power level than the single tone case:

$$P_1 = P_{1dBCP} - 2P_2 \tag{5.15}$$

As a matter of fact, as the power of the interferer increases, the power of the desired signal decreases by twice the amount of the interferer power, eventually leading the desired signal to be completely blocked as will be discussed in the next section.

5.2 Signal Desensitization and Blocking

When a weak desired signal is present along with a strong dominant interferer, the latter, due to the compressive behavior of the circuit, tends to reduce its overall average gain, thus serving to further weaken the desired signal. This process is known as *desensitization*. To illustrate this effect, consider the two-tone scenario described above where the desired tone at ω_1 and the interfering tone is at ω_2. The output at the desired frequency can be written as:

$$\hat{y}(t) = \left\{ \beta_1\alpha_1 + \frac{3}{2}\beta_3\left(\frac{1}{2}\alpha_1^3 + \alpha_1\alpha_2^2 \right) \right\} \cos(\omega_1 t) \tag{5.16}$$

As the interfering tone level gets larger in comparison to the desired tone level, the output at the desired signal becomes:

$$\hat{y}(t)\Big|_{\alpha_2 \gg \alpha_1} = \left\{ \beta_1\alpha_1 + \frac{3}{2}\beta_3\alpha_1\left(\frac{\alpha_1^2}{2} + \alpha_2^2 \right) \right\} \cos(\omega_1 t) = \underbrace{\left\{ \beta_1\alpha_1 + \frac{3}{2}\beta_3\alpha_1\alpha_2^2 \right\}}_{\text{Gain}} \cos(\omega_1 t)$$

$$\tag{5.17}$$

As discussed earlier, for compressive behavior where $\beta_3 < 0$, the gain expression in (5.17) diminishes as the magnitude of the interfering tone α_2 increases. As the desired signal gain approaches zero, the signal is said to be blocked, and the interferer is often referred to as a blocker. Signal desensitization will be further addressed in Section 5.5.

5.3 Intermodulation Distortion

Intermodulation distortion generally specifies the level of in-band nonharmonic distortion due to second and third order nonlinearity. The intermodulation (IM) product is a spur typically specified in response to two strong narrowband signals or tones at the input of the receiver (or nonlinear device). IM products due to second and third order distortion are the most common types encountered. In practical RF systems, intermodulation could result from two strong interferers such as 802.11a signals producing a significant IM product in band of a desired UWB signal. One method of alleviating the problem is proper filtering before the LNA so the offending signals are sufficiently attenuated. However, in most cases, filtering has to be accompanied by designing nonlinear components with sufficient linearity to further minimize distortion.

Second order nonlinearity is particularly important in direct conversion architecture where signals are mixed down from RF to baseband and vice versa via a single mixing stage. Given a nonlinear system, as illustrated in Figure 5.2, a second order nonlinearity due to a two-tone input could manifest itself as an IM product at the output. This IM product could fall into the desired signal band. For example, transistor mismatches in a single balanced mixer coupled with the deviation of the LO duty cycle from 50% produces asymmetry in the circuitry, allowing for certain signals before the mixer to feed through without mixing into the baseband, as shown in Figure 5.3 [1].

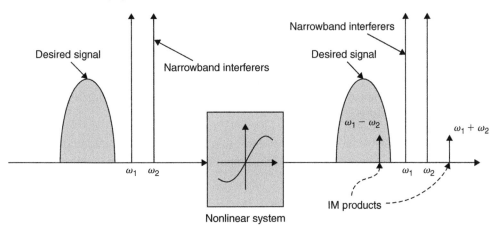

Figure 5.2 Intermodulation distortion due to second order nonlinearity

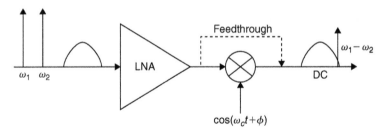

Figure 5.3 Effect of mixer feedthrough on the received desired baseband signal

This phenomenon manifests itself as follows. Suppose a received signal exhibits a certain amount of AM variation due to either transmit filtering or channel variations under fading conditions. Then the received signal can be conceptually thought of as:

$$x(t) = \{\Gamma + \gamma\cos(\omega_m t)\}\{A\cos(\omega_c t) - B\sin(\omega_c t)\} \tag{5.18}$$

where Γ is a slow-varying signal assumed to be constant over an extended period of time, and $\gamma\cos(\omega_m t)$ represents a low frequency AM signal as stated in [1]. After the received signal undergoes second order nonlinearity, then:

$$x^2(t) = \{\Gamma + \gamma\cos(\omega_m t)\}^2 \{A\cos(\omega_c t) - B\sin(\omega_c t)\}^2$$

$$x^2(t) = \left\{\Gamma^2 + \frac{\gamma^2}{2} + 2\Gamma\gamma\cos(\omega_m t) + \frac{\gamma^2}{2}\cos(2\omega_m t)\right\}$$

$$\times \left\{\frac{A^2 + B^2}{2} + \frac{A^2}{2}\cos(2\omega_c t) - \frac{B^2}{2}\cos(2\omega_c t) - AB\sin(2\omega_c t)\right\} \tag{5.19}$$

The term of interest in (5.19) is the product $(A^2 + B^2)\Gamma\gamma\cos(\omega_m t)$ which, when fed through the mixer, appears at baseband. This demodulation of AM components serves to further corrupt the received signal.

Second order nonlinearity can be characterized using the second order intercept point. Define the input RMS power of a single tone centered at ω_i to a nonlinear system or device as:

$$P_i = \frac{1}{2}\frac{\alpha_i^2}{R_s}, \quad i = 1, 2 \tag{5.20}$$

where R_s is the source resistor. The RMS output power of the nonlinear system or device at the fundamental frequency is:

$$P_o = \frac{1}{2} \frac{\beta_1^2 a_i^2}{R_L}, \quad i = 1, 2 \tag{5.21}$$

where R_L is the load resistor. Let the distortion power of the IM product delivered to a load resistor R_L be:

$$P_d = \frac{(\beta_2 \alpha_1 \alpha_2)^2}{2R_L} \tag{5.22}$$

The IM signal in question is due to the $|\omega_1 - \omega_2|$ product, which is assumed to fall in band as shown in Figure 5.3. For simplicity's sake, assume that the input and output powers to be normalized to their respective source and load resistors. Furthermore, assume that the two tones have equal power at the input of the nonlinear device, that is $P_1 = P_2 = P$ or simply $\alpha = \alpha_1 = \alpha_2$. Define the signal amplitude due to the fundamental frequency as:

$$\Psi_{\omega_{1,2}} = |\beta_1| \alpha \tag{5.23}$$

and the signal amplitude due to the IM product as:

$$\Psi_{\omega_1 - \omega_2} = |\beta_2| \alpha_1 \alpha_2 = |\beta_2| \alpha^2 \tag{5.24}$$

The absolute values are used to simplify the mathematical assumptions. Define the second order input-referred intercept point (IIP2) as the point for which the IM product magnitude equals that of the fundamental signal magnitude, as shown in Figure 5.4. That is,

$$\Psi_{\omega_{1,2}} = \Psi_{\omega_1 - \omega_2} \Rightarrow |\beta_1| \alpha = |\beta_2| \alpha^2 \Rightarrow IIP2 \equiv \alpha = \frac{|\beta_1|}{|\beta_2|} \tag{5.25}$$

Consider the ratio:

$$\frac{\Psi_{\omega_{1,2}}}{\Psi_{\omega_1 - \omega_2}} = \frac{|\beta_1| \alpha}{|\beta_2| \alpha^2} = \frac{|\beta_1|}{|\beta_2|} \frac{1}{\alpha} = \frac{IIP2}{\alpha} \tag{5.26}$$

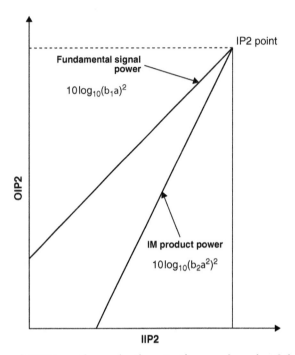

Figure 5.4 IIP2 and OIP2 are the projections to the x and y axis of the second order intercept point for which the IM product amplitude equals that of the fundamental output signal amplitude

Then the input-referred IM product—that is, the value of the IM product referred to the input of the system or the device—is:

$$IM = \frac{\Psi_{\omega_1 - \omega_2}}{|\beta_1|} = \frac{\alpha^2}{IIP2} \qquad (5.27)$$

Expressing the relation in (5.27) in terms of power we obtain:

$$P_{IM} = \frac{1}{2}(IM)^2 = \frac{1}{2}\frac{\alpha^4}{(IIP2)^2} = \frac{4p_i^2}{2(IIP2)^2} = \frac{p_i^2}{P_{IIP2}} \qquad (5.28)$$

Expressing (5.28) in dBm, we obtain the well-known relation:

$$P_{IM,dB} = 2P_{i,dBm} - P_{IIP2,dBm} \qquad (5.29)$$

Observe that if input tone power is given as a certain Δ_{dB} in dB below $P_{IIP2,dBm}$, say $P_{i,dBm} = P_{IIP2,dBm} - \Delta_{dB}$, then:

$$P_{IM,dB} = 2P_{i,dBm} - P_{IIP2,dBm} = 2(P_{IIP2,dBM} - \Delta_{dB}) - P_{IIP2,dBm} = P_{IIP2,dBm} - 2\Delta_{dB} \tag{5.30}$$

That is, the input referred IM power is $2\Delta_{dB}$ below $P_{IIP2,dBm}$.

In the event where the two tones have different powers, it can be easily shown that (5.29) can be expressed in terms of the two-tone powers as:

$$P_{IM,dB} = P_{i,1,dBm} + P_{i,2,dBm} - P_{IIP2,dBm}$$
$P_{i,1,dBm}$ is the input power of the first tone
$P_{i,2,dBm}$ is the input power of the second tone
$$\tag{5.31}$$

An alternative way of arriving at (5.29) is as follows. The linear power ratio of the output IM power to the output power of the fundamental is given as:

$$\frac{P_{OIM}}{P_o} = \frac{0.5\beta_2^2\alpha^4}{0.5\beta_1^2\alpha^2} = \frac{\alpha^2}{(IIP2)^2} = \frac{P_i}{P_{IIP2}} \Rightarrow P_{IM} = \frac{P_{OIM}}{\beta_1^2} = \frac{P_i^2}{P_{IIP2}} \tag{5.32}$$

This is true since $P_o = 0.5\beta_1^2\alpha^2 = \beta_1^2 P_i$. Expressing the relation (5.32) in dBm, we obtain the relation in (5.29).

When designing a system, it is important to keep in mind that IIP2 is a fictitious point that serves to only specify second order nonlinearity. A system output power would long compress before the input power reaches the IIP2$_{dBm}$ power.

As mentioned earlier, second order distortion is problematic in direct conversion receivers. DC and low frequency product at the LNA are usually filtered out via AC coupling and bandpass filtering. However, the key contributors due to second order distortion in such architecture are the RF to baseband I/Q downconverter as well as the baseband gain stages. Note that in direct conversion architecture most of the gain is in the baseband after downconversion. Such a transceiver usually requires a high system IIP2.

Example 5-1: System *IIP2* of MBOA UWB

The system NF of a MBOA UWB device is 6.5 dB. The occupied signal bandwidth is 528 MHz. In order to decode the highest data rate of 480 Mbps, a CNR of approximately 9.0 dB is required. Suppose that there are two interfering 802.11a signals. Each interferer signal power is −29 dBm. Assume that we anticipate that the noise floor will increase by 0.5 dB only. What is the required IIP2?

Recall that the noise floor is defined as:

$$P_{noisefloor} = 10\log_{10}(kTB) + NF = -174 \text{ dBm/Hz} + NF + 10\log_{10}(B)$$
$$P_{noisefloor} = -174 + 9 + 10\log_{10}(528 \times 10^6) = -80.27 \text{ dBm} \tag{5.33}$$

The degradation due to the IM will increase the noise floor by 0.5 dB to −79.77 dBm from −80.27 dBm. The IM product power can then be computed as:

$$P_{IM} = 10\log_{10}(10^{-79.77/10} - 10^{-80.27/10}) = -89.41 \text{ dBm} \tag{5.34}$$

Recall that:

$$P_{IM,dB} = 2P_{i,dBm} - IIP2_{dBm} \Rightarrow P_{IIP2,dBm} = 2P_{i,dBm} - P_{IM,dB}$$
$$P_{IIP2,dBm} = 2P_{i,dBm} - P_{IM,dB} = 2 \times (-29) - 89.41 = 31.41 \text{ dBm} \tag{5.35}$$

These results are summarized in Table 5.2.

Table 5.2 Computation of system IIP2

Required IIP2 (dBm)	Value
System NF	6.50
Bandwidth (MHz)	528
Required CNR @ 480 Mbps (dB)	9.00
Noise floor	−80.27366
Increase to noise floor due to IM (dB)	0.5
Degraded noise floor (dB)	−79.77366
IM signal power (dBm)	−89.40941
Interferer signal power (dBm)	−29
Required IIP2 (dBm)	**31.40941**

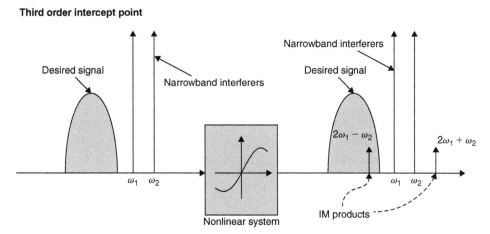

Figure 5.5 Intermodulation distortion due to third order nonlinearity

The third order intercept point is an important performance measure that specifies the amount of distortion inflicted on the desired signal due to the intermodulation product of two interfering signals. This is illustrated in Figure 5.5 where the input to a nonlinear system consists of the desired input signal and two interfering tones. The nonlinearity of the system produces an IM product which could occur in the desired signal band. Consider the interfering signals in (5.2). Assume that the intermodulation product due to $2\omega_1 - \omega_2$ falls within the desired signal bandwidth. Furthermore, assume that the magnitude of the two tones α_1 and α_2 impinging on the receiver is sufficiently small such that terms due to higher order nonlinearities can be ignored. Typically, the effect of third order nonlinearity can be minimized at the input of the system by minimizing the bandwidth of the band select or band definition filter (BDF).[1] However, decreasing the bandwidth of this filter increases its cost, complexity, size, and insertion loss. In designing a system, a trade-off is made between the BDF filter parameters and system IIP3.

Consider the output of a nonlinear system:

$$y(t) = \left\{ \beta_1\alpha_1 + \frac{3}{2}\beta_3\left[\frac{1}{2}\alpha_1^3 + \alpha_1\alpha_2^2\right]\right\}\cos(\omega_1 t)$$
$$+ \frac{3}{4}\beta_3\alpha_1^2\alpha_2\cos\left[(2\omega_1 - \omega_2)t\right] + \dots \qquad (5.36)$$

---

[1] This filter is often referred to as a preselector filter.

Assume that the two interfering tones have equal input powers, that is $\alpha = \alpha_1 = \alpha_2$. Furthermore, assume that $|\beta_1| \gg |\beta_3|$; then the relation in (5.36) can be re-expressed as:

$$y(t) = \underbrace{\beta_1 \alpha \cos(\omega_1 t)}_{\text{Linear Term}} + \frac{9}{4} \beta_3 \alpha^3 \cos(\omega_1 t) + \frac{3}{4} \beta_3 \alpha^3 \cos\left[(2\omega_1 - \omega_2)t\right] + \dots \qquad (5.37)$$

Define the input IP3 (IIP3) point as the magnitude for which the linear amplitude of $\cos(\omega_1 t)$ is equal to that of the intermodulation product term associated with $\cos\left[(2\omega_1 - \omega_2)t\right]$. Mathematically, this can be expressed as:

$$|\beta_1| \alpha = \frac{3}{4} |\beta_3| \alpha^3 \qquad (5.38)$$

which implies that the IIP3 point is:

$$IIP3 \equiv \alpha|_{\alpha = IP3} = \sqrt{\frac{4}{3} \frac{|\beta_1|}{|\beta_3|}} \qquad (5.39)$$

This phenomenon is illustrated in Figure 5.6 where the projection of the intercept point onto the x-axis is known as the IIP3 point, whereas the projection of the same point onto the y-axis is the OIP3 point. Therefore, as the amplitude of the input tones α increases, the amplitude of the IM product increases in a cubic proportion, respectively. On a logarithmic scale, as shown in Figure 5.6, the IM product magnitude increases three times faster than the amplitude due to the fundamental frequency. Again, as was expressed earlier concerning IIP2, IIP3 is a fictitious point used in specifying third order nonlinearity. The system output power will long compress before the input power reaches the $IIP3_{dBm}$ power. Nonetheless, IIP3 and OIP3 are excellent specification measures by which one can judge and compare system and device nonlinearities.

To specify IIP3 and OIP3, consider the ratio:

$$\frac{\Psi_{\omega_{1,2}}}{\Psi_{2\omega_1 - \omega_2}} = \frac{|\beta_1| \alpha}{\frac{3}{4} |\beta_3| \alpha^3} = \frac{(IIP3)^2}{\alpha^2}$$

$$\Psi_{2\omega_1 - \omega_2} = \frac{3}{4} |\beta_3| \alpha^3 \qquad (5.40)$$

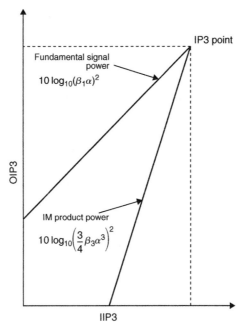

Figure 5.6 IIP3 and OIP3 are the projections to the x and y axis of the third order intercept point for which the IM product amplitude equals that of the fundamental output signal amplitude

where:

$\Psi_{\omega_{1,2}} = |\beta_1|\alpha$ is the linear amplitude of the fundamental at the output of the system, and

$\Psi_{2\omega_1 - \omega_2} = \frac{3}{4}|\beta_3|\alpha^3$ is the amplitude of the IM product. Further manipulations of (5.40)

show that:

$$\Psi_{2\omega_1 - \omega_2} = \frac{|\beta_1|\alpha^3}{(IIP3)^2} \tag{5.41}$$

Referring the IM product to the input, the relation in (5.41) becomes:

$$IM = \frac{\Psi_{2\omega_1 - \omega_2}}{|\beta_1|} = \frac{\alpha^3}{(IIP3)^2} \tag{5.42}$$

Re-expressing the result in (5.42) in dBm, we obtain the well-known relation:

$$P_{IM,dB} = 3P_{in,dBm} - 2P_{IIP3,dBm} \tag{5.43}$$

In the event where the two tones are not of equal power, the IM product power at the IM frequency $f_{IM} = \pm 2f_1 \pm f_2$, where f_1 and f_2 are the frequencies of the first and second tones, can be rewritten as:

$$P_{IM,dB} = 2P_{in,1,dBm} + P_{in,2,dBm} - 2P_{IIP3,dBm}$$

$P_{in,1,dBm}$ is the input power of the first tone

$P_{in,2,dBm}$ is the input power of the second tone

$$(5.44)$$

Finally, it is interesting to compare IIP3 to the 1-dB-CP:

$$20\log_{10}\left(\frac{\alpha_{1dBCP}}{IIP3}\right) = 20\log_{10}\left(\frac{\sqrt{0.145\left|\dfrac{\beta_1}{\beta_3}\right|}}{\sqrt{\dfrac{4}{3}\dfrac{|\beta_1|}{|\beta_3|}}}\right) = -9.6357 \text{ dB} \qquad (5.45)$$

The relation presented in (5.45) is true for memoryless systems only.

Example 5-2: System IIP3 of MBOA UWB

Assume that the band definition filter required at the front end of an MBOA UWB system has a minimum rejection of 20 dB in the UNII[2] band. Furthermore, assume that two 802.11a interfering signals impinge on the receiver with the average power of −16.0103 dBm each. Let the UWB system operate in FFI mode[3] with center frequency at 4488 MHz. The transmit signal power is −41 dBm/ MHz. In the 528-MHz channel bandwidth, there are 122 bins that carry information. Each bin is 4.125 MHz wide. For this design, we require the IM product power to be 6 dB less than the received desired signal power. What is the required IIP3 of the system? To simplify the analysis, assume that the signals undergo free space path loss. The distance between the transmitter and receiver of the MBOA signal is 1.2 m. The antenna gains at both transmitter and receiver are 0 dBi.

The transmit signal power in FFI mode is given as −41 dBm/MHz. Given that the number of carriers (or bins) that contain energy in an MBOA OFDM signal is 122, with each bin having a bandwidth of 4.125 MHz, the total transmit power in this mode is:

$$P_{tx} = -41 \text{ dBm/MHz} + 10\log_{10}(122 \text{ bins} \times 4.125 \text{ MHz/bin}) \approx -14 \text{ dBm} \qquad (5.46)$$

At the carrier frequency of 4488 MHz, the wavelength is:

$$\lambda = \frac{c}{f_c} = \frac{299,800,000}{4488e6} = 0.0668004 \text{ mete} \qquad (5.47)$$

The free space path loss is given as:

$$PL = 10\log_{10}\left(\frac{4\pi d}{\lambda}\right)^2 = 10\log_{10}\left(\frac{4\pi \times 1.2}{0.0668004}\right)^2 \approx 47.07 \text{ dB} \qquad (5.48)$$

Hence, the received signal power:

$$P_{Rx} = P_{tx} - PL = -14 \text{ dBm} - 47.07 \text{ dB} \approx -61.05 \text{ dBm} \qquad (5.49)$$

The received interferer power at the output of the BDF is:

$$P_i = -16.01 \text{ dBm} - 20 \text{ dB} = -36.01 \text{ dBm} \qquad (5.50)$$

Since the requirement is to design for a system IIP3 which produces an IM product of 6 dB less than the received signal power at the given above parameters, the IIP3 is given as:

$$P_{IIP3,\text{dBm}} = \frac{3P_i - P_{IM}}{2} = \frac{3 \times (-36.0103) - (-61.0544 - 6)}{2} = -20.4883 \text{ dBm} \qquad (5.51)$$

The results of this exercise are summarized in Table 5.3.

Table 5.3 Computation of IIP3 based on interference power and IM product power

IIP3 (out of band)	Value
Rx 802.11 interference at input of BDF (dBm)	−16.0103
BDF rejection	20
Rx interference power after BDF (dBm)	−36.0103
Ptx FFI UWB (dBm/MHz)	−41
Single bin bandwidth (MHz)	4.125
Number of occupied bins	122
Occupied signal bandwidth (MHz)	503.25
Transmit signal power in FFI mode (dBm)	−13.98216
Carrier frequency UWB (MHz)	4488
Speed of light	299,800,000
Lambda UWB	0.0668004
Transmitter to receiver distance (m)	1.2
Path loss (dB)	47.072247
Rx desired signal power (dBm)	−61.05441
Desired IM signal power (dBm)	−67.05441
Required IIP3 (dBm)	−20.48825

Example 5-3: System IIP3 for CDMA IS95 Handset

The sensitivity level of a CDMA IS95 handset is specified in IS98 to be −104 dBm with a minimum acceptable CNR of −1 dB. Given the receiver noise figure to be 6.8 dB, what is the receiver IIP3 such that the third order IM degrades (increases) the receiver noise floor by 3.5 dB? Recall that the bandwidth of the receiver is 1.23 MHz. The received interferer power is −43 dBm.

The increased receiver noise floor referred to the input of the system is defined as:

$$P_{degraded\ noise\ floor} = P_{noisefloor} + 3.5 = \underbrace{-174\,dBm/Hz + 10\log_{10}(B) + NF}_{Receiver\ noise\ floor} + 3.5$$

$$= -174\,dBm/Hz + NF + 10\log_{10}(B) = -103.3\,dBm \tag{5.52}$$

The noise power due to the IM product is given as:

$$P_{IM} = 10\log_{10}(10^{P_{degraded\ noise\ floor}/10} - 10^{P_{noise\ floor}/10})$$
$$= 10\log_{10}(10^{(-174+NF+10\log_{10}(B)+3.5)/10} - 10^{(-174+NF+10\log_{10}(B))/10})$$
$$= -106.32\,dBm \tag{5.53}$$

The IIP3 of the system is then given as:

$$P_{IIP3,dBm} = \frac{3P_i - P_{IM}}{2} = \frac{3 \times (-43) - (-106.32)}{2} = -11.34\,dBm \tag{5.54}$$

5.4 Cascaded Input-Referred Intercept Points

As with noise, the systems IIP2 and IIP3 depend largely on the receiver line-up. That is, the overall linearity of the system, and hence its intermodulation characteristics, is determined by the various blocks' linearity and gain as well as their distribution (location) in the receiver line-up.

Consider a receiver line-up made-up of N-blocks. The cascaded input-referred intercept point has the general form:

$$\left(\frac{1}{IIPx_{system}}\right)^{\frac{1}{2}(x-1)} = \left(\frac{1}{IIPx_1}\right)^{\frac{1}{2}(x-1)} + \sum_{n=2}^{N}\left(\frac{\prod_{i=1}^{n-1}g_i}{IIPx_n}\right)^{\frac{1}{2}(x-1)} \tag{5.55}$$

where x designated the order of the linearity—that is, for $x = 2$, it is the various block and system *IIP2* that we are referring to. For a second order system, the relation in (5.55) may be expressed as:

$$IIP2_{system} = \left(\frac{1}{\left(\frac{1}{IIP2_1}\right)^{\frac{1}{2}} + \left(\frac{g_1}{IIP2_2}\right)^{\frac{1}{2}} + \cdots + \left(\frac{g_1 g_2 \cdots g_{N-1}}{IIP2_N}\right)^{\frac{1}{2}}} \right)^2 \qquad (5.56)$$

Likewise, the third order system intercept point may be expressed in terms of the various gains and block IIP3s in the system:

$$IIP3_{system} = \frac{1}{\frac{1}{IIP3_1} + \frac{g_1}{IIP3_2} + \cdots + \frac{g_1 g_2 \cdots g_{N-1}}{IIP3_N}} \qquad (5.57)$$

In certain architectures, such as IF-sampling and super heterodyne, the third order intermodulation product due to the IF amplifier can be reduced by increasing the selectivity of the first IF filter, for example, as shown in Figure 5.7. To account for the selectivity in (5.56) and (5.57), these relations can be modified respectively as:

$$IIP2_{system} = \left(\frac{1}{\left(\frac{1}{IIP2_1}\right)^{\frac{1}{2}} + \left(\frac{g_1}{IIP2_2 S_1^2}\right)^{\frac{1}{2}} + \cdots + \left(\frac{g_1 g_2 \cdots g_{N-1}}{IIP2_N S_1^2 S_2^2 \ldots S_{N-1}^2}\right)^{\frac{1}{2}}} \right) \qquad (5.58)$$

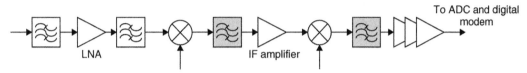

Figure 5.7 Simplified diagram of IF sampling receiver

And,

$$IIP3_{system} = \cfrac{1}{\cfrac{1}{IIP3_1} + \cfrac{g_1}{IIP3_2 \, S_1^{3/2}} + \cdots + \cfrac{g_1 g_2 \cdots g_{N-1}}{IIP3_N \, S_1^{3/2} S_2^{3/2} \cdots S_{N-1}^{3/2}}} \tag{5.59}$$

where $\{S_i, i = 1, \ldots, N-1\}$ designates the selectivity of the filters at the various stages.

5.4.1 Generalization of Intermodulation Distortion

The relationships presented above can be extended to higher order nonlinearities. Consider the Nth order memoryless polynomial:

$$y(t) = \sum_{n=0}^{N} \beta_n x^n(t) \tag{5.60}$$

Let

$$g_n = 10 \log_{10}(\beta_n) \tag{5.61}$$

Then it can be shown that:

$$P_{IMn} = nP_i - (n-1)P_{IIPn} \tag{5.62}$$

and,

$$P_{OIPn} = g_1 + P_{IIPn} = g_n + nP_{IIPn} \tag{5.63}$$

Recall that the relationships expressed above are only applicable for weak signal nonlinearities under memoryless conditions.

5.5 Cross Modulation Distortion

Cross modulation is the amount of AM modulation that transfers from an undesired strong signal to a weak desired signal after passing through a nonlinear system or device. Consider the input to a nonlinear device $x(t)$ comprised of a desired weak unmodulated signal and an undesired strong modulated signal:

$$x(t) = \underbrace{\alpha_1 \cos(\omega_1 t)}_{\substack{\text{unmodulated desired} \\ \text{signal}}} + \underbrace{\alpha_2 \{1 + m(t)\} \cos(\omega_2 t)}_{\text{modulated interferer signal}} \tag{5.64}$$

Passing through a memoryless nonlinear channel as described in (5.3), we obtain:

$$y(t) = \left\{ \underbrace{\beta_1\alpha_1}_{\substack{\text{linear gain} \\ \text{term}}} + \underbrace{\frac{3}{4}\beta_3\alpha_1^3 + \frac{3}{2}\beta_3\alpha_1\alpha_2^2}_{\substack{\text{compression and desensitization} \\ \text{term}}} + \underbrace{\frac{3}{2}\beta_3\alpha_1\alpha_2^2\left(2m(t) + m^2(t)\right)}_{\text{cross modulation term}} \right\} \cos(\omega_1 t) + \ldots$$

(5.65)

The cross modulation term in (5.65) shows that, due to third order nonlinearity or any odd order nonlinearity for that matter, the amplitude modulation on the undesired signal has transferred to the desired signal. Further manipulation of (5.65) reveals:

$$y(t) = \beta_1\alpha_1 \left\{ 1 + \frac{3}{4}\frac{\beta_3}{\beta_1}\alpha_1^2 + \frac{3}{2}\frac{\beta_3}{\beta_1}\alpha_2^2 + \frac{3}{2}\frac{\beta_3}{\beta_1}\alpha_2^2\left(2m(t) + m^2(t)\right) \right\} \cos(\omega_1 t) + \ldots$$

$$= \beta_1\alpha_1 \left\{ 1 + \frac{\alpha_1^2}{(IIP3)^2} + \frac{2\alpha_2^2}{(IIP3)^2} + \frac{2\alpha_2^2}{(IIP3)^2}\left(2m(t) + m^2(t)\right) \right\} \cos(\omega_1 t) + \ldots \quad (5.66)$$

An interesting observation from (5.66) reveals that the higher the IIP3 value, the lower the cross modulation and other interfering terms.

Cross modulation can also degrade the receive channel close to the single tone interferer in a full duplex system such as IS95-CDMA. A transmitted OQPSK signal, for example, exhibits certain AM after filtering. In the event where the handset is far away from the base station, the weak received signal is susceptible to cross modulation distortion [2]. A strong single tone interferer, for example, would amplify the transmit signal leaking through the duplexer and passing through the LNA as illustrated in Figure 5.8.

To illustrate this, consider a full duplex CDMA system with a baseband signal $m(t)$ modulated on a transmit carrier ω_c. The transmit signal path filtered by the duplexer leaks through to the receive path and gets amplified by the LNA. This leakage signal does not constitute a significant degradation to the receiver on its own. However, coupled with a strong narrowband interfering signal, the transmit leakage signal is transferred to the interferer, as will be demonstrated next, thus causing degradation to the receiver sensitivity.

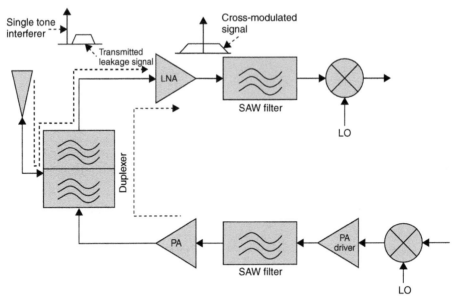

Figure 5.8 Cross-modulation distortion due to a strong single tone (or narrowband) interferer and transmit signal leaking through the duplexer

Consider a strong single tone interferer, or possibly a narrowband AMPS signal, centered at ω_i that is also received and amplified via the LNA. Upon entering the LNA, the leakage and interferer signals can be expressed as:

$$x(t) = \underbrace{\alpha_1 \cos(\omega_i t)}_{\text{single tone interferer}} + \underbrace{\alpha_2 m(t) \cos(\omega_c t)}_{\substack{\text{modulated-transmit} \\ \text{leakage signal}}} \tag{5.67}$$

where $E\{m(t)\} = 0$. At the output of the LNA, and considering only third order nonlinearity due to cross modulation, the amplified signal modulated on the single tone interferer's frequency is given as

$$y_{CM}(t) = \left\{ \underbrace{\beta_1 \alpha_1}_{\substack{\text{linear gain} \\ \text{term}}} + \underbrace{\frac{3}{2} \beta_3 \alpha_1 \alpha_2^2 m^2(t)}_{\substack{\text{cross modulation} \\ \text{term}}} \right\} \cos(\omega_i t) \tag{5.68}$$

The relation in (5.68) reveals the effects of cross modulation. Substituting the relation given in (5.39) into (5.68) we obtain:

$$
\begin{aligned}
y_{CM}(t) &= \beta_1 \left\{ \alpha_1 + \frac{3}{2}\frac{\beta_3}{\beta_1}\alpha_1\alpha_2^2 m^2(t) \right\} \cos(\omega_i t) \\
&= \beta_1\alpha_1 \left\{ 1 + \frac{2}{(IIP3_{LNA})^2}\alpha_2^2 m^2(t) \right\} \cos(\omega_i t)
\end{aligned}
\tag{5.69}
$$

The voltage ratio of the amplitude of the single-tone interferer to the amplitude of the cross modulation product is:

$$
\Delta = \frac{P_{\text{single tone}}}{P_{\text{cross modulation}}} = \frac{(IIP3_{LNA})^2}{2\alpha_2^2 m^2(t)}
\tag{5.70}
$$

Equivalently, the voltage ratio in (5.70) can be expressed in terms of power in dB as:

$$
\begin{aligned}
\Delta_{dB} &= P_{\text{single tone}} - P_{\text{cross modulation}} = 2(IIP3_{LNA})_{dBm} - 2P_{\text{TX leakage}} - 2P_{m,dBm} - 6 \\
P_{\text{cross modulation}} &= P_{\text{single tone}} - \Delta_{dB} = P_{\text{single tone}} + 2(P_{\text{TX leakage}} + P_{m,dBm} + 3) \\
&\quad - 2(IIP3_{LNA})_{dBm}
\end{aligned}
\tag{5.71}
$$

where

- $\Delta_{dB} = 10\log_{10}(\Delta)^2$

- $P_{\text{single tone}} = 10\log_{10}(\rho_{\text{single tone}})^2$

- $P_{\text{cross modulation}} = 10\log_{10}(\rho_{\text{cross modulation}})^2$

- $(IIP3_{LNA})_{dBm}$ is the input-referred intercept point of the LNA

- $P_{\text{TX leakage}}$ is the transmit leakage power at the input of the LNA after passing through the duplexer, and

- $P_{m,dBm}$ is the power of the variance of the baseband signal in dBm. It defines the power variations in the baseband signal due to AM.

At this point the following observations can be made:

- After passing through the LNA, the transmitted signal leaking through the duplexer has been squared and modulated on the interferer's in-band carrier frequency. Recall that squaring a signal in the time domain is equivalent to performing a convolution in the frequency domain, that is $m^2(t) \Leftrightarrow m(f) * m(f)$. The resulting squared signal's bandwidth $m^2(t)$ is twice as wide as that of the original signal $m(t)$.

- The amplitude of the single tone interferer has also served to amplify the baseband signal $m(t)$.

- The higher the IIP3 the lower the cross modulation term.

Depending on the interferer's frequency, the cross modulation signal may or may not overlap with the desired signal since the LNA's bandwidth is typically much wider than the desired signal's bandwidth. In this case, in order to properly estimate the degradation caused by cross modulation on the desired signal, the cross modulation product term in (5.71) must be adjusted by the ratio of the bandwidth affected. This phenomenon is depicted in Figure 5.9. The amount of overlap is

$$f_i + B - (f_{crx} - B/2) = (f_i - f_{crx}) + 3B/2$$

where f_i is the single-tone interferer frequency and f_{crx} is the center frequency of the received signal.

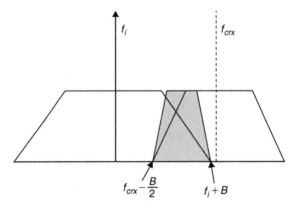

Figure 5.9 The percentage of degraded spectrum of the desired receive signal is a function of the receive carrier frequency and the single tone interferer frequency

From the above discussion, an approximate adjustment to (5.71) is

$f_i + B - (f_{crx} - B/2) = (f_i - f_{crx}) + 3B/2$. In general, the adjustment can be expressed as:

$$A_{dB} = 10 \log_{10} \left[\frac{-|f_i - f_{crx}| + 3B/2}{2B} \right] \tag{5.72}$$

Therefore, modifying (5.71) by A_{dB}, it becomes:

$$P_{\text{cross modulation}} = P_{\text{single tone}} + 2 \left(P_{\text{TX leakage}} + P_{m,dBm} + 3 \right) - 2(IIP3_{LNA})_{dBm} + A_{dB} \tag{5.73}$$

Cross modulation is very problematic in SDR design. Radios that operate multiple waveforms simultaneously suffer from cross modulation effects in the form of coexistence (cosite) interference.

Example 5-4: Cross Modulation Distortion in CDMA IS95 Handset [2]

IS98 dictates the single tone desensitization test in the cellular band at 900 kHz. It places an exceptional linearity requirement on the LNA. Assume that the single tone interferer power at the antenna is −30 dBm. The maximum transmit power at the antenna $P_{\text{TX leakage}} + P_{m,dBm}$ is +23 dBm. Let the duplexer insertion loss at the transmitter to antenna and antenna to receiver be $L_{il} = 2.7\,dB$, and the duplexer Tx-Rx isolation in the Tx band is $LTx - Rx = 53\,dB$. Assume that the forward link CDMA signal power at the antenna is $\hat{I}_{or} = -104\,dBm$ (or at sensitivity level). The traffic channel power to the total received signal power $E_c/I_{or} = -15.6\,dB$. In order to guarantee a FER of 0.5%, the required E_b/N_o should be 4.5 dB. The spreading gain is 64 (18.062 dB) and the gain due to the Viterbi decoder in AWGN is approximately 3 dB, bringing the total processing gain to 128 (21.07 dB). Assume the degradation due to phase noise to be −72.9 dBc. In the presence of a single tone interferer due to, say, an AMPS or TDMA IS136 signal, the FER is allowed to degrade to 1% when $\hat{I}_{or} = -101\,dBm$, thus resulting in a E_b/N_o of 4.3 dB. What is the required IIP3? Note that for this scenario $|f_i - f_{crx}|$ is specified as 900 MHz.

The maximum allowed thermal noise at the antenna is given as:

$$\begin{aligned} P_{thermal} &= \hat{I}_{or} + PG + G_{Viterbi} + E_c/I_{or} - E_b/N_0 \\ P_{thermal} &= -104 + 10 \log(128) - 15.6 - 4.5 = -103\,dBm \end{aligned} \tag{5.74}$$

Adding 3 dB of margin to the figure above results in a thermal noise floor of −106 dBm

In the presence of a single tone interferer, however, the maximum allowed thermal noise at the antenna becomes:

$$P_{thermal} = \hat{I}_{or} + PG + G_{Viterbi} + E_c/I_{or} - E_b/N_0$$
$$P_{thermal} = -101 + 10\log(128) - 15.6 - 4.3 = -99.83 \text{ dBm} \qquad (5.75)$$

The degraded performance due to phase noise caused by reciprocal mixing (see Section 5.7) is:

$$P_{PN} = -72.9 \text{ dBc/Hz} + P_{single\ tone} = -72.9 \text{ dBc/Hz} - 30 = -102.9 \text{ dBm} \qquad (5.76)$$

Therefore, the combined phase noise, thermal noise, and cross modulation degradation must not exceed -99.83 dBm, resulting in a cross modulation product no bigger than:

$$P_{cross\ modulation} = 10\log_{10}(10^{-99.8/10} - 10^{-106/10} - 10^{-102.9/10}) = -105.5 \text{ dBm} \qquad (5.77)$$

The correction factor in this case is given as:

$$A_{dB} = 10\log_{10}\left[\frac{-|f_i - f_{crx}| + 3B/2}{2B}\right]$$

$$A_{dB} = 10\log_{10}\left[\frac{-900e3 + 1.5 \times 1.23e6}{2 \times 1.23e6}\right] = -4.1550 \text{ dB} \qquad (5.78)$$

Note that since the transmit power is expected to be 23 dBm at the antenna, this implies that transmit power at the input of the duplexer is 23 dBm plus the insertion loss of $L_{il} = 2.7$ dB resulting in a transmit power of $+23$ dBm $+ 2.7$ dB $= 25.7$ dBm. Furthermore, the single-tone interferer power has been reduced by the insertion loss of the duplexer. After the transmit signal leaks through the duplexer, the signal strength is reduced by Tx to Rx isolation in the Tx band, resulting in:

$$P_{TX\ leakage} + P_{m,dBm} = P_{tx\ power\ at\ input\ of \atop duplexer} - L_{Tx-Rx} = 25.7 - 53 = -27.3 \text{ dBm} \qquad (5.79)$$

Hence, the IIP3 of the system can be computed based on (5.73) as:

$$(IIP3_{LNA})_{dBm} = \frac{P_{single\ tone} + 2\left(P_{TX\ leakage} + P_{m,dBm} + 3\right) + A_{dB} - P_{cross\ modulation}}{2}$$

$$(IIP3_{LNA})_{dBm} = \frac{-32.7 + 2(-27.3 + 3) - 4.15 + 105.5}{2} = 10.02 \text{ dBm} \qquad (5.80)$$

Note that, in a superheterodyne receiver, the leakage (transmit) signal is typically attenuated by the image reject filter following the LNA. Furthermore, cross-modulation distortion is negligible in the mixer compared to intermodulation products, especially due to third order nonlinearity distortion.

5.6 Harmonics

Harmonics are signals whose center frequencies are integer multiples of the fundamental's center frequency. Assume, for simplicity's sake, that the input signal in (5.2) is comprised of only a single tone $x(t) = \alpha_1 \cos(\omega_1 t)$ (that is $\alpha_2 = 0$), then from (5.3) and Table 5.1, the output $y(t)$ of a nonlinear device can be arranged in terms of the first harmonic as:

$$y(t) = \frac{1}{2}\beta_2\alpha_1^2 + \left(\beta_1\alpha_1 + \frac{3}{4}\beta_3\alpha_1^3\right)\cos(\omega_1 t) + \frac{1}{2}\beta_2\alpha_1^2\cos(2\omega_1 t)$$
$$+ \frac{1}{4}\beta_3\alpha_1^3\cos(3\omega_1 t) + \cdots \tag{5.81}$$

From the relation in (5.81), we observe the following:

1 If the input signal is centered around ω_1 then the output of the nonlinear system will produce harmonic signals, among others, whose center frequencies are DC, $2\omega_1$, $3\omega_1$, ... , $N\omega_1$. Generally speaking, for a two-tone input signal centered around ω_1 and ω_2 the output signals are centered at $\pm m\omega_1 \pm n\omega_2$ where m and n are integers.

2 From (5.81) we observe that terms with even order harmonics $k\omega_1$ and $k\omega_2$ where k is even are proportional to their respective input tone-magnitudes raised to an even power (that is, α_1^k and α_2^k) and amplified by β_k. Even order harmonics become insignificant in a fully differential system.

3 Generally speaking, the relationship in (5.81) shows that harmonics up to the nth term are proportional to α_1^n and α_2^n.

5.6.1 Second Order Harmonic

Express the relation in (5.3) in response to (5.2) as:

$$y(t) = \beta_1\left\{\alpha_1\cos(\omega_1 t) + \alpha_2\cos(\omega_2 t)\right\} + \beta_2\left\{\frac{\alpha_1^2 + \alpha_2^2}{2} + \frac{\alpha_1^2}{2}\cos(2\omega_1 t) + \right.$$
$$\left. \frac{\alpha_2^2}{2}\cos(2\omega_2 t) + \alpha_1\alpha_2\cos\left[(\omega_1 - \omega_2)t\right] + \alpha_1\alpha_2\cos\left[(\omega_1 + \omega_2)t\right]\right\} + \cdots \tag{5.82}$$

In a two-tone test where the tones are of equal amplitude, that is $\alpha = \alpha_1 = \alpha_2$, then the harmonic terms become

$$\frac{\alpha^2}{2} \cos(\omega_i t) \quad \text{for } i = 1, 2$$

and the intermodulation product terms become $\alpha^2 \cos[(\omega_1 \pm \omega_2)t]$. In order to determine the harmonic power as a function of IIP2 power, consider the output power due to the harmonic term $\beta_2 \frac{\alpha_1^2}{2} \cos(2\omega_1 t)$ as $P_{ho2} = \frac{1}{8} \beta_2^2 \alpha_1^4$

and the output power due to the fundamental term $\beta_1 \alpha_1 \cos(\omega_1 t)$ as $P_o = \frac{1}{2} \beta_1^2 \alpha_1^2$; then the ratio becomes:

$$\frac{P_{ho2}}{P_o} = \frac{\frac{1}{8} \beta_2^2 \alpha_1^4}{\frac{1}{2} \beta_1^2 \alpha_1^2} \tag{5.83}$$

Recall that IIP2 was derived as $IIP2 = |\beta_1|/|\beta_2|$, so the ratio in (5.83) becomes:

$$\frac{P_{ho2}}{P_o} = \frac{\alpha_1^2}{4(IIP2)^2} = \frac{P_i}{4P_{IIP2}} \tag{5.84}$$

where P_i was previously defined as $P_i = \frac{1}{2} \alpha_1^2$ and P_{IIP2} as $P_{IIP2} = \frac{1}{2}(IIP2)^2$.

The relationship in (5.84) becomes:

$$\frac{P_{ho2}}{P_o} = \frac{\beta_1^2 P_{hi2}}{\beta_1^2 P_i} = \frac{P_{hi2}}{P_i} = \frac{P_i}{4P_{IIP2}} \Rightarrow P_{hi2} = \frac{P_i^2}{4P_{IIP2}} \tag{5.85}$$

where P_{hi} is the harmonic signal power referred to the input. In dBm, we can express (5.85) as:

$$P_{hi2,dBm} = 2P_{i,dBm} - P_{IIP2,dBm} - 6 \tag{5.86}$$

Note that from (5.86) the input referred harmonic power is 6 dB below that of the input referred IM product power due to second order nonlinearity.

5.6.2 Third Order Harmonic

In a similar manner to the analysis presented in the previous subsection, the third order harmonic can be related to IIP3 power. Assuming that the two-tone signals have equal power at the input of the system or device, then the third order harmonic signal $\frac{1}{4}\beta_3\alpha_i^3 \cos(3\omega_i t)$ for $i = 1, 2$ has an average power of $P_{ho3} = \frac{1}{32}\beta_3^2\alpha^6$.

The linear power ratio of the output harmonic power to the output power of the fundamental is:

$$\frac{P_{ho3}}{P_o} = \frac{\frac{1}{32}\beta_3^2\alpha^6}{\frac{1}{2}\beta_1^2\alpha^2} = \frac{1}{16}\frac{\beta_3^2}{\beta_1^2}\alpha^4 \tag{5.87}$$

Substituting the IIP3 power $P_{IIP3} = \frac{2}{3}\frac{|\beta_1|}{|\beta_3|}$ into (5.87), we obtain:

$$\frac{P_{ho3}}{P_o} = \frac{\frac{1}{32}\beta_3^2\alpha^6}{\frac{1}{2}\beta_1^2\alpha^2} = \frac{1}{9}\frac{P_i^2}{P_{IIP3}^2} \tag{5.88}$$

Further manipulation of (5.88) allows us to express the harmonic signal power referred to the input in terms of input signal power and IIP3 power as:

$$\frac{P_{hi3}}{P_i} = \frac{1}{9}\frac{P_i^2}{P_{IIP3}^2} \Rightarrow P_{hi3} = \frac{1}{9}\frac{P_i^3}{P_{IIP3}^2} \tag{5.89}$$

Expressing the relationship in (5.89) in dBm, we obtain:

$$P_{hi3,dBm} = 3P_{i,dBm} - 2P_{IIP3} - 9.54 \tag{5.90}$$

In a similar manner to the analysis presented above, harmonics related to higher order nonlinearities can also be developed.

5.7 Phase Noise and Reciprocal Mixing

The phase noise profile at the output of the LO is due to the crystal oscillator and VCO phase noise profiles shaped by the PLL and other mechanisms such as frequency multipliers and dividers used in generating the LO frequency output. This phase noise gets transferred to any incoming signal, wanted or unwanted, thus degrading the desired signal's SNR. We will discuss phase noise generation and analysis in detail in the coming chapters. For the purposes of this chapter, it will be sufficient to specify the phase noise level at a certain given frequency offset.

In the absence of an out of band interferer, phase noise raises the noise floor in the desired signal's band. In the presence of, say, a strong narrowband interferer outside the signal bandwidth, the phase noise raises the signal noise floor, even further degrading the receiver's sensitivity. Figure 5.10 depicts a generic GMSK signal accompanied by a narrowband interferer offset at a certain frequency away from the desired-signal center frequency. After mixing with the LO, the phase noise transfers onto both the desired signal and narrowband interferer as shown in Figure 5.11. A quick comparison between Figure 5.10 and Figure 5.11 shows the effect of phase noise on the desired signal noise floor. The amount of degradation due to the interferer alone can be estimated as:

$$P_{pn} = P_{\text{interferer}} + 10\log_{10}\left(\int_{f_L}^{f_H} L(f)df\right) \tag{5.91}$$

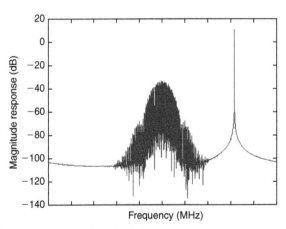

Figure 5.10 Normalized spectrum of desired GMSK signal and strong out of band interferer before mixing by the LO signal

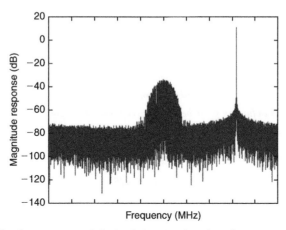

Figure 5.11 Normalized spectrum of desired GMSK signal and strong out of band interferer after mixing by the LO signal

where L(*f*) is the SSB phase noise power spectral density in dBc/Hz, and f_H and f_L are the integration limits equal to the channel filter's noise equivalent bandwidth as defined in Section 5.9.2, and $P_{\text{interferer}}$ is the interferer signal power in dBm. In the event where the phase noise is constant in the desired channel band (5.91) can be expressed as:

$$P_{pn} = P_{\text{interferer}} + L_{dBc/Hz}(f) + 10\log_{10}(B) \tag{5.92}$$

Often the integrated phase noise is expressed in degrees, that is:

$$\Delta\theta_{RMS,\text{deg}} = \frac{180° \times \sqrt{\int_{f_1}^{f_2} L(f)df}}{\pi} \tag{5.93}$$

Example 5-5: Receiver Desensitization Due to Reciprocal Mixing

Given the specifications of the following receiver:

- System noise figure is 7.2 dB
- Channel bandwidth is 3 MHz

- Narrowband jammer-signal power is 12 dBm

- LO broadband noise is −160 dBc/Hz

Assume that the jammer's center frequency is spaced away from the desired signal center frequency such that the phase noise density in the desired signal band due to the jammer is flat. Compute the rejection due to the band definition filter before the receiver's front-end such that the desensitization impact on the system sensitivity is less than 0.5 dB.

The received signal noise floor can be computed as

$$P_{noisefloor} = -174\,\text{dBm/Hz} + 10\log_{10}(B) + NF$$
$$= -174\,\text{dBm/Hz} + 10\log_{10}(3 \times 10^6\,\text{Hz}) + 7.2 = -102.03\,\text{dBm} \qquad (5.94)$$

The total noise floor due to thermal noise and reciprocal mixing is

$$P_T = P_{noisefloor} + 0.5\,\text{dB} = -101.53\,\text{dBm} \qquad (5.95)$$

The noise due to reciprocal mixing can then be estimated as

$$P_{pn} = 10\log_{10}(10^{P_T/10} - 10^{P_{noisefloor}/10}) = -111.16\,\text{dBn} \qquad (5.96)$$

Recall that from (5.92) we can write

$$P_{interferer} = P_{pn} - L_{dBc/Hz}(f) - 10\log_{10}(B)$$
$$P_{interferer} = -111.16 + 160 - 10\log_{10}(3 \times 10^6) = -15.94\,\text{dBm} \qquad (5.97)$$

where $P_{interferer}$ is the jammer power after the band definition filter. Therefore, the rejection required by the band definition filter is

$$R = P_{interferer} - P_{jammer} = -15.94\,\text{dBm} - 12\,\text{dBm} = -28\,\text{dB} \qquad (5.98)$$

5.8 Spurious Signals

Spurs can be generated by numerous mechanisms in the transceiver. They can be classified into various categories:

- Modulated or unmodulated. Unlike unmodulated spurs, which consist of a single tone, modulated spurs are generated by a single tone spur modulating a signal. For example, a spur could modulate the transmit signal onto an out-of-band frequency, thus causing signal degradation to transceivers operating in that band.

Spurs could also produce demodulated output in the receiver, thus degrading the quality of the received signal. This is particularly troubling in wideband receivers such as software radios.

- Single-frequency or multi-frequency spurs. Unlike IM products, which require at least two signals to occur, single-frequency spurs transpire when the transceiver is tuned to a certain frequency band.

- Internal spurs occur due to the internal mechanisms of the transceiver, such as harmonics.

- Spurs can be dependent on the LO or interferer power levels and are expressed in dBc. Or they can be independent of the LO and other signals; in this case, they are expressed typically in dBm.

A more detailed treatment of spurs will be provided in the next chapters.

5.9 Appendix

5.9.1 Two-Tone Analysis of the Nonlinear Memoryless Channel

5.9.1.1 Analysis

Let $x(t)$ be the input signal to a memoryless nonlinear device consisting of two tones at two distinct frequencies:

$$x(t) = \alpha_1 \cos(\omega_1 t) + \alpha_2 \cos(\omega_2 t) \tag{5.99}$$

and let $y(t)$ be the output of a nonlinear time variant device, limited for simplicity to third order nonlinearity, in response to $x(t)$:

$$y(t) = \beta_1 x(t) + \beta_2 x^2(t) + \beta_3 x^3(t) \tag{5.100}$$

The fundamental terms from (5.100) are:

$$y_1(t) = \beta_1 \left\{ \alpha_1 \cos(\omega_1 t) + \alpha_2 \cos(\omega_2 t) \right\} \tag{5.101}$$

The effect of second order nonlinearity on the signal is:

$$y_2(t) = \beta_2 x^2(t) = \beta_2 \left\{ \alpha_1 \cos(\omega_1 t) + \alpha_2 \cos(\omega_2 t) \right\}^2$$
$$y_2(t) = \beta_2 \left\{ \alpha_1^2 \cos^2(\omega_1 t) + \alpha_2^2 \cos^2(\omega_2 t) + 2\alpha_1\alpha_2 \cos(\omega_1 t)\cos(\omega_2 t) \right\} \tag{5.102}$$

Using the trigonometric identity in (5.112) and (5.114), the relation in (5.102) can be rewritten as:

$$y_2(t) = \beta_2 \left\{ \alpha_1^2 \frac{1 + \cos(2\omega_1 t)}{2} + \alpha_2^2 \frac{1 + \cos(2\omega_2 t)}{2} \right.$$

$$\left. + \alpha_1 \alpha_2 \left\{ \cos((\omega_1 + \omega_2)t) + \cos((\omega_1 - \omega_2)t) \right\} \right\}$$

$$y_2(t) = \underbrace{\beta_2 \frac{\alpha_1^2 + \alpha_2^2}{2}}_{\text{DC Term}} + \underbrace{\beta_2 \frac{\alpha_1^2 \cos(2\omega_1 t) + \alpha_2^2 \cos(2\omega_2 t)}{2}}_{\text{Second Harmonic Terms}} \tag{5.103}$$

$$\underbrace{+\beta_2 \alpha_1 \alpha_2 \cos\left((\omega_1 + \omega_2)t\right) + \beta_2 \alpha_1 \alpha_2 \cos\left((\omega_1 - \omega_2)t\right)}_{\text{Intermodulation Terms}}$$

Next, consider the effect of third order nonlinearity:

$$y_3(t) = \beta_3 x^3(t) = \beta_3 \left\{ \alpha_1 \cos(\omega_1 t) + \alpha_2 \cos(\omega_2 t) \right\}^3$$

$$y_3(t) = \beta_3 \left\{ \alpha_1^3 \cos^3(\omega_1 t) + 3\alpha_1^2 \alpha_2 \cos^2(\omega_1 t)\cos(\omega_2 t) \right. \tag{5.104}$$

$$\left. + 3\alpha_1 \alpha_2^2 \cos(\omega_1 t)\cos^2(\omega_2 t) + \alpha_2^3 \cos^3(\omega_2 t) \right\}$$

Using the relationships expressed in (5.112) through (5.115), the various identities in (5.104) can be rewritten as:

$$\alpha_1^3 \cos^3(\omega_1 t) = \alpha_1^3 \frac{\cos 3\omega_1 t + 3\cos \omega_1 t}{4}, \alpha_2^3 \cos^3(\omega_2 t)$$

$$= \alpha_2^3 \frac{\cos 3\omega_2 t + 3\cos \omega_2 t}{4}$$

$$3\alpha_1^2 \alpha_2 \cos^2(\omega_1 t)\cos(\omega_2 t) = 3\alpha_1^2 \alpha_2 \left[\frac{\cos(2\omega_1 t) + 1}{2} \right] \cos(\omega_2 t)$$

$$= 3\alpha_1^2 \alpha_2 \left\{ \frac{\cos(2\omega_1 t)\cos(\omega_2 t) + \cos(\omega_1 t)}{2} \right\}$$

$$= 3\alpha_1^2 \alpha_2 \left\{ \frac{\cos\left((2\omega_1 + \omega_2)t\right) + \cos\left((2\omega_1 - \omega_2)t\right)}{4} + \frac{\cos(\omega_2 t)}{2} \right\}$$

$$3\alpha_1 \alpha_2^2 \cos(\omega_1 t)\cos^2(\omega_2 t) = 3\alpha_1 \alpha_2^2 \left\{ \frac{\cos\left((2\omega_2 + \omega_1)t\right) + \cos\left((2\omega_2 - \omega_1)t\right)}{4} + \frac{\cos(\omega_2 t)}{2} \right\}$$

$$\tag{5.105}$$

The relation in (5.104) can now be expressed as:

$$y_3(t) = \beta_3 \left(\frac{3}{4}\alpha_1^3 + \frac{3}{2}\alpha_1\alpha_2^2 \right) \cos \omega_1 t + \beta_3 \left(\frac{3}{4}\alpha_2^3 + \frac{3}{2}\alpha_1^2\alpha_2 \right) \cos \omega_2 t +$$

$$\underbrace{\phantom{\beta_3 \left(\frac{3}{4}\alpha_1^3 + \frac{3}{2}\alpha_1\alpha_2^2 \right) \cos \omega_1 t + \beta_3 \left(\frac{3}{4}\alpha_2^3 + \frac{3}{2}\alpha_1^2\alpha_2 \right) \cos \omega_2 t}}_{\text{Gain Compression (Blocking and cross-modulation) Terms}}$$

$$+ \underbrace{\frac{1}{4}\beta_3\alpha_1^3 \cos 3\omega_1 t + \frac{1}{4}\beta_3\alpha_2^3 \cos 3\omega_2 t}_{\text{Third Harmonic Terms}}$$

$$+ \underbrace{\frac{3}{4}\beta_3\alpha_1^2\alpha_2 \left\{ \cos \left((2\omega_1 + \omega_2)t \right) + \cos \left((2\omega_1 - \omega_2)t \right) \right\}}_{\text{Intermodulation Term}}$$

$$+ \underbrace{\frac{3}{4}\beta_3\alpha_1^2\alpha_2 \left\{ \cos \left((2\omega_2 + \omega_1)t \right) + \cos \left((2\omega_2 - \omega_1)t \right) \right\}}_{\text{Intermodulation Term}} \qquad (5.106)$$

5.9.1.2 Observations

Based on De Moivre's theorem of complex numbers, which states that:

$$\{ \cos(\omega t) + j \sin(\omega t) \}^n = e^{jn\omega t} = \cos(n\omega t) + j \sin(n\omega t) \qquad (5.107)$$

we note that the trigonometric expansion of $\cos^n(\omega t)$ can be expressed as[1]:

$$\cos^n(\omega t) = \cos(n\omega t) + \binom{n}{2}\cos^{n-2}(\omega t)\sin^2(\omega t) - \binom{n}{4}\cos^{n-4}(\omega t)\sin^4(\omega t) + \dots \quad (5.108)$$

For even n, $\cos^n(\omega t)$ will only produce even harmonics of the fundamental ranging from DC to the nth harmonic, whereas for odd n, only odd harmonics of the fundamental will be produced ranging from the fundamental up to the nth harmonic. The strength of a given harmonic in (5.100) depends on various terms in the polynomial and gets smaller as n increases. For example, the nth harmonic is smaller than $(n-1)$th harmonic and the $(n-1)$th harmonic is smaller than $(n-2)$th harmonic and so on. Note that these rules apply only to small signal analysis, that is, for a small input signal. These rules do not apply to mixers, for example, where the device is built with a certain enhanced nonlinearity in mind.

---

[1] Recall that the combination $C_{(k)}^{(n)}$ of k elements taken from n elements is $C_{(k)}^{(n)} = \binom{n}{k} = \dfrac{n!}{(n-k)!k!}$

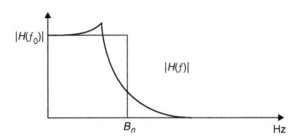

Figure 5.12 Equivalent noise bandwidth filter

5.9.2 Definition of Noise Equivalent Bandwidth

The noise equivalent bandwidth, illustrated in Figure 5.12, for a lowpass filter is defined as the bandwidth of an ideal filter such that the power at the output of this filter, if excited by white Gaussian noise, is equal to that of the real filter given the same input signal; that is:

$$B_n = \frac{1}{|H(f_0)|^2} \int_0^\infty |H(f)|^2 \, df \qquad (5.109)$$

In reality, however, the integration limit is set by the dynamic range of the ADC. In other words, the integration limit is at the point where the noise is less than or equal to ½ LSB.

The estimation of the noise equivalent bandwidth allows us to compute the amount of in-band noise and its effect on the received signal SNR regardless of the filter's transfer function. Therefore, when specifying sensitivity, it is essential to estimate the noise equivalent bandwidth, and use this bandwidth figure when computing the final sensitivity of the receiver[2].

5.9.3 Useful Mathematical Identities

The majority of these relations can be found in [3]. The following trigonometric identities are used in the memoryless polynomial expansion:

$$\cos(a + b) = \cos a \cos b - \sin a \sin b$$
$$\cos(a - b) = \cos a \cos b + \sin a \sin b$$

---

[2]Furthermore, note that the filter response determines the amount of aliasing noise that folds onto the desired signal band after sampling

$$\sin(a + b) = \sin a \cos b + \sin b \cos a$$
$$\sin(a - b) = \sin a \cos b - \sin b \cos a \tag{5.110}$$

Furthermore, the sin relation can be expanded as:

$$\sin(a + b) = \sin a \cos b + \sin b \cos a = \{\sin a + \tan b \cos a\}\cos b \tag{5.111}$$

From the relationship in (5.110), we can deduce:

$$\cos a \cos b = \frac{1}{2}\left(\cos(a + b) + \cos(a - b)\right)$$
$$\sin a \sin b = \frac{1}{2}\left(\cos(a - b) - \cos(a + b)\right) \tag{5.112}$$

The function of double angles can be expressed as:

$$\cos 2a = \cos^2 a - \sin^2 a = 2\cos^2 a - 1 = 1 - 2\sin^2 a$$
$$\sin 2a = 2\sin a \cos a \tag{5.113}$$

Consequently, we can re-express the function of double angle as:

$$\cos^2 a = \frac{\cos 2a + 1}{2}$$
$$\sin^2 a = \frac{1 - \cos 2a}{2} \tag{5.114}$$

The sine and cosine functions of multiple angles can be expressed as:

$$\cos 3a = 4\cos^3 a - 3\cos a \Rightarrow \cos^3 a = \frac{\cos 3a + 3\cos a}{4}$$
$$\sin 3a = 3\sin a - 4\sin^3 a \Rightarrow \sin^3 a = \frac{3\sin a - \sin 3a}{4} \tag{5.115}$$

The relation in (5.115) can be generalized as:

$$\cos^n(x) = \left(\frac{1}{2}\right)^{n-1}\left\{\cos(nx) + \binom{n}{1}\cos\left((n - 2)x\right) + \binom{n}{2}\cos\left((n - 4)x\right)\right.$$

$$\left. + \cdots + \binom{n}{\frac{n-1}{2}}\cos(x)\right\} \text{ for } n \text{ odd}$$

$$\cos^n(x) = \left(\frac{1}{2}\right)^{n-1} \left[\cos(nx) + \binom{n}{1}\cos\big((n-2)x\big) + \binom{n}{2}\cos\big((n-4)x\big) \right.$$

$$\left. + \cdots + \left(\frac{n}{\frac{n-2}{2}}\right)\cos(x) \right\} \text{ for } n \text{ even} \tag{5.116}$$

A polynomial expansion can be expressed in terms of a multinomial series as:

$$(x_1 + x_2 + \cdots + x_k)^n = \sum_{\substack{n_1,n_2,\ldots,n_k \geq 0 \\ n_1+n_2+\ldots+n_k=n}} \frac{n!}{n_1!n_2!\ldots n_k!} x_1^{n_1} x_2^{n_2} \ldots x_k^{n_k} \tag{5.117}$$

As an example, consider the polynomial:

$$(x_1 + x_2 + x_3)^3 = x_1^3 + 3x_1^2 x_2 + 3x_1 x_2^2 + x_2^3 + 3x_1^2 x_3 + 6x_1 x_2 x_3 +$$
$$3x_2^2 x_3 + 3x_1 x_3^2 + 3x_2 x_3^2 + x_3^3 \tag{5.118}$$

References

[1] Razavi B. Design considerations for direct-conversion receivers. IEEE Transactions on Circuit and Systems-II June 1997;44(6).

[2] Aparin V, Zeisel E, Gazerro P. Highly linear BiCMOS LNA and mixer for cellular CDMA/AMPS applications. IEEE Radio Frequency Integrated Circuits Symposium 2002;4:129–32.

[3] Korn G, Korn T. Mathematical Handbook for Scientists and Engineers. Mineola, NY: Dover Publications; 1968.

CHAPTER 6

Transceiver System Analysis and Design Parameters

Thus far, Chapters 4 and 5 have discussed important system design parameters and analysis pertinent to sensitivity and system linearity. It was shown how a system's noise figure and second and third order input intercept points are dependent on the linearity, noise figure, and gain of the individual blocks. Analysis and design tools were provided to evaluate the performance. This chapter presents a continuation of the previous two chapters. In Section 6.1, we discuss the selectivity of the receiver that is its ability to reject unwanted signals such as blocker and other types of interferers. The contribution due to phase noise and spurs is also discussed. Another major set of design parameter that are of great concern to the system performance are the various parameters that define its dynamic range. The 1-dB compression dynamic range, the desensitization dynamic range, the spur-free dynamic range, and the dynamic range obtained through the noise power ratio are all presented in Section 6.2. AM/AM and AM/PM distortions presented in the context of memoryless systems are studied in Section 6.3. Frequency bands, accuracy, and tuning are briefly discussed in Section 6.5. Modulation accuracy and EVM due to various anomalies such as IQ imbalance and DC offsets, for example, are presented in Section 6.5. Adjacent channel power ratio and transmitter broadband noise are considered in Sections 6.6 and 6.7, respectively.

6.1 Receiver Selectivity

Selectivity is the ability of the receiver to reject unwanted signals, mostly in the adjacent and alternate channels, thus allowing proper demodulation of the desired signal. Selectivity is *mainly* determined by the IF and channel filter (analog and digital) rejection. In the super-heterodyne architecture, the filter at IF presents a certain amount of selectivity to aid the channel filter(s)[1] before down-conversion to baseband. There are several factors that determine the amount of selectivity needed, namely spur-levels, phase noise, bandwidth, and the amount of cochannel rejection required. The ratio between the wanted signal and the unwanted blocker or interferer is commonly referred to as cochannel rejection, or in FM terms *capture ratio*.[2]

Define the maximum allowed degradation as:

$$P_D = 10 \log_{10} \left(10^{\frac{P_{desired} - CNR}{10}} - 10^{\frac{P_{noisefloor}}{10}} \right) \tag{6.1}$$

where $P_{desired}$ is the desired signal power in the presence of interference, and $P_{noisefloor}$ is the noise floor as defined previously:

$$P_{noisefloor} = -174\,\text{dBm/Hz} + NF + 10\log_{10}(B) \tag{6.2}$$

The mixing of the narrowband interferer or blocker modulates with the LO to produce inband phase noise as well as a potential in-band spur. Assuming the phase noise to be flat in the desired band, the noise floor due to phase noise is:

$$P_{PN,LO,dBm} = P_{\text{intermodulation tone},dBm} + L(f = \Delta f)\text{dBc/Hz} + 10\log_{10}(B)$$

$$P_{PN,LO,linear} = 10^{\frac{P_{PN,LO,dBm}}{10}} = 10^{\frac{P_{\text{intermodulation tone},dBm} + L(f=\Delta f)\text{dBc/Hz} + 10\log_{10}(B)}{10}} \tag{6.3}$$

where $P_{\text{intermodulation tone},dBm}$ is the intermodulation signal power in dBm, $L(f = \Delta f_1)$ is the phase noise density at $f = \Delta f$.[3] The spur signal is expressed in dBm:

$$P_{spur,dBm} = P_{\text{intermodulation tone},dBm} + P_{spur,dBc}$$

$$P_{spur,lin} = 10^{\frac{P_{spur,dBm}}{10}} = 10^{\frac{P_{\text{intermodulation tone},dBm} + P_{spur,dBc}}{10}} \tag{6.4}$$

[1]Recall that channel filtering in SDR is split between analog IF and digital baseband.
[2]The capture ratio is associated with the capture effect in FM. If two simultaneous signals, one strong and one weak, present at the input of a limiting amplifier, the "limiting operation" further strengthens the strong signal at the expense of weakening the weak signal. Thus, the strong signal "captures" the weak signal.
[3]In order to simplify the analysis, the phase noise density $L(f)$ is assumed to be constant at Δf.

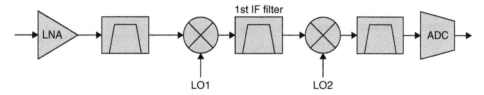

Figure 6.1 Conceptual SDR receiver with dual conversion architecture

Signal-dependent spurs are byproducts of an incoming blocker at mixing with an LO spur to produce an inband spur. In a dual conversion SDR transceiver, depicted in Figure 6.1, the first IF filter provides some rejection of the blocker signal, then the phase noise contribution due to the second LO is:

$$P_{PN,LO2,dBm} = P_{\text{blocker, dBm}} + L(f = \Delta f_2)\text{dBc/Hz} + 10\log_{10}(B) - R$$

$$P_{PN,LO2,lin} = 10^{\frac{P_{PN,LO2,dBm}}{10}} = 10^{\frac{P_{\text{blocker, dBm}}}{10}} 10^{\frac{L(f=\Delta f_2)\text{dBc/Hz}+10\log_{10}(B)-R}{10}} \qquad (6.5)$$

where R is the first IF filter rejection in dB of the blocking signal. In a similar manner, a spur due to blocking signal mixed with the second LO is expressed as:

$$P_{spur\ 1,dBm} = P_{\text{blocker,dBm}} + P_{spur\ 1,dBc} - R$$

$$P_{spur\ 1,lin} = 10^{\frac{P_{spur\ 1,dBm}}{10}} = 10^{\frac{P_{\text{blocker, dBm}}}{10}} 10^{\frac{P_{spur\ 1,dBc}-R}{10}} \qquad (6.6)$$

The blocking signal power can then be determined from (6.1) through (6.6) as:

$$10^{\frac{P_{\text{blocker,dBm}}}{10}} \left(10^{\frac{L(f=\Delta f_1)\text{dBc/Hz}+10\log_{10}(B)}{10}} + 10^{\frac{P_{spur\ 1,dBc}}{10}} + 10^{\frac{L(f=\Delta f_2)\text{dBc/Hz}+10\log_{10}(B)-R}{10}} + 10^{\frac{P_{spur\ 1,dBc}-R}{10}} \right) = 10^{\frac{P_D}{10}}$$

$$P_{\text{blocker,dBm}} = 10\log_{10}\left(\frac{10^{\frac{P_D}{10}}}{10^{\frac{L(f=\Delta f_1)\text{dBc/Hz}+10\log_{10}(B)}{10}} + 10^{\frac{P_{spur\ 1,dBc}}{10}} + 10^{\frac{L(f=\Delta f_2)\text{dBc/Hz}+10\log_{10}(B)-R}{10}} + 10^{\frac{P_{spur\ 1,dBc}-R}{10}}} \right) \qquad (6.7)$$

The selectivity due to this architecture can then be defined as:

$$S = P_{blocker,dBm} - P_{signal} \qquad (6.8)$$

Example 6-1: Receiver Selectivity in WCDMA [1]

Given the architecture depicted in Figure 6.1, the reference sensitivity level for a WCDMA handset is −106.7 dBm per 3.84 MHz of bandwidth. In the presence of a blocking signal, WCDMA allows the desired signal level to be 14 dB higher than the sensitivity level. Let the receiver noise figure be 7.9 dB, and let the phase noise density due to LO_1 be constant in the desired band at −100 dBc. Let the LO spur due to the first LO be −85 dBc. Similarly, let the phase noise density due to the second LO be −90 dBc/Hz and the spur level be −70 dBc. The first IF filter selectivity is assumed to be 30 dB. Compute the selectivity of this system given that the required CNR is −7.9 dB. How does the estimated blocker level compare with the required blocker level of −52 dBm as specified in the standard?

According to the WCDMA specification, the signal under the selectivity test can be as high as 14 dB above sensitivity, that is −106.7 dBm + 14 dB = −92.7 dBm. The noise floor of the receiver is given as:

$$P_{noisefloor} = -174\,dBm/Hz + NF + 10\log_{10}(B)$$

$$P_{noisefloor} = -174\,dBm/Hz + 7.9 + 10\log_{10}(3.84 \times 10^6) = -100.26\,dBm \qquad (6.9)$$

Using the relation expressed in (6.1), the maximum allowed degradation is given as:

$$P_D = 10\log_{10}\left(10^{\frac{-92.7+7.3}{10}} - 10^{\frac{-100.26}{10}}\right) = -85.54\,dB \qquad (6.10)$$

Next, using the relation in (6.7), we can estimate the maximum allowable blocker the receiver can handle:

$$P_{blocker,dBm} = 10\log_{10}\left(\frac{10^{\frac{-85.54}{10}}}{10^{\frac{-100+10\log_{10}(3.84\times10^6)}{10}} + 10^{\frac{-85}{10}} + 10^{\frac{-90+10\log_{10}(3.84\times10^6)-30}{10}} + 10^{\frac{-70-30}{10}}}\right)$$

$$= -51.43\,dBm \qquad (6.11)$$

The result in (6.11) is comparable to the blocker level as specified in the standard. Hence, the selectivity at IF becomes:

$$S = -51.43 - (-92.7) = 41.27 \, dB \tag{6.12}$$

The results are summarized in Table 6.1.

Table 6.1 Selectivity requirement for WCDMA

Parameter	Value
Sensitivity (dBm)	−106.7
Desired signal level (dBm)	−92.7
CNR (dB)	−7.3
NF (dB)	7.9
Bandwidth (MHz)	3.84
Noise floor (dBm)	−100.26
Maximum noise floor (dBm)	−85.40
Allowed degradation (dBm)	−85.54
IF selectivity (dB)	30
Phase noise density 1st LO (dBc/Hz)	−100
Phase noise density 2nd LO (dBc/Hz)	−90
Spur level after LO1 (dBc)	−85
Spur level after LO2 (dBc)	−70
Estimated blocker level (dBm)	−51.43
Selectivity at IF (dB)	41.27
Allowable blocker (dBm)	−52

6.2 Receiver Dynamic Range

At this point, we have discussed the various parameters that influence the receiver dynamic range, save those dictated by the AGC, ADC, and DAC, which will be discussed in later chapters. The receiver dynamic range specifies the range of the various signal power levels that the receiver can process consistently.

6.2.1 1-dB Compression Dynamic Range

The 1-dB compression point is defined as the difference between the 1-dB compression point and the receiver noise floor [2],[3]:

$$
\begin{aligned}
\Delta_{\text{1-dB dynamic range}} &= P_{ICP,dB} - P_{noisefloor} \\
&= P_{ICP,dB} + 174\,\text{dBm/Hz} - NF - 10\log_{10}(B)
\end{aligned}
\tag{6.13}
$$

The 1-dB compression dynamic range is only meaningful in the presence of the desired signal only. It is not applicable in the presence of interference.

6.2.2 Desensitization Dynamic Range

The desensitization dynamic range is concerned with the receiver's ability to process a desired signal in the presence of a strong single-tone out of band interferer. As seen earlier, a strong interferer can have detrimental effects on receiver performance due to phase noise in the form of reciprocal mixing and spurs. The desensitization dynamic range is defined for a given desired signal received with a certain CNR as the difference between the strongest interferer in the presence of which the desired signal can still be demodulated with a certain CNR to the noise floor:

$$
\begin{aligned}
\Delta_{\text{desensitization dynamic range}} &= P_i - P_{noisefloor} \\
&= P_i + 174\,\text{dBm/Hz} - NF - 10\log_{10}(B)
\end{aligned}
\tag{6.14}
$$

where P_i is the blocking signal power in dBm.

6.2.3 Spur-Free Dynamic Range

The spur-free dynamic range (SFDR) is defined as the difference between the receiver's IIP2 or IIP3 and its respective noise floor.

$$
\begin{aligned}
\Delta_{SFDR-IIP2} &= \frac{1}{2}\left(IIP2 + 174 - NF - 10\log_{10}(B)\right) \\[2mm]
\Delta_{SFDR-IIP3} &= \frac{2}{3}\left(IIP3 + 174 - NF - 10\log_{10}(B)\right)
\end{aligned}
\tag{6.15}
$$

SFDR is a popular figure of merit that allows us to compare the performance of various receivers in a well-defined manner. Although SFDR is in general a better measure of dynamic range than the desensitization dynamic range, it neglects certain key factors that influence the dynamic range of the receiver. The limitation is due to the fact that most interfering signals in a real environment are not necessarily narrowband, limited in number to two, and of equal amplitude.

6.2.4 Noise Power Ratio Dynamic Range

The noise power ratio dynamic range (NPR-DR) is a more comprehensive figure of merit than what was previously discussed. The basic premise is to stimulate the signal and interference environment with high power white noise.[4] The noise is filtered at the receiver's center frequency via a notch filter. At the output of the receiver, the signal power present in the notched-frequency is measured. This degradation is mainly due to the anomalies in the receiver due to IM products, spurs, and the like.

6.3 AM/AM and AM/PM

The purpose of this section is to provide a certain insight into the processes of amplitude (AM/AM) and phase (AM/PM) distortion and their adverse effects on the desired signal. First, the effect of a narrowband interferer in terms of AM/AM and AM/PM on a desired transmit or receive signal is discussed. Finally, under certain assumptions, we discuss the relationships that exist between the coefficients of the memoryless nonlinearity model and AM/AM and AM/PM.

6.3.1 AM/AM and AM/PM Distortion Due to Narrowband Interference

In this subsection, the process of AM/AM and AM/PM as it manifests itself in narrowband systems is described. Suppose the nonlinear IM contribution of a certain device due to $2\omega_1 \pm \omega_2$ falls within the carrier frequency range, say $2\omega_c + \Delta\omega$ of a given signal $A \sin(\omega_c t)$. Then at the output of the device, the desired signal and interference can be expressed as

$$y(t) = A(t)\sin(\omega_c t) + \Gamma\cos(\omega_c t + \Delta\omega t) \tag{6.16}$$

where A is the magnitude of the desired signal, and Γ is the magnitude of the IM product.

---

[4]The noise is expected to be spectrally flat across the band.

Expand $y(t)$:

$$y(t) = A\sin(\omega_c t) + \Gamma\{\cos(\omega_c t)\cos(\Delta\omega t) - \sin(\omega_c t)\sin(\Delta\omega t)\}$$

$$y(t) = (A - \Gamma\sin(\Delta\omega t))\sin(\omega_c t) + \Gamma\cos(\Delta\omega t)\cos(\omega_c t)$$

$$y(t) = (A - \Gamma\sin(\Delta\omega t))\left\{\sin(\omega_c t) + \frac{\Gamma\cos(\Delta\omega t)}{(A - \Gamma\sin(\Delta\omega t))}\cos(\omega_c t)\right\} \tag{6.17}$$

Let

$$\tan(\phi(t)) = \frac{\Gamma\cos(\Delta\omega t)}{(A - \Gamma\sin(\Delta\omega t))} \tag{6.18}$$

Using the relation expressed in the trigonometric identities given in the appendix in Chapter 5, the relation in (6.17) becomes:

$$y(t) = \frac{(A - \Gamma\sin(\Delta\omega t))}{\cos(\phi(t))}\{\sin(\omega_c t) + \tan(\phi(t))\cos(\omega_c t)\}\cos(\phi(t))$$

$$y(t) = A\frac{\left[1 - \dfrac{\Gamma}{A}\sin(\Delta\omega t)\right]}{\cos(\phi(t))}\sin(\omega_c t + \phi(t)) \tag{6.19}$$

It is obvious from (6.19) that the AM/AM distortion due to $\Gamma(t)\cos(\omega_c t + \Delta\omega t)$ is:

$$D_{AM/AM}(t) = \frac{\cos(\phi(t))}{\left[1 - \dfrac{\Gamma}{A}\sin(\Delta\omega t)\right]} \tag{6.20}$$

The average power of the denominator of (6.20) is

$$P_{D,AM/AM} = 1 + \frac{\Gamma^2}{2A^2} \tag{6.21}$$

And likewise the AM/PM distortion is:

$$\phi(t) = \tan^{-1}\left(\frac{\Gamma\cos(\Delta\omega t)}{A - \Gamma\sin(\Delta\omega t)}\right) \tag{6.22}$$

6.3.2 AM/AM and AM/PM Distortion in Nonlinear Memoryless Systems

Consider the memoryless polynomial:

$$y(t) = \sum_{n=0}^{\infty} \beta_n x^n(t) = \beta_0 + \beta_1 x(t) + \beta_2 x^2(t) + \beta_3 x^3(t) + \cdots \tag{6.23}$$

For an infinite series polynomial such as the one expressed in (6.23), the AM/AM and AM/PM contribution can be obtained for an input signal $x(t) = \alpha\cos(\omega t)$ using the trigonometric expansion of $\cos^n(x)$ relation given in the appendix in Chapter 5 as

$$y(t) = \sum_{\substack{k=1 \\ k\text{ odd}}}^{n} \left(\frac{1}{2}\right)^{k-1}\binom{k}{\frac{k-1}{2}}\beta_k\alpha^{k-1}, \quad n \text{ is odd} \tag{6.24}$$

$$y(t) = \beta_1\alpha\left\{1 + \frac{3}{4}\frac{\beta_3}{\beta_1}\alpha^2 + \frac{5}{8}\frac{\beta_5}{\beta_1}\alpha^4 + \frac{35}{64}\frac{\beta_7}{\beta_1}\alpha^6 + \cdots + \left(\frac{1}{2}\right)^{n-1}\binom{n}{\frac{n-1}{2}}\frac{\beta_n}{\beta_1}\alpha^{n-1}\right\}\cos(\omega t)$$

For a third order nonlinearity, the output signal due to AM/AM and AM/PM is

$$\begin{aligned}y(t) &= \left\{\beta_1\alpha + \frac{3}{4}\beta_3\alpha^3\right\}\cos(\omega t) \\ &= \beta_1\alpha\left\{1 + \frac{3}{4}\frac{\beta_3}{\beta_1}\alpha^2\right\}\cos(\omega t)\end{aligned} \tag{6.25}$$

So far we have assumed that the coefficients of the memoryless system are real. This assumption is a fair assumption when it comes to analyzing worst-case degradations due to, say, IM products, harmonics, etc. In reality, however, most systems exhibit a complex behavior, and as such affect the output of the device in both amplitude and phase. For the sake of this analysis, assume that $\beta_1 = |\beta_1| e^{j\varphi_1}$ and $\beta_3 = |\beta_3| e^{j\varphi_3}$; then the relation in (6.41) becomes

$$y(t) = |\beta_1| e^{j\varphi_1} \alpha \left\{ 1 + \frac{3}{4} \frac{|\beta_3|}{|\beta_1|} e^{j\theta} \alpha^2 \right\} \cos(\omega t) \tag{6.26}$$

where $\theta = \varphi_3 - \varphi_1$. Expanding (6.26) in terms of real and imaginary components, we obtain

$$y(t) = |\beta_1| e^{j\varphi_1} \alpha \left\{ 1 + \frac{3}{4} \frac{|\beta_3|}{|\beta_1|} \alpha^2 \cos(\theta) + j \frac{3}{4} \frac{|\beta_3|}{|\beta_1|} \alpha^2 \sin(\theta) \right\} \cos(\omega t) \tag{6.27}$$

The phase $\phi(t)$ due to AM/PM in $y(t)$ in (6.27) is

$$\phi(t) = \tan^{-1} \left\{ \frac{\frac{3}{4} \frac{|\beta_3|}{|\beta_1|} \alpha^2 \sin(\theta)}{1 + \frac{3}{4} \frac{|\beta_3|}{|\beta_1|} \alpha^2 \cos(\theta)} \right\} \tag{6.28}$$

For a small value of $\phi(t)$, we can approximate $\tan(\phi(t)) \approx \phi(t)$ and the AM/PM becomes

$$\phi(t) \approx \frac{\frac{3}{4} \frac{|\beta_3|}{|\beta_1|} \alpha^2 \sin(\theta)}{1 + \frac{3}{4} \frac{|\beta_3|}{|\beta_1|} \alpha^2 \cos(\theta)} \tag{6.29}$$

The relation (6.29) is *similar* to Saleh's model for AM/PM [4]:

$$\phi(\alpha) = \frac{\Omega \alpha^2}{1 + \Delta \alpha^2} \tag{6.30}$$

where Ω and Δ are determined from empirical data. The coefficients in Saleh's model however, are not the same as the coefficients expressed in (6.25). Note that, for simplicity's sake, we used a third order model to derive our AM/PM model. In reality, a true AM/PM model must accommodate higher-order odd terms. In the Saleh model, the coefficients are determined via curve fitting of the AM/PM as well as the AM/AM data rather than series expansion of the polynomial. Next, substituting $IIP3 \equiv \alpha|_{\alpha=IP3} = \sqrt{(4/3)(|\beta_1|/|\beta_3|)}$ into (6.29), we obtain

$$\phi(t) \approx \frac{\alpha^2 \sin(\theta)}{IIP3^2 + \alpha^2 \cos(\theta)} \tag{6.31}$$

Note from (6.31) that the higher the $IIP3$ point, the lower the AM/PM degradation to the output signal.

The AM/AM contribution due to third order nonlinearity can be derived from the magnitude $|A|$ of $y(t) = (|A|\angle A)\cos(\omega t)$ from (6.43):

$$A = |\beta_1|\alpha\sqrt{\left(1 + \frac{3}{4}\frac{|\beta_3|}{|\beta_1|}\alpha^2 \cos(\theta)\right)^2 + \left(\frac{3}{4}\frac{|\beta_3|}{|\beta_1|}\alpha^2 \sin(\theta)\right)^2}$$

$$A = |\beta_1|\alpha\sqrt{\left(1 + \left(\frac{\alpha}{IIP3}\right)^2 \cos(\theta)\right)^2 + \left(\left(\frac{\alpha}{IIP3}\right)^2 \sin(\theta)\right)^2} \tag{6.32}$$

where $\angle$ is the phase operator. The AM/AM contribution is given as:

$$P_{D,AM/AM} = \sqrt{\left(1 + \left(\frac{\alpha}{IIP3}\right)^2 \cos(\theta)\right)^2 + \left(\left(\frac{\alpha}{IIP3}\right)^2 \sin(\theta)\right)^2} \tag{6.33}$$

Note that the Saleh model for AM/AM is expressed as:

$$A(\alpha) = \frac{\Gamma\alpha}{1 + \Psi\alpha^2} \tag{6.34}$$

where Γ and Ψ are computed from empirical data to fit the AM/AM function. The relation in (6.34) can be expressed in terms of a series expansion as:

$$A(\alpha) = \frac{\Gamma\alpha}{1 + \Psi\alpha^2} = \Gamma\alpha\{1 - \Psi\alpha^2 + \Psi^2\alpha^4 - \Psi^3\alpha^6 + \Psi^4\alpha^8 + \cdots\} \tag{6.35}$$

In the event where the imaginary values (6.24) are zeros or negligible, the degradation is purely due to AM/AM. In this case, a direct comparison between (6.24) and (6.35) reveals that the polynomial coefficients are related to Γ and Ψ as:

$$A(\alpha) = \beta_1 \alpha \left\{ 1 + \frac{3}{4} \frac{\beta_3}{\beta_1} \alpha^2 + \frac{5}{8} \frac{\beta_5}{\beta_1} \alpha^4 + \frac{35}{64} \frac{\beta_7}{\beta_1} \alpha^6 + \cdots + \left(\frac{1}{2}\right)^{n-1} \left(\begin{array}{c} n \\ \frac{n-1}{2} \end{array}\right) \frac{\beta_n}{\beta_1} \alpha^{n-1} \right\} \quad (6.36)$$

and consequently,

$$\beta_1 = \Gamma$$

$$\beta_3 = -\frac{4}{3} \Gamma \Psi$$

$$\beta_5 = -\frac{8}{5} \Gamma \Psi^2$$

$$\beta_7 = -\frac{64}{35} \Gamma \Psi^3$$

$$\vdots$$

$$\beta_n = (-1)^{\frac{n-1}{2}} \frac{2^{n-1}}{\left(\begin{array}{c} n \\ \frac{n-1}{2} \end{array}\right)} \Gamma \Psi^{\frac{n-1}{2}} \quad (6.37)$$

For a third order nonlinearity, the output can be described as:

$$y(t) = \Gamma \alpha \{ 1 - \Psi \alpha^2 \} \cos(\omega t) \quad (6.38)$$

It is important to keep in mind that AM/AM and AM/PM are interrelated and not necessarily independent processes. The results above are at best approximations that aid us in understanding these phenomena.

6.4 Frequency Bands, Accuracy, and Tuning

The purpose of this section is to discuss the high-level requirements pertinent to frequency that are typical in an SDR application. The intent is not to go into the details, but rather to enumerate the various parameters and high-level requirements that need to be specified. A more complete discussion and analysis of frequency generation mechanisms and anomalies will be presented in detail in the coming chapters. It is worthy to note, however, that in most modern communication systems, an AFC algorithm is usually employed to compensate for the various frequency errors inflicted on the signal. This is done by estimating the frequency error or offset between the LO and the received desired signal.

6.4.1 Frequency Bands

Defining the frequency bands of operation of a given radio is one of the most important design specifications. This is even more important for SDR radios that support multiple wireless applications and must operate in multiple bands and with varying bandwidths. Furthermore, the architecture of the radio is greatly dictated by the frequency plan that covers the various desired bands of operation, which in turn dictates the performance of the reference oscillator and the VCO(s). The frequency band is typically defined in terms of a lower frequency limit f_l and an upper frequency limit f_h such that one or more channel bandwidths are contained within these limits. Often, a guard band is used to separate various information bands.

6.4.2 Frequency Accuracy and Drift

Frequency accuracy can be specified in terms of reference oscillator frequency accuracy or synthesizer's output frequency accuracy. The specification is usually expressed in terms of range or frequency offset from the desired center frequency. Frequency drift is a long-term effect. For example, the reference-oscillator frequency may be offset from the desired carrier frequency due to change in temperature and/or voltage supply and aging. Frequency accuracy is expressed as a deviation from the center frequency in parts per million or ppm.[5] The acceptable frequency range around the carrier is

$$f = f_c \pm \frac{\partial f \text{ (in ppm)}}{10^6} f_c \qquad (6.39)$$

Oscillator aging is a long term accuracy parameter expressed in ppm or Hz as frequency offset from the reference oscillator's frequency. Initially, when the oscillator is first used, the reference frequency changes or ages rapidly over a certain period of time, usually expressed in weeks. After this initial period, the crystal changes more slowly until it is stable over the course of possibly several months. Aging is a process independent of temperature and supply voltage variations.

---

[5]1 ppm is 0.0001%.

Example 6-2: VCTCXO Accuracy in CDMA IS-95

The VCTCXO commonly employed in IS95 CDMA handsets is 19.2 MHz. The frequency temperature stability is ±2 ppm. The control range is specified as ±12 ppm. The aging stability over a period of five years is ±4 ppm. Convert the parameters expressed in ppm to Hz.

Using the relationship expressed in (6.39) the temperature stability is given as:

$$f = f_c \pm \frac{\partial f(\text{in ppm})}{10^6} f_c = 19.2 \text{ MHz} \pm \frac{2}{10^6} \times 19.2 \times 10^6 = 19.2 \text{ MHz} \pm 38.4 \text{ Hz}$$

In a similar manner, it can be found that the control range parameter is 19.2 MHz ± 230.4 Hz and the aging range is 76.8 Hz.

6.4.3 Short-Term Frequency Stability and Phase Noise

The intention of this section is to present the reader with an overview of the two most common types of techniques in specifying frequency stability and accuracy. Time domain techniques rely on time-domain measurement and time jitter statistics such as the Allan Variance (see [4] and [5]). The other technique relies on frequency domain statistics, otherwise known as phase noise. Phase noise is defined either in terms of phase noise density in dBc/Hz at a certain frequency offset from the carrier or integrated in dBc or degrees over a given bandwidth. The subject of phase noise and time jitter will be discussed in depth in a future chapter and won't be discussed further in this section. The Allan variance is a measure of short-term frequency stability. It is particularly suitable, and more reliable than phase noise measurements, for measuring frequency stability at small offsets from the carrier. For example, UFO satellites reference oscillators are required to have a phase noise density of −46 dBc/Hz at 10 Hz due to vibration in airborne terminals. Spectrum analyzers cannot reliably measure close in-phase noise at 10 Hz offset. Test engineers resort to time domain techniques to perform these measurements. For a finite number of measurements N of adjacent frequency counts of time interval τ, the Allan variance can be approximated as

$$\sigma_y^2(\tau) \approx \frac{1}{2(N-1)f_{avg}^2} \sum_{n=1}^{N-1} \Delta_n^2(\tau) \tag{6.40}$$

where f_{avg} is the average frequency and $\Delta_n(\tau)$ is the difference in frequency between contiguous (or adjacent) averaging periods.

6.4.4 Tuning Speed and Settling Time

Tuning speed is defined as the time it takes the synthesizer to reach 90% or more of its final commanded frequency from the time a tuning step is applied. This time is most critical in TDMA and frequency-hopping applications. It is mostly defined in the form of transmit to transmit settling time and transmit to receive settling time, especially when the transceiver is designed to frequency-hop or time division multiplex in various frequency bands. The tuning speed is dictated by the architecture of the synthesizer and is mainly influenced by other requirements like phase noise, which in some cases can limit the tuning speed.

6.5 Modulation Accuracy: EVM and Waveform Quality Factor

Modulation accuracy is a measure of the transmitter's precision in transmitting a certain waveform. There are various ways to measure the modulation accuracy of the transmitter, error vector magnitude (EVM) and waveform quality factor being the most common.

6.5.1 EVM

The error vector magnitude is defined as the square root of the mean square of the difference (error) between the ideal and actual transmitted signal divided by the mean square of the actual signal, that is [6]:

$$EVM = \sqrt{\frac{\int |e(t)|^2 \, dt}{\int |y(t)|^2 \, dt}} \approx \sqrt{\frac{\sum_k |e(k)|^2}{\sum_k |y(k)|^2}} \tag{6.41}$$

where $y(k) \cong y(kT) = y(t)|_{t=kT}$ is the sampled desired output signal of $y(t)$ and $e(k) \cong e(kT) = e(t)|_{t=kT}$ is the sampled error signal. The error signal, as mentioned above, is defined as:

$$e(t) = y(t) - y_a(t) \tag{6.42}$$

where $y_a(t)$ is the actual signal. From (6.42), it is obvious that the EVM is related to "*transmit*" SNR as

$$SNR_{dB} = 10 \log_{10} \left(\frac{1}{EVM^2} \right) \tag{6.43}$$

Actual EVM measurements typically entail the use of test equipment that demodulates the transmitted waveform. The soft symbol decisions are then compared to the ideal symbol decisions, and an EVM measure is obtained according to (6.41) as depicted in Figure 6.2.

EVM is intended as an overall measure of the transmitter accuracy. Anomalies such as IQ imbalance, LO leakage, phase noise, nonlinear degradation, filter passband ripple and group delay, and thermal noise are all factors that contribute to EVM. The total EVM attributed to the various degradation effects is the sum of the square of EVMs due to each degradation:

$$EVM = \sqrt{\sum_k EVM_k^2} \tag{6.44}$$

For both TD-SCDMA and WCDMA, the required EVM for QPSK is less than or equal to 17.5%.

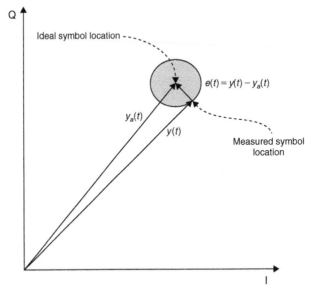

Figure 6.2 EVM is the vector designating the difference between the actual and measured symbols

6.5.2 Effect of Phase Noise on EVM

Assume that the desired output signal is $y(t) = A \sin(\omega t)$ and the actual output signal $y_a(t) = A \sin(\omega t + \rho(t))$. The error $e(t)$ is given as:

$$
\begin{aligned}
e(t) &= y(t) - y_a(t) \\
&= A\{\sin(\omega t) - \sin(\omega t + \phi(t))\} \\
&= A\{\sin(\omega t) - \sin(\omega t)\cos(\phi(t)) - \cos(\omega t)\sin(\phi(t))\} \\
&= A\{[1 - \cos(\phi(t))]\sin(\omega t) - \cos(\omega t)\sin(\phi(t))\}
\end{aligned}
\tag{6.45}
$$

In turn, using the trigonometric identities presented in Chapter 5, the relation in (6.45) may be expressed as:

$$
\begin{aligned}
e(t) &= A\left\{\left[1 - 1 + 2\sin^2\left(\frac{\phi(t)}{2}\right)\right]\sin(\omega t) - 2\sin\left(\frac{\phi(t)}{2}\right)\cos\left(\frac{\phi(t)}{2}\right)\cos(\omega t)\right\} \\
&= 2A\sin\left(\frac{\phi(t)}{2}\right)\left\{\sin\left(\frac{\phi(t)}{2}\right)\sin(\omega t) - \cos\left(\frac{\phi(t)}{2}\right)\cos(\omega t)\right\} \\
&= 2A\sin\left(\frac{\phi(t)}{2}\right)\sin\left(\omega t + \frac{\phi(t)}{2} + \frac{\pi}{2}\right)
\end{aligned}
\tag{6.46}
$$

Then, the EVM defined in (6.41) becomes:

$$
EVM = \sqrt{\frac{\int |e(t)|^2}{\int |y(t)|^2}} = \sqrt{\frac{\int \left|2A\sin\left(\frac{\phi(t)}{2}\right)\sin\left(\omega t + \frac{\phi(t)}{2} + \frac{\pi}{2}\right)\right|^2 dt}{\int |A\sin(\omega t)|^2 dt}}
\tag{6.47}
$$

where $\phi(t)$ is slowly varying and can be approximated by a constant that is $\phi \approx \phi(t)$, the relation in (6.47) becomes:

$$
\begin{aligned}
EVM &= \sqrt{\left|2E\left\{\sin\left(\frac{\varphi(t)}{2}\right)\right\}\right|^2} \\
&\approx \sqrt{2 - 2\cos(\varphi)}
\end{aligned}
\tag{6.48}
$$

And hence the transmit SNR:

$$SNR_{dB} = 10 \log_{10} \left(\frac{1}{\left| 2E\left\{ \sin\left(\frac{\varphi(t)}{2} \right) \right\} \right|^2} \right)$$

$$\approx 10 \log_{10} \left(\frac{1}{2 - 2\cos(\varphi)} \right) \tag{6.49}$$

The SNR degradation due to phase noise is depicted in Figure 6.3 and can be as low as $-6.02\,dB$.

Note that for a small value of $\phi(t)$, the EVM relationship in (6.48) can be approximated as

$$EVM = \sqrt{E\left\{ \phi^2(t) \right\}} = \sqrt{\int_{-\infty}^{\infty} L(f)\,df} \tag{6.50}$$

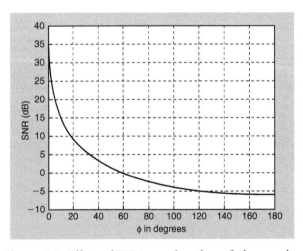

Figure 6.3 Effect of EVM as a function of phase noise

Example 6-3: Impact of Phase Noise on EVM

Assume the average SSB phase noise of a certain transmitter to be $-80\,dBc/Hz$. What is the EVM due to phase noise within the IF bandwidth of 300 kHz? Furthermore, assume that dual upconversion is used to bring the digital signal to RF, and that the DSB integrated phase noise of the first or VHF synthesizer to be $-27\,dBc$. What is the integrated phase noise of the second or UHF synthesizer?

According to (6.50), the EVM can be estimated as:

$$EVM = \sqrt{\int_{-\infty}^{\infty} L(f)\,df} = \sqrt{2 \times 10^{-80/10} \times 75{,}000} = 0.077 \text{ or } 7.7\% \qquad (6.51)$$

Then the integrated phase noise of the second synthesizer is:

$$\sqrt{10^{-27/10} + 10^{\Phi/10}} = 0.077$$
$$\Phi = 10\log_{10}(0.077^2 - 10^{-27/10}) = -24\,dBc \qquad (6.52)$$

Example 6-4: Contribution of Various PLL Components to Phase Noise

Consider a PLL used in a WCDMA transmitter, as shown in Figure 6.4. Given the phase noise of the various PLL components at given frequency offsets as shown in Table 6.2, compute the DSB integrated phase noise using a simple piece-wise integration technique. What is the impact on EVM and SNR?

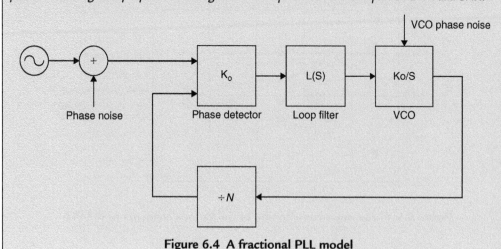

Figure 6.4 A fractional PLL model

Table 6.2 PLL phase noise of various PLL blocks at given frequency offsets

Frequency Offset (Hz)	VCO (dBc/Hz)	Ref (dBc/Hz)	Detector (dBc/Hz)	Loop Filter (dBc/Hz)
100	−216.7	−91.22	−101.3	−197.8
1000	−196.7	−111.2	−101.3	−177.8
10,000	−176.8	−130.6	−101.3	−157.8
100,000	−157.7	−137.5	−99.65	−138.6
1,000,000	−163.5	−152.5	−114.3	−142.9

Table 6.3 Total phase noise at various frequency offsets

Frequency Offset (Hz)	Total PN (dBc/Hz)
100	−90.8
1000	−100.9
10,000	−101.3
100,000	−99.6
1,000,000	−114.3

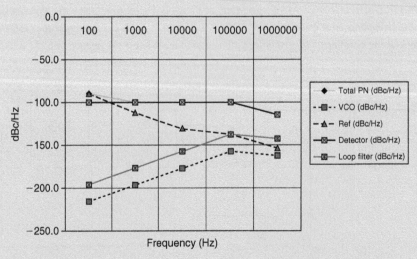

Figure 6.5 Phase noise contribution of the various elements of the PLL

The total phase noise of the various PLL components at the various frequency offsets is shown in Table 6.3 and depicted in Figure 6.5. The DSB integrated phase noise, EVM, and SNR are shown in Table 6.4.

Table 6.4 DSB integrated phase noise, EVM, and SNR

Parameter	Value
Integrated DSB Phase Noise (dBc/Hz)	-38.97554742
EVM (%)	1.125181619
SNR (dB)	19.48777371

6.5.3 Effect of Carrier Leakage on EVM

Although carrier leakage largely depends on the design and architecture of the transmitter, it is worthwhile to describe its effect on EVM from a traditional analog quadrature-modulation sense. The results can be easily modified to apply to IF sampling architectures. DC offsets at the analog I and Q can manifest themselves as an IF signal or carrier leakage. Let

$$s(t) = \underbrace{A_I(t)\cos(\theta(t))\cos(\omega_c t) - A_Q(t)\sin(\theta)(t))\sin(\omega_c t)}_{\text{desired signal}}$$

$$+ \underbrace{d_I \cos(\omega_c t) - d_Q \sin(\omega_c t)}_{\text{carrier leakage term due to I and Q DC offset}} \tag{6.53}$$

be the transmitted signal with carrier leakage, where $A_I(t)$ and $A_Q(t)$ are the in-phase and quadrature amplitudes of the I and Q signal and $\theta(t)$ is the corresponding phase angle. Furthermore, let d_I and d_Q be the DC offsets at the output of the I and Q DACs. The relationship in (6.53) can be re-expressed as:

$$s(t) = \underbrace{\sqrt{A_I^2(t)\cos^2(\theta(t)) + A_Q^2(t)\sin^2(\theta(t))}\cos(\omega_c t + \phi(t))}_{\text{desired signal}}$$

$$+ \underbrace{\sqrt{d_I^2 + d_Q^2}\cos(\omega_c t + \delta)}_{\text{carrier leakage}} \tag{6.54}$$

where:

$$\phi(t) = \tan^{-1}\left(\frac{A_Q(t)\sin(\theta(t))}{A_I(t)\cos(\theta(t))}\right) \tag{6.55}$$

and:

$$\delta = \tan^{-1}\left(\frac{d_Q}{d_I}\right) \tag{6.56}$$

Define the carrier suppression as the average carrier leakage P_{cl} to the average signal power P_s:

$$R = 10\log_{10}\left(\frac{P_{cl}}{P_s}\right) = 20\log_{10}\left(\frac{V_{cl}}{V_s}\right) \tag{6.57}$$

where V_{cl} and V_s are the average (RMS) voltage levels of the carrier leakage and desired signal respectively. The degradation in terms of EVM is then given as:

$$EVM = \sqrt{10^{R/10}} \tag{6.58}$$

This EVM is pertinent only to DC offset on the I and Q data paths. However, DC offset is not the only contributor to LO leakage, as will be seen in the following example.[6]

Example 6-5: Impact of Carrier Leakage on EVM in a Direct Conversion Transmitter

The crest factor or peak to average power ratio (PAR) of a WCDMA base station transmitter depends on the number of code channels transmitted, as shown in Table 6.5 [7]. Assume, for the sake of simplicity, that the degradation due to the analog blocks of the transmitter is negligible and that this quantity can be used to estimate the PAR of the envelope signal at baseband. Theoretically, after quadrature upconversion, the PAR at RF is 3 dB higher than the envelope PAR. The PAR of

---

[6]Note that in HSPA and LTE standards, origin offset or LO leakage is no longer counted in the definition of EVM. The reason is that smart base stations can, within reason, dial-out origin offsets, hence they are no longer part of Tx SNR.

Table 6.5 Envelope PAR versus number of code channels in a WCDMA transmitter system

Number of Code Channels	PAR (dB)
1	4.5
4	9
16	10
128	11

the envelope signal at baseband is also different than the PAR defined for the in-phase or quadrature signals defined at the DACs. Note that, when referring to the PAR of a signal, most engineers refer to the PAR of the envelope of the signal at baseband.

Consider the isolation between the LO and the upconverter to be I = −26 dBc. In order to maintain an EVM degradation of 6.2% due to LO leakage and DC offset, find the maximum allowable DC offset for the various number of code channels as listed in Table 6.5. For this example, let the peak-to-peak voltage of the discrete I and Q signals be 1.0 V.

The carrier suppression can be computed from the EVM and LO leakage as:

$$EVM = \sqrt{10^{I/10} + 10^{R/10}} \Rightarrow R = 10 \log_{10}(EVM^2 - 10^{I/10})$$

$$R = 10 \log_{10}\left(\left(\frac{6.2}{100}\right)^2 - 10^{-26/10}\right) = -29.63 \, dBc \tag{6.59}$$

The PAR is defined as

$$PAR = 10 \log_{10}\left(\frac{P_{peak}}{P_s}\right) = 10 \log_{10}\left(\frac{\max|\hat{S}(nT)|^2}{E\left\{\frac{1}{N}\|\hat{S}(nT)\|^2\right\}}\right) \tag{6.60}$$

where $\hat{S}(nT)$ is the sampled discrete baseband envelope signal. In terms of voltage ratios, the relation in (6.60) can be written as:

$$PAR = 10 \log_{10}\left(\frac{P_{peak}}{P_s}\right) = 20 \log_{10}\left(\frac{V_{peak}}{V_s}\right) \tag{6.61}$$

In this case, and accounting for the 3-dB difference between PAR of the envelope signal and that of, say, the in-phase or quadrature signal (see Chapter 7, for the underlying assumptions), the RMS voltage can be computed from (6.61) as:

$$V_s = 10^{-(PAR+3)/20} V_{peak} \tag{6.62}$$

The carrier leakage voltage due to DC offset can then be derived from the carrier suppression via the relation

$$R = 10\log_{10}\left(\frac{P_{cl}}{P_s}\right) = 20\log_{10}\left(\frac{V_{cl}}{V_s}\right) \tag{6.63}$$

Finally, the carrier leakage voltage due to DC offset voltages only is:

$$V_{cl} = 10^{R/20} V_s \tag{6.64}$$

and the maximum allowed DC offset on the I or Q is:

$$d_{I,Q} = \sqrt{2}V_{cl} = \sqrt{2} \times 10^{R/20} \times V_s \tag{6.65}$$

The results for the maximum allowable DC offsets for the various numbers of code channels are summarized in Table 6.6.

Table 6.6 Maximum allowable DC offset on the I and Q channels for various numbers of code channels

Number of Code Channels	PAR (dB)	V_s (volt)	DC offset (mVolt)
1	4.5	0.420697571	21.71478972
4	9	0.250593617	51.6161519
16	10	0.223341796	11.52804407
128	11	0.199053585	10.27438009

6.5.4 Effect of IQ Imbalance on EVM

Consider the transmitter shown in Figure 6.6. Assume the transmitted signal on the I and Q to be a simple sinusoid:

$$I(t) + jQ(t) = Ae^{j\omega t} \tag{6.66}$$

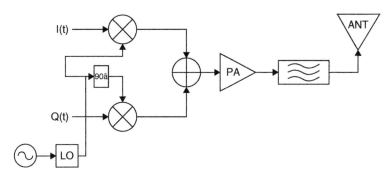

Figure 6.6 Conceptual direct conversion transmitter

where A is the signal's amplitude. The I and Q signals are mixed with the LO signals:

$$LO_I(t) = A_{LO} \cos(\omega_c t)$$
$$LO_Q(t) = (A_{LO} + \delta) \sin(\omega_c t + \theta)$$

(6.67)

where A_{LO} is LO voltage signal amplitude, δ is the amplitude imbalance, and θ is the phase imbalance. The transmitted signal at RF is then expressed as the sum of the desired signal plus the image signal:

$$x(t) = x_{desired}(t) + x_{image}(t)$$
$$x_{desired}(t) = \frac{AA_{LO}}{2} \cos(\omega_c t + \omega t) + \frac{A(A_{LO} + \delta)}{2} \cos(\omega_c t + \omega t + \theta)$$
$$x_{image}(t) = \frac{AA_{LO}}{2} \cos(\omega_c t - \omega t) - \frac{A(A_{LO} + \delta)}{2} \cos(\omega_c t - \omega t + \theta)$$

(6.68)

The image suppression is the ratio of the image signal power to the desired signal power (or the inverse of image rejection):

$$IS = \frac{1}{IR} = \frac{A^2 A_{LO}^2 - 2A^2 A_{LO}(A_{LO} + \delta) \cos\theta + A^2(A_{LO} + \delta)^2}{A^2 A_{LO}^2 + 2A^2 A_{LO}(A_{LO} + \delta) \cos\theta + A^2(A_{LO} + \delta)^2}$$

(6.69)

To simplify the analysis, we can normalize the expression (6.69) by letting $A = A_{LO} = 1$; the image suppression then becomes:

$$IS = \frac{1}{IR} = \frac{1 - 2(1 + \delta)\cos\theta + (1 + \delta)^2}{1 + 2(1 + \delta)\cos\theta + (1 + \delta)^2} \qquad (6.70)$$

The EVM due to image suppression is

$$EVM = \sqrt{10^{IS/10}} \qquad (6.71)$$

Example 6-6: Impact of IQ Imbalance on EVM in a Direct Conversion Transmitter

Given that the amplitude imbalance of a direct conversion receiver is 0.5 dB and the phase imbalance is 5 degrees, what is the impact of this IQ imbalance on EVM? Recall that the amplitude imbalance is $10\log_{10}(1 + \delta)$ dB.

Using the relations expressed in (6.70) and (6.71), the IQ imbalance and its impact on EVM are summarized in Table 6.7.

Table 6.7 Impact of IQ imbalance on EVM

Parameter	Value
Amplitude imbalance (dB)	0.5
Phase imbalance (degrees)	5
Image suppression (dB)	−25.63163702
EVM (%)	5.228994065

6.5.5 Effect of Channel Noise on EVM

The impact of the in-band channel noise on EVM is most profound at low transmit power and is almost insignificant when the transmitter is transmitting at high power. The impact on EVM can simply be stated as

$$EVM = \sqrt{10^{\frac{P_{TX,noisefloor} - P_{TX}}{10}}} \qquad (6.72)$$

where $P_{TX,noisefloor}$ is the transmitter noise floor and P_{TX} is the transmit signal power. This effect is best illustrated with a practical example.

Example 6-7: Impact of In-Band Channel Noise on EVM

In WCDMA, the minimum transmit power is $-50\,dBm$. In order to maintain the impact of channel noise on EVM to less than 8%, compute the maximum integrated noise power.

Using the relation in (6.72), the integrated noise power can be estimated as shown in Table 6.8.

Table 6.8 Impact of in-band channel noise on EVM

Parameter	Value
EVM (%)	8
P_{TX} (dBm)	-50
Integrated noise power (dBm)	-71.93820026

6.5.6 Effect of Nonlinearity on EVM

The EVM due to nonlinearity in a wireless transmitter can be defined with reference to the 1-dB output compression point $P_{output,1\,dB}$. Define the instantaneous transmit signal distortion in terms of excess amplitude level above the 1-dB OCP as:

$$D(t) = P_{Tx}(t) - P_{output,1\,dB} + 1\,dB \tag{6.73}$$

where $P_{TX}(t)$ is the desired signal's instantaneous transmit power. Assume that the phase distortion is minimal,[7] then the actual transmit signal $y_a(t)$ can be expressed as:

$$y_a(t) = 10^{\frac{P_{TX}(t)+D(t)}{20}} e^{j\varphi(t)} \tag{6.74}$$

where $\phi(t)$ is the ideal phase response. The desired signal, on the other hand, is given as:

$$y(t) = 10^{\frac{P_{TX}(t)}{20}} e^{j\varphi(t)} \tag{6.75}$$

---

[7]This assumption breaks down at high transmit power and especially when dealing with nonmemoryless systems.

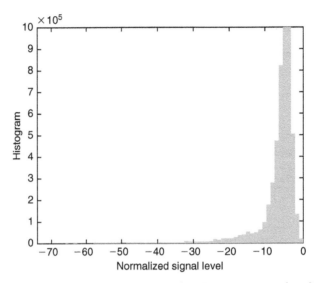

Figure 6.7 Histogram of transmitted TD-SCDMA signal

The resulting EVM, in discrete notation, must be weighed with the linear amplitude PDF p_k of the desired transmit signal as:

$$EVM = \sqrt{\frac{\sum_k |e(k)|^2}{\sum_k |y(k)|^2}} = \sqrt{\frac{10^{\frac{P_{TX}(k)}{20}} \sum_k p_k^2 \left|10^{\frac{D(k)}{20}} - 1\right|^2}{10^{\frac{P_{TX}(k)}{20}}}} = \sqrt{\sum_k p_k^2 \left|10^{\frac{D(k)}{20}} - 1\right|^2} \quad (6.76)$$

In order to compute the EVM, a histogram of the transmitted power is needed. In TD-SCDMA, for example, this is not a concern if the modulation used is QPSK. However, this becomes a big contributor to EVM for higher order constellations. A simulation of a TD-SCDMA signal is shown in Figure 6.7. A characterization of the PA based on this distribution is needed to estimate the impact on EVM.

6.5.7 Effect of Filtering on EVM

Depending on the modulation scheme, a nonideal filter can potentially cause a significant degradation to the EVM, mainly due to the filter's passband ripple and group delay. Pulse-shaping filters, however, like the root raised cosine filters used in WCDMA and

TD-SCDMA, are usually not considered in estimating the EVM. The reason is that a transmit pulse-shaping filter is matched at the receiver with an identical receive pulse-shaping filter to form, as in the case of WCDMA and TD-SCDMA, for example, a raised cosine filter. A raised cosine filter is a Nyquist filter and consequently it is ISI-free as discussed in Chapter 2. In general, it can be shown that any filter pair that results in a Nyquist filter does not affect the EVM. In most standards, EVM is defined assuming an ISI-free detection. To show this, consider the data stream:

$$x(t) = \sum_n d(n)\delta(t - nT_s) \tag{6.77}$$

where $d(n)$ is the data stream and T_s is the symbol rate. In general, passing the data through a shaping filter (analog or digital) $h(t)$, the data stream presented in (6.77) becomes:

$$x(t) = \sum_n d(n)\delta(t - nT_s) \oplus h(t) = \sum_n d(n)\delta(t - nT_s) \oplus h(t)$$
$$= \sum_n d(n)h(t - nT_s) \tag{6.78}$$

where $\oplus$ denotes the convolution operator. On the receive side, and neglecting the effect of the channel to simplify the analysis, the data is matched with a shaping filter, $h^*(-t)$ where * denotes complex conjugation:

$$y(t) = x(t) \oplus h^*(-t) = \sum_n d(n)h(t - nT_s) \oplus h^*(-t)$$
$$= \sum_n d(n) \int_{-\infty}^{\infty} h(\tau - nT_s)h^*(\tau - t)d\tau \tag{6.79}$$

However, when sampled at $t = l\,T_s$ the convolution integral in (6.79) becomes:

$$\int_{-\infty}^{\infty} h(\tau - nT_s)h^*(\tau - lT_s)d\tau = C\delta(l - n) = \begin{cases} C = \text{constant} & l = n \\ 0 & l \neq n \end{cases} \tag{6.80}$$

where C is a constant. Substituting (6.80) into (6.79) we obtain:

$$y(n) \cong y(t)\big|_{t=nT_s} = \sum_n d(n) \int_{-\infty}^{\infty} h(\tau - nT_s)h^*(\tau - lT_s)d\tau = d(n) \tag{6.81}$$

This is true due to the Nyquist characteristic of the resulting filter. Therefore, when computing the EVM of the transmitter, the effect of the shaping filter will be ignored. Next, consider the EVM due to a filter $h(k)$. The filtered signal is given as:

$$y_a(k) = h(k) \oplus y(k) = \sum_{n=-\infty}^{\infty} y(k-n)h(n) = y(k)h(0) + \sum_{\substack{n=-\infty \\ n\neq 0}}^{\infty} y(k-n)h(n) \quad (6.82)$$

where $y(k)$ is approximated as a zero mean Gaussian process and variance $E\{y^2(k)\} = \sigma_y^2$. Next, consider the error:

$$e(k) = y(k) - y_a(k) = [1 - h(0)]y(k) - \sum_{\substack{n=-\infty \\ n\neq 0}}^{\infty} y(k-n)h(n) \quad (6.83)$$

Taking the expected value of the error square in (6.83) we obtain

$$E\{e^2(k)\} = E\left\{\left(\left[1 - h(0)\right]y(k) - \sum_{\substack{n=-\infty \\ n\neq 0}}^{\infty} y(k-n)h(n)\right)^2\right\}$$

$$= [1 - h(0)]^2 E\{y^2(k)\} - 2[1 - h(0)] \sum_{\substack{n=-\infty \\ n\neq 0}}^{\infty} E\{y(k)y(k-n)\}h(n)$$

$$+ \sum_{\substack{n=-\infty \\ n\neq 0}}^{\infty} \sum_{\substack{l=-\infty \\ l\neq 0}}^{\infty} E\{y(k-n)y(k-l)\}h^2(n) \quad (6.84)$$

The autocorrelation function satisfies the relation:

$$E\{y(k-n)y(k-l)\} = \begin{cases} \sigma_y^2 & l = n \\ 0 & l \neq n \end{cases} \quad (6.85)$$

Then (6.84) becomes:

$$E\{e^2(k)\} = [1 - h(0)]^2 \sigma_y^2 + \sum_{\substack{n=-\infty \\ n\neq 0}}^{\infty} \sigma_y^2 h^2(n) \quad (6.86)$$

and the EVM due to filtering is finally given as:

$$EVM = \sqrt{\frac{[1 - h(0)]^2 \sigma_y^2 + \sum_{\substack{n=-\infty \\ n \neq 0}}^{\infty} \sigma_y^2 h^2(n)}{\sigma_y^2}} = \sqrt{[1 - h(0)]^2 + \sum_{\substack{n=-\infty \\ n \neq 0}}^{\infty} h^2(n)} \qquad (6.87)$$

Therefore, the EVM due to filtering can be estimated purely by its coefficients. Following this method, the analog filter must then be transformed to an equivalent digital IIR filter. The IIR digital filter is then approximated with an FIR filter. A simpler nonanalytic approach will involve simulating an AWGN data stream passing through the filter and computing the impact on EVM.

6.5.8 Relationship between EVM and Waveform Quality Factor

The waveform quality factor ρ,[8] like EVM, is a measure of the accuracy of the *distorted* modulated waveform at RF compared to the *ideal* waveform at digital baseband or IF. More specifically, ρ is a correlation measure between the actual waveform and the ideal waveform defined as [9]:

$$\rho = \frac{E^2\{y(t)y_a^*(t)\}}{E\{|y(t)|^2\}E\{|y_a(t)|^2\}} \qquad (6.88)$$

To establish a relationship between ρ and SNR, let the actual signal be written as the sum of the ideal signal plus an error signal:

$$y_a(t) = y(t) + e(t) \qquad (6.89)$$

Then the cross-correlation between the actual and ideal signal is:

$$E\{y(t)y_a^*(t)\} = E\{y(t)(y(t) + e(t))^*\} = E\{y(t)y^*(t)\} + E\{y(t)e^*(t)\} \qquad (6.90)$$

Assuming that the data is zero-mean, (i.e., $E\{y(t)\} = 0$) and that the error function is uncorrelated to the desired signal (i.e., $E\{y(t)e^*(t)\} = 0$) then let $E\{|y(t)|^2\} = \sigma_y^2$ and

---

[8]The waveform quality factor ρ is defined in J-STD-019 [8].

$E\{|e(t)|^2\} = \sigma_e^2$ where the error signal is also zero-mean. Furthermore, consider the relation:

$$E\{|y_a(t)|^2\} = E\{|y(t)|^2\} + 2E\{y(t)e^*(t)\} + E\{|e(t)|^2\} = \sigma_y^2 + \sigma_e^2 \qquad (6.91)$$

Substituting (6.91) into (6.88), we obtain:

$$\rho = \frac{\sigma_y^4}{\sigma_y^2(\sigma_y^2 + \sigma_e^2)} = \frac{\sigma_y^2}{\sigma_e^2\left(\dfrac{\sigma_y^2}{\sigma_e^2} + 1\right)} = \frac{SNR}{SNR + 1} \qquad (6.92)$$

Recall that the SNR is inversely proportional to the square of the EVM; this implies that a relationship between EVM and ρ can be derived as:

$$\rho = \frac{1}{EVM^2 + 1} \qquad (6.93)$$

Recall that (6.92) and hence (6.93) are approximations, since it is assumed that the cross-correlation between the ideal signal and the error is negligible. This is particularly true when the source of the error is thermal noise. This approximation becomes less and less reliable in the presence of nonlinear degradation.

Example 6-8: EVM versus ρ for IS95 Mobile Station

The transmit waveform of a CDMA mobile station must correlate 94.4% to the ideal transmitted signal. What are the EVM and SNR values as the correlation coefficient ρ increases from 94.4% to 100%?

From (6.92) and (6.93), we can express EVM and the corresponding SNR as

$$SNR = 10\log_{10}\left(\frac{\rho}{1 - \rho}\right)$$

$$EVM = \sqrt{\frac{1}{\rho} - 1} \qquad (6.94)$$

The EVM and SNR are depicted in Figure 6.8 and Figure 6.9. As one might expect, as the correlation coefficient increases between the actual and the ideal signal, the EVM decreases and the SNR increases.

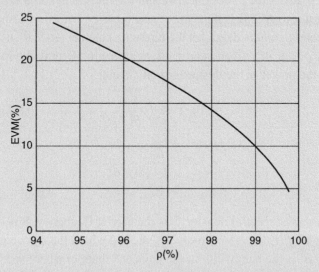

Figure 6.8 EVM as a function of ρ of the transmitted CDMA signal

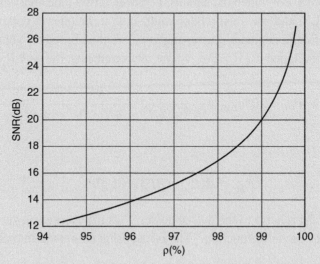

Figure 6.9 SNR as a function of ρ of the transmitted CDMA signal

6.6 Adjacent Channel Leakage Ratio (ACLR)

ACLR is a figure of merit that describes the maximum allowable spectral regrowth in the adjacent band. It is attributed to nonlinearities in the PA and in some cases the PA-driver. The main contributor to ACLR is third order nonlinearity distortion compared to, say, fifth order nonlinearity, which degrades the alternate channel (ACLR2). ACLR is defined as the ratio of the power in a given bandwidth at a certain offset in the lower or upper adjacent band to the power in the desired signal's band:

$$ACLR = \frac{\displaystyle\int_{f_c+f_0}^{f_c+f_0+B'} S(f)df}{\displaystyle\int_{f_c-B/2}^{f_c+B/2} S(f)df} \tag{6.95}$$

where $S(f)$ is the power spectral density, f_c is the carrier frequency, B is the bandwidth of the desired signal, f_o is the offset frequency around which the adjacent channel power is integrated, and B' is the bandwidth in the adjacent channel where spectral regrowth is measured. Traditionally, two-tone analysis was employed to characterize ACLR for narrowband systems. For wideband signals, however, a two-tone analysis is not sufficient to determine ACLR, and one must rely on multi-tone techniques to better evaluate the transmitter out-of-band degradation. An analytical method based on Volterra-Wienner techniques was developed in [10]. For N-tone measurements, the ACLR is given as:

$$ACLR = P_{IM} - 10\log_{10}\left(\frac{3N^3}{4N^3 - 3N^2 - 4N + 3(1 - \text{mod}_2(N))}\right) \tag{6.96}$$

where

$$P_{IM} = 2(P_i - OIP3) - 6\,\text{dB} \tag{6.97}$$

is the third order IM relative to the output power, P_i is the average power of a single tone, and $\text{mod}_2(N)$ is the modulus function.[9] The relationship expressed in (6.96) implies that

---

[9] The modulus function $\text{mod}_M(N) = N - xM$ where $x = \text{floor}(N/M)$.

all the tones used in determining the ACLR are coherent and that the adjacent channel occupies the same bandwidth as the desired signal channel. In the event where the signaling channel bandwidth is different from that of the adjacent channel bandwidth, a correction factor of $10 \log_{10}(B'/B)$ needs to be added to (6.96). A further adjustment to (6.96) includes the statistics of the transmitted signal in the form of PAR. Since (6.96) utilizes the average power of the various tones to compute ACLR, a correction by the PAR is necessary when estimating the behavior of actual signals. Finally, the relationship in (6.96) can be adjusted to include both bandwidth considerations and ACLR as:

$$ACLR = P_{IM} - 10 \log_{10} \left(\frac{3N^3}{4N^3 - 3N^2 - 4N + 3(1 - \mathrm{mod}_2(N))} \right) + PAR$$
$$+ 10 \log_{10} \left(\frac{B'}{B} \right) \tag{6.98}$$

Finally, if the signal is best approximated by a bandlimited white Gaussian noise, the ACLR was found to approximate

$$ACLR = P_{IM} + 10 \log_{10} \left(\frac{3}{4} \right) = P_{IM} - 1.25 \, dB \tag{6.99}$$

Example 6-9: ACLR for TD-SCDMA

The PAR for a TD-SCDMA waveform is 4.5 dB. Based on the maximum allowable transmit average power of 24 dBm, let the single tone average power be 21 dBm. Using two-tone analysis and (6.98), compute the ACLR of the transmitter given that OIP3 = 45 dBm. How do the results vary for 3, 4, 5, and 6 tones?

From the relation in (6.98), the two-tone ACLR is computed as:

$$ACLR = P_{IM} - 10 \log_{10} \left(\frac{3N^3}{4N^3 - 3N^2 - 4N + 3(1 - \mathrm{mod}_2(N))} \right)$$
$$+ PAR + 10 \log_{10} \left(\frac{B'}{B} \right)$$
$$ACLR = -54 - 10 \log_{10} (1.6) + 4.5 = -51.54 \, dBc \tag{6.100}$$

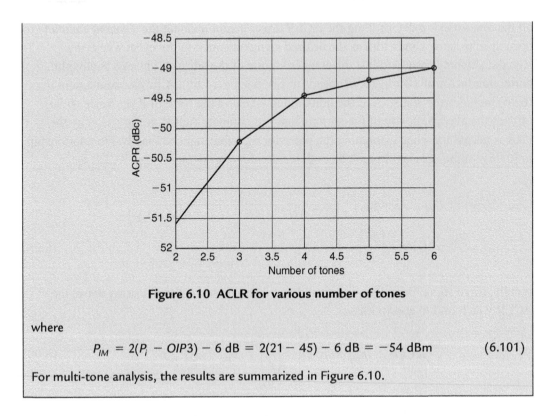

Figure 6.10 ACLR for various number of tones

where

$$P_{IM} = 2(P_i - OIP3) - 6\,dB = 2(21 - 45) - 6\,dB = -54\,dBm \qquad (6.101)$$

For multi-tone analysis, the results are summarized in Figure 6.10.

6.7 Transmitter Broadband Noise

Broadband noise generated by the PA can degrade the sensitivity of the receiver. This type of noise can be mitigated by designing better PAs, which may prove to be expensive and not so easily accomplished, or via filtering and transmit-receive isolation.

Example 6-10: PA Noise in UHF Follow-On (UFO) SATCOM

Assume that in a military airborne platform, a UV application is transmitting while a UFO SATCOM is receiving. Both applications are running on a single JTRS SDR platform. The UV PA is transmitting at 50 Watts. Its broadband noise in the UFO SATCOM receive band is −165 dBc/Hz. Assume that the sensitivity of the UFO SATCOM receiver is −126 dBm in 30 kHz band (occupied signal bandwidth is 25 kHz). What is the required isolation between the Tx and Rx chain such that the

SATCOM terminal noise floor is not degraded by more than 0.7 dB? The required C/KT for the 25-kHz channel is 47.6 dB-Hz. The signaling rate is 16 Kbps.

The E_b/N_o of the SATCOM terminal can be estimated as:

$$CNR_{dB} = 10\log_{10}\left(\frac{E_b}{N_0}\frac{R_b}{B}\right) = SNR_{dB} + 10\log_{10}\left(\frac{R_b}{B}\right)$$

$$\left.\frac{E_b}{N_0}\right|_{dB} = \left.\frac{C}{KT}\right|_{dB} - 10\log_{10}(R_b) = 47.6 - 10\log_{10}(16,000) = 5.56 \, dB \qquad (6.102)$$

where R_b is the data rate in bits per second. The required CNR is then given as:

$$CNR = \left.\frac{E_b}{N_0}\right|_{dB} + 10\log_{10}\left(\frac{R_b}{B}\right)$$

$$CNR = 5.56 + 10\log_{10}\left(\frac{16,000}{30,000}\right) = 2.83 \, dB \qquad (6.103)$$

From the sensitivity and the CNR of the SATCOM terminal, we can determine the noise floor of the receiver:

$$P_{noisefloor} = P_{dBm} - CNR_{dB} = -126 - 2.83 = -128.83 \text{ dBm} \qquad (6.104)$$

The allowable degradation is given as the difference between the total noise floor minus the ideal noise floor

$$P_D = 10\log_{10}\left(10^{P_T/10} - 10^{P_{noisefloor}/10}\right)$$

$$P_D = 10\log_{10}(10^{(-128.83+0.7)/10} - 10^{-128.83/10}) = -136.4 \text{ dBm} \qquad (6.105)$$

Again, recall that the total noise floor is the ideal noise floor plus the allowable degradation. Finally, the requirement states that the system isolation must be such that the degraded noise floor due to integrated broadband PA noise does not exceed -136.4 dBm; that is,

$$P_{TX,dBm} + P_{BBN,dBc.Hz} + 10\log_{10}(B) - Isolation = P_D$$

$$Isolation = -P_D + P_{TX,dBm} + P_{BBN,dBc.Hz} + 10\log_{10}(B)$$

$$Isolation = 136.4 + 47 - 165 + 10\log_{10}(30,000 \text{ Hz}) = 63.17 \text{ dB} \qquad (6.106)$$

where the transmit power in dBm is $P_{TX,dBm} = 10\log_{10}(50e3) \approx 47$ dbm.

References

[1] Ramachandran B, et al. Key specifications and implementations of WCDMA receiver. IEEE proceedings on VLSI Technology and Applications 2001:49–52.

[2] Watson R. Receiver Dynamic Range: Part 1. Watkins-Johnson Technical Note, WJ Communications, San Jose, CA.

[3] Watson R. Receiver Dynamic Range: Part 2. Watkins-Johnson Technical Note, WJ Communications, San Jose, CA.

[4] Saleh A. Frequency-independent and frequency independent nonlinear models of TWT amplifiers. IEEE Trans. Commun. November 1981;COM-29(11):1715–20.

[5] Allan DW. The measurement of frequency and frequency stability of precision oscillators. Proc. 6th PTTI Planning Meeting 1975:109.

[6] Mashour A, Borjak A. A method for computing error vector magnitude in GSM EDGE systems—simulation results. IEEE Commun. Letters March 2001;5(3):88–91.

[7] Pretl H, et al. Linearity considerations of W-CDMA front-ends for UMTS. IEEE Microwave Symposium Digest June 2000;1:433–6.

[8] ANSI J-STD-019: Recommended Minimum Performance Requirements for Base Stations Supporting 1.8 to 2.0 GHz Code Division Multiple Access (CDMA) Personal Stations (1996).

[9] Gharaibeh KM, et al. Accurate estimation of digital communication system metrics—SNR, EVM and ρ in a nonlinear amplifier environment. IEEE ARTFG Microwave Measurements Conference Dec. 2004:41–4.

[10] Pedro JC, Borges de Carvalho N. On the use of multitone techniques for assessing RF components' intermodulation distortion. IEEE Trans. on Microwave Theory and Techniques December 1999;47(12).

CHAPTER 7

Uniform Sampling of Signals and Automatic Gain Control

Uniform sampling of both bandlimited lowpass signals and bandpass signals is studied in detail. The conditions for sampling a signal without aliasing are reviewed. Special scenarios relating to the IF frequency and the sampling rate are explained. Signal reconstruction as it relates to the sampling theorem is also discussed. In the context of data conversion, it is imperative to the performance of the modem to control the long-term changes in the received signal strength, thus preserving the overall dynamic range. This must be done without introducing significant clipping if the signal becomes large, or under-representation by the data converter if the signal becomes small. Either case would degrade the signal quality and would have adverse effects on the desired SNR. The automatic gain control algorithm discussed in this chapter, one of several ways of implementing gain control, serves to regulate the signal strength in order to preserve optimal performance.

This chapter is divided into four major sections. In Sections 7.1 and 7.2, the uniform sampling theorem is discussed in some detail as it relates to both lowpass and bandpass sampling. That is, uniform sampling as applied to quadrature baseband signaling as well as complex sampling at IF or RF is discussed. The AGC algorithm is considered in Section 7.3. This section also addresses the implication of the peak to average power ratio of a signal and its impact on the AGC. Section 7.4 contains the appendix.

7.1 Sampling of Lowpass Signals

The sampling theorem requires that a lowpass signal be sampled at least at twice the highest frequency component of the analog band-limited signal. This in essence ensures that the spectral replicas that occur due to sampling do not overlap and the original signal can be reconstructed from the samples with *theoretically* no distortion.

7.1.1 Signal Representation and Sampling

Given a band-limited analog lowpass signal $x_a(t)$—that is, the highest frequency component of $x_a(t)$ is strictly less than a given upper bound, say $B/2$—then $x_a(t)$ can be suitably represented by a discrete signal $x(n)$ made up of uniformly spaced samples collected at a minimum rate of B samples per second. B is known as the Nyquist rate. Inversely, given the discrete samples $x(n)$ sampled at least at the Nyquist rate, the original analog signal $x_a(t)$ can then be reconstructed without loss of information. In the remainder of this section, we will discuss the theory behind lowpass sampling.

The Fourier transform of an analog signal $x_a(t)$ is expressed as

$$X_a(F) = \int_{-\infty}^{\infty} x_a(t)e^{-j2\pi Ft}\,dt \tag{7.1}$$

The analog time domain signal can be recovered from $X_a(F)$ via the inverse Fourier transform as:

$$x_a(t) = \int_{-\infty}^{\infty} X_a(F)e^{j2\pi Ft}\,dF \tag{7.2}$$

Next, consider sampling $x_a(t)$ periodically every T_s seconds to obtain the discrete sequence:

$$x(n) = x_a(t)\big|_{t=nT_s} \equiv x_a(nT_s) \tag{7.3}$$

The spectrum of $x(n)$ can then be obtained via the Fourier transform of discrete aperiodic signals as:

$$X(f) = \sum_{n=-\infty}^{\infty} x(n)e^{-j2\pi fn} \tag{7.4}$$

Similar to the analog case, the discrete signal can then be recovered via the inverse Fourier transform as:

$$x(n) = \int_{-1/2}^{1/2} X(f)e^{j2\pi fn}\,df \tag{7.5}$$

To establish a relationship between the spectra of the analog signal and that of its counterpart discrete signal, we note from (7.2) and (7.3) that:

$$x(n) = x_a(nT_s) = \int_{-\infty}^{\infty} X_a(F)e^{j2\pi\frac{F}{F_s}n} dF \tag{7.6}$$

Note that from (7.6), periodic sampling implies a relationship between analog and discrete frequency $f = F/F_s$, thus implying that when comparing (7.5) and (7.6) we obtain:

$$\int_{-1/2}^{1/2} X(f) e^{j2\pi fn} df \Bigg|_{\substack{f=F/F_s \\ df=dF/F_s}} = \frac{1}{F_s}\int_{-F_s/2}^{F_s/2} X\left(\frac{F}{F_s}\right)e^{j2\pi\frac{F}{F_s}n} dF = \int_{-\infty}^{\infty} X_a(F)e^{j2\pi\frac{F}{F_s}n} dF \tag{7.7}$$

The integral on the right hand side of (7.7) can be written as the infinite sum of integrals:

$$\int_{-\infty}^{\infty} X_a(F)e^{j2\pi\frac{F}{F_s}n} dF = \sum_{l=-\infty}^{\infty} \int_{(l-1/2)F_s}^{(l+1/2)F_s} X_a(F)e^{j2\pi\frac{F}{F_s}n} dF \tag{7.8}$$

Recognizing that $X_a(F)$ in the interval $((l-1/2)F_s,(l+1/2)F_s)$ is equivalent to $X_a(F+lF_s)$ in the interval $(-F_s/2, F_s/2_s)$, then the summation term on the right hand side of (7.8) becomes:

$$\sum_{l=-\infty}^{\infty} \int_{(l-1/2)F_s}^{(l+1/2)F_s} X_a(F)e^{j2\pi\frac{F}{F_s}n} dF = \sum_{l=-\infty}^{\infty} \int_{-F_s/2}^{F_s/2} X_a(F+lF_s)e^{j2\pi\frac{F+lF_s}{F_s}n} dF \tag{7.9}$$

Swapping the integral and the summation sign on the right hand side of (7.9), and noting that:

$$e^{j2\pi\frac{F+lF_s}{F_s}n} = e^{j2\pi\frac{lF_s}{F_s}n}e^{j2\pi\frac{F}{F_s}n} = e^{j2\pi\frac{F}{F_s}n} \tag{7.10}$$

we obtain:

$$\sum_{l=-\infty}^{\infty} \int_{-F_s/2}^{F_s/2} X_a(F+lF_s)e^{j2\pi\frac{F+lF_s}{F_s}n} dF = \int_{-F_s/2}^{F_s/2} \sum_{l=-\infty}^{\infty} X_a(F+lF_s)e^{j2\pi\frac{F_s}{F_s}n} dF \tag{7.11}$$

Comparing (7.7) and (7.11) we obtain:

$$\frac{1}{F_s} \int_{-F_s/2}^{F_s/2} X\left(\frac{F}{F_s}\right) e^{j2\pi\frac{F}{F_s}n} dF = \int_{-F_s/2}^{F_s/2} \sum_{l=-\infty}^{\infty} X_a(F + lF_s) e^{j2\pi\frac{F_s}{F_s}n} dF \quad (7.12)$$

And hence, from (7.12), one can deduce the relation:

$$X\left(\frac{F}{F_s}\right) = F_s \sum_{l=-\infty}^{\infty} X_a(F + lF_s) \text{ or}$$

$$X(f) = F_s \sum_{l=-\infty}^{\infty} X_a\left[(f + l)\right] F_s \quad (7.13)$$

The relation in (7.13) implies that the spectrum of $X(f)$ is made up of replicas of the analog spectrum $X_a(F)$ periodically shifted in frequency and scaled by the sampling frequency F_s as shown in Figure 7.1.

The relationship in (7.13) expresses the association between the analog spectrum and its discrete counterpart. The discrete spectrum is essentially made up of a series of periodic replicas of the analog spectrum. If the sampling frequency F_s is selected such that $F_s \geq B$, where B is the IF bandwidth not to be confused with the baseband bandwidth, which is only the positive half of the spectrum, the analog signal can then be reconstructed without loss of information due to aliasing from the discrete signal via filtering scaled by F_s. Note that the minimum sampling frequency allowed to reconstruct the analog signal from its discrete-time counterpart is $F_s = B$. This frequency is known as the Nyquist rate. However, if F_s is chosen such that $F_s < B$, then the reconstructed analog signal suffers from loss of information due to aliasing, as shown in Figure 7.1(c) and (d), and an exact replica of the original analog signal cannot be faithfully reproduced. In this case, as can be seen from Figure 7.1(c), the discrete spectrum is made up of scaled overlapped replicas of the original spectrum $X_a(F)$. The reconstructed signal is corrupted with aliased spectral components coexisting on top of the original spectral components as can be seen in Figure 7.1(d), thus preventing us from recreating the original analog signal.

7.1.2 Out-of-Band Energy

A band-limited lowpass signal is theoretically defined as a signal that has no frequency components above a certain upper frequency. Similarly, a bandpass band-limited signal is a

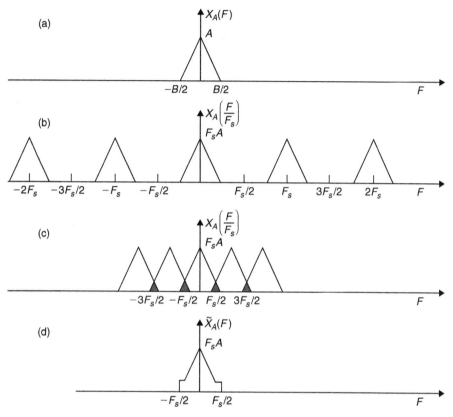

Figure 7.1 Lowpass sampling of analog signals: **(a)** spectrum of analog bandlimited signal, **(b)** spectrum of discrete-time sampled analog signal scaled by F_s and replicated, **(c)** aliased spectrum of discrete-time sampled signal, and **(d)** spectrum of reconstructed analog signal derived from aliased discrete-time signal in **(c)**

signal that is bounded by a certain lower frequency limit and a certain upper frequency limit. In reality, however, a received signal at an antenna is not strictly band-limited per the definition just given. Certain out-of-band frequency components are attenuated via filtering but not entirely eliminated. Thermal noise, which is white Gaussian in nature, is also always present at the input of the ADC. This out-of-band energy at the input of the ADC must be managed properly in order to minimize the distortion imposed on the desired signal by aliasing. Therefore, one key question the designer must ask is: how much SNR degradation is tolerable?

In order to answer the question properly, the analysis must take into account the entire receive-chain, and consider the sampling frequency and the location of the aliased signal components. What is meant by the latter is whether the aliased distortion components lie within the signal bandwidth or not. In the event where the aliased distortion does not overlap the desired signal after sampling, digital filtering can be used to recover the desired signal; otherwise the designer must solely rely on the performance of the antialiasing filter. The anti-aliasing filter must be designed such that performance distortion due to spectral overlap is less than the largest ADC spur appearing between DC and the Nyquist rate, as well as less than ½ LSB. These requirements tend to be very stringent on the filter design and can be alleviated by oversampling the signal and spreading the spectral replicas farther apart. Oversampling enables less complex anti-aliasing filter design with gradual roll-off at the expense of higher complexity and power consumption at the ADC. Oversampling requires that the ADC run at a faster clock rate, thus increasing the power consumption and complexity of the circuitry.

7.1.3 Reconstruction of Lowpass Signals

The Fourier transform of the analog signal $x_a(t)$ is:

$$x_a(t) = \int_{-\infty}^{\infty} X_a(F)e^{j2\pi Ft}dF \tag{7.14}$$

Assume for simplicity's sake that the discrete-time signal is sampled at the Nyquist rate $F_s = B$, that is:

$$X_a(F) = \begin{cases} \dfrac{1}{F_s} X\left(\dfrac{F}{F_s}\right) & -\dfrac{F_s}{2} < F < \dfrac{F_s}{2} \\ 0 & \text{otherwise} \end{cases} \tag{7.15}$$

Substituting the relation in (7.15) into (7.14), we obtain:

$$\hat{x}_a(t) = \frac{1}{F_s} \int_{-F_s/2}^{F_s/2} X\left(\frac{F}{F_s}\right)e^{j2\pi Ft}dF \tag{7.16}$$

where $\hat{x}_a(t)$ is the reconstructed analog signal. Recall that the Fourier transform of $X\left(\dfrac{F}{F_s}\right)$ is given as:

$$X\left(\frac{F}{F_s}\right) = \sum_{n=-\infty}^{\infty} x(nT_s)e^{-j2\pi \frac{F}{F_s}n} \tag{7.17}$$

where T_s is the sampling rate. Substituting (7.17) into (7.16), it becomes:

$$\hat{x}_a(t) = \frac{1}{F_s} \int_{-F_s/2}^{F_s/2} \left[\sum_{n=-\infty}^{\infty} x(nT_s) e^{-j2\pi \frac{F}{F_s} n} \right] e^{j2\pi Ft} dF \qquad (7.18)$$

Rearranging the order of the summation and integral in (7.18):

$$\hat{x}_a(t) = \sum_{n=-\infty}^{\infty} x(nT_s) \left\{ \frac{1}{F_s} \int_{-F_s/2}^{F_s/2} e^{j2\pi F\left(t - \frac{n}{F_s}\right)} dF \right\} \qquad (7.19)$$

The integral in (7.19) is the sinc function:

$$\frac{1}{F_s} \int_{-F_s/2}^{F_s/2} e^{j2\pi F\left(t - \frac{n}{F_s}\right)} dF = \frac{1}{F_s} \frac{e^{j2\pi \frac{F_s}{2}\left(t - \frac{n}{F_s}\right)} - e^{-j2\pi \frac{F_s}{2}\left(t - \frac{n}{F_s}\right)}}{j2\pi\left(t - \frac{n}{F_s}\right)} = \frac{1}{F_s} \frac{\sin\left(2\pi \frac{F_s}{2}\left(t - \frac{n}{F_s}\right)\right)}{\pi\left(t - \frac{n}{F_s}\right)}$$

$$\left. \frac{1}{F_s} \int_{-F_s/2}^{F_s/2} e^{j2\pi F\left(t - \frac{n}{F_s}\right)} dF \right|_{T_s = \frac{1}{F_s}} = \frac{\sin\left(\frac{\pi}{T_s}(t - nT_s)\right)}{\frac{\pi}{T_s}(t - nT_s)} \qquad (7.20)$$

Hence, substituting (7.20) into (7.19), it becomes:

$$\hat{x}_a(t) = \sum_{n=-\infty}^{\infty} x(nT_s) \frac{\sin\left(\frac{\pi}{T_s}(t - nT_s)\right)}{\frac{\pi}{T_s}(t - nT_s)} \qquad (7.21)$$

The reconstructed analog signal in (7.21) involves weighing each individual discrete sample by a *sinc* function shifted-in-time by the sampling period. The *sinc* function:

$$p(t) = \frac{\sin\left(\frac{\pi}{T_s}(t - nT_s)\right)}{\frac{\pi}{T_s}(t - nT_s)} \qquad (7.22)$$

is known as the ideal interpolation filter expressed in the frequency domain as:

$$P(F) = \begin{cases} 1 & |F| < F_s/2 \\ 0 & |F| \geq F_s/2 \end{cases} \qquad (7.23)$$

Applying the ideal interpolation filter in (7.23) to the spectrum of a nonaliased discrete-time signal results in recovering the original analog signal without any loss of information, as shown in Figure 7.2.

Therefore, another simple way to arrive at the relationship expressed in (7.21) is to start with filtering the spectrum of the discrete-time signal by the interpolation filter (7.23), that is:

$$\hat{X}\left(\frac{F}{F_s}\right) = X\left(\frac{F}{F_s}\right) P(F) \qquad (7.24)$$

Multiplication in the frequency domain is convolution in the time domain,

$$\hat{x}_a(t) = x_a(t) * p(t) = \sum_{n=-\infty}^{\infty} x(nT_s) \frac{\sin\left(\dfrac{\pi}{T_s}(t - nT_s)\right)}{\dfrac{\pi}{T_s}(t - nT_s)} \qquad (7.25)$$

which is none other than the relationship developed in (7.21). Note that certain modulation schemes require a flat response at the output of the DAC. Hence, the inband distortion due to the weighing function of the *sinc* must be corrected for. This is typically done with an equalizer mimicking an inverse *sinc* filter.

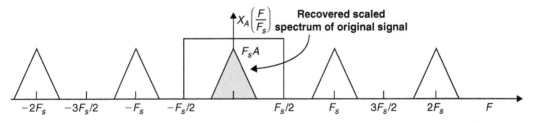

Figure 7.2 Recovered sampled spectrum of original analog signal

7.2 Sampling of Bandpass Signals

The sampling process is critical for radio receivers that digitize signals at RF or IF. Sampling an analog signal at IF or RF results in replicas of the signal's spectrum repeated at uniform intervals. The choice of the sampling rate of such signals is dependent on the signal's bandwidth and the IF or RF center frequency.

Bandpass sampling does not require the use of analog quadrature mixing, thus eliminating certain DC offsets due to the carrier feedthrough in the mixers, VGA gain stages, and filters. Furthermore, it does not require analog phase and amplitude compensation due to IQ imbalance. Bandpass sampling requires only one ADC, as shown in Figure 7.3, allowing for the final IF (or low IF) to baseband conversion to occur in the digital domain. Note that the second DAC feeding off the phase-to-accumulator block is not a transmit DAC, but rather generates a sinusoid as part of the direct digital synthesis (DDS) block. A DDS system is a mechanism for generating a sinusoid digitally and passing the signal through a DAC to be used for mixing.

On the other hand, bandpass sampling is sensitive to carrier or IF frequency variations, as well as sampling frequency and jitter. In this case, the ADC tends to consume more power due to a faster sample and hold (S/H) and digital circuitry and the performance of the system becomes more prone to degradations due to mixed-signal circuit imperfections. Furthermore, the requirements imposed on the bandpass filter at IF before the ADC become much more stringent and much more difficult to build compared to the more benign lowpass filters used in analog quadrature downconversion. Note that this IF filter also performs the function of an antialiasing filter used in the lowpass case.

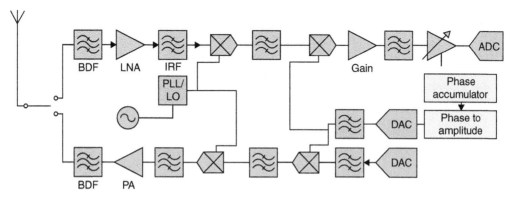

Figure 7.3 IF sampling architecture showing the use of only one ADC and one DAC

7.2.1 Representation of Bandpass Signals

Consider the conceptual modulator depicted in Figure 7.4. The in-phase $I(t)$ and quadrature $Q(t)$ signals are obtained from discrete-time signals converted to analog by two DACs. The signal at IF, or RF in the case of a direct conversion transmitter, can be expressed as:

$$x_a(t) = I(t)\cos(2\pi F_c t) - Q(t)\sin(2\pi F_c t) \tag{7.26}$$

again, where $I(t)$ and $Q(t)$ are the real and complex components of the complex analog baseband signal $s_a(t) = I(t) - jQ(t)$. The bandpass signal can then be related to the complex baseband signal as:

$$x_a(t) = \text{Re}\left\{s_a(t)e^{j2\pi F_c t}\right\} = \text{Re}\left\{(I(t) + jQ(t))e^{j2\pi F_c t}\right\} \tag{7.27}$$

Furthermore, the relation in (7.26) can be expressed as:

$$x_a(t) = \sqrt{I^2(t) + Q^2(t)}\,\cos\left(2\pi F_c t + \tan^{-1}\left(\frac{-Q(t)}{I(t)}\right)\right) \tag{7.28}$$

7.2.2 Sampling of Bandpass Signals — Integer Positioning

The fractional bandwidth is typically referred to as the fractional number of bandwidths separating the origin to the lower edge of the passband [1]. Integer band positioning is a

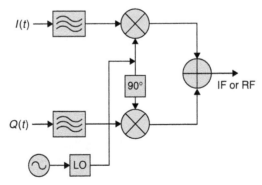

Figure 7.4 Conceptual upconversion of baseband signal to IF or RF frequency

special case where the lower or upper edge of the band is located at an integer multiple of bandwidths from the origin, that is:

$$F_c \pm \frac{B}{2} = lB \qquad l \text{ is a positive integer} \qquad (7.29)$$

An example for integer band positioning for $l = 5$ for $F_c + \frac{B}{2} = 5B$ is depicted in Figure 7.5.

Note that one important consideration concerning the bandwidth of the sampled signal implies that the signal is bounded by a lower frequency component F_l and an upper frequency component F_u such that B is equal to the interval (F_l, F_u) such that there are no signal components at either F_l or F_u. This assumption will hold true throughout this discussion and will be elaborated upon when discussing the general bandpass sampling case.

Assume that the carrier frequency (or IF frequency before the ADC) is chosen such that:

$$F_c + \frac{B}{2} = lB \qquad l \text{ is a positive integer} \qquad (7.30)$$

Sampling (7.26) at the rate $T_s = 1/2B$ and substitute $F_c = (2l - 1)\frac{B}{2}$, we obtain:

$$x_a(nT_s) = I(nT_s)\cos\left(2\pi(2l-1)\frac{B}{2}nT_s\right) - Q(nT_s)\sin\left(2\pi(2l-1)\frac{B}{2}nT_s\right)$$

$$x_a(nT_s) = I(nT_s)\cos\left(\frac{\pi}{2}(2l-1)n\right) - Q(nT_s)\sin\left(\frac{\pi}{2}(2l-1)n\right) \qquad (7.31)$$

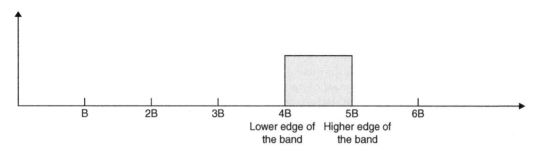

Figure 7.5 Example of integer band positioning for *l* = 4

Let's consider the two cases where n is even and n is odd. For n even, that is $n = 2m$, the relation in (7.31) becomes:

$$x_a(2mT_s) = I(2mT_s)\cos((2l - 1)m\pi) - Q(nT_s)\sin((2l - 1)m\pi)$$
$$x_a(2mT_s) = (-1)^m I(2mT_s) \text{ or } x_a(nT_s) = (-1)^{n/2} I(nT_s) \text{ for } n \text{ even} \qquad (7.32)$$

For n odd, that is $n = 2m - 1$, the relation in (7.31) becomes:

$$x_a((2m - 1)T_s) = I((2m - 1)T_s)\cos\left(2\pi(2l - 1)\frac{B}{2}(2m - 1)T_s\right)$$
$$- Q((2m - 1)T_s)\sin\left(2\pi(2l - 1)\frac{B}{2}(2m - 1)T_s\right)$$
$$x_a((2m - 1)T_s) = I((2m - 1)T_s)\cos\left(\frac{\pi}{2}(4lm - 2l - 2m + 1)\right)$$
$$- Q((2m - 1)T_s)\sin\left(\frac{\pi}{2}(4lm - 2l - 2m + 1)\right)$$
$$x_a((2m - 1)T_s) = (-1)^{l+m+1}Q((2m - 1)T_s) \text{ or } x_a(nT_s) = (-1)^{l+\frac{n+1}{2}+1}Q(nT_s) \text{ for } n \text{ odd} \qquad (7.33)$$

Let $T_s' = 2T_s = 1/B$ and using the relations expressed in (7.32) and (7.33) we can represent the sampled signal in yet another form:

$$x_a(mT_s') = (-1)^m I(mT_s') \qquad \text{for } n \text{ even}$$
$$x_a\left(mT_s' - \frac{T_s'}{2}\right) = (-1)^{l+m+1}Q\left(mT_s' - \frac{T_s'}{2}\right) \quad \text{for } n \text{ odd} \qquad (7.34)$$

Observation

With this choice of relationship between the sampling rate and the bandwidth of the IF signal, we note that the even-numbered samples are related to the in-phase baseband signal $I(t)$ whereas the odd-numbered samples are related to the quadrature baseband signal $Q(t)$. Therefore, at the output of the ADC sampling at IF, the quadrature conversion to baseband is rather simple. To obtain the in-phase digital component of the sampled signal, it is sufficient to digitally multiply the signal at the output of the ADC by $(-1)^{n/2}$ for even n and by 0 for odd n. Similarly, in order to obtain the digital quadrature component of the received signal, it is sufficient to digitally multiply the output of the ADC by $(-1)^{l+\frac{n+1}{2}+1}$ for n odd and by 0 for n even.

7.2.3 Reconstruction of Bandpass Signal — Integer Positioning

The discrete samples of the in-phase and quadrature components of (7.34), that is

$I(mT_s')$ and $Q\left(mT_s' - \dfrac{T_s'}{2}\right)$, can be used to reconstruct the equivalent analog lowpass

signals according to the relationship developed in (7.21) as:

$$I(t) = \sum_{m=-\infty}^{\infty} I(2mT_s) \frac{\sin\left(\dfrac{\pi}{2T_s}(t - 2mT_s)\right)}{\dfrac{\pi}{2T_s}(t - 2mT_s)}$$

$$Q(t) = \sum_{m=-\infty}^{\infty} Q(2mT_s - T_s) \frac{\sin\left(\dfrac{\pi}{2T_s}(t - 2mT_s + T_s)\right)}{\dfrac{\pi}{2T_s}(t - 2mT_s + T_s)} \tag{7.35}$$

Substituting (7.34) and (7.35) into (7.26), we can express the analog bandpass signal as:

$$x_a(t) = I(t)\cos(2\pi F_c t) - Q(t)\sin(2\pi F_c t)$$

$$x_a(t) = \sum_{m=-\infty}^{\infty} \left\{ (-1)^m x_a(2mT_s) \frac{\sin\left(\dfrac{\pi}{2T_s}(t - 2mT_s)\right)}{\dfrac{\pi}{2T_s}(t - 2mT_s)} + \right.$$

$$\left. (-1)^{l+m} x_a(2mT_s - T_s) \frac{\sin\left(\dfrac{\pi}{2T_s}(t - 2mT_s + T_s)\right)}{\dfrac{\pi}{2T_s}(t - 2mT_s + T_s)} \right\} \tag{7.36}$$

The relationship in (7.36) can be further expressed as:

$$x_a(t) = \sum_{m=-\infty}^{\infty} x_a(mT_s) \frac{\sin\left(\dfrac{\pi}{2T_s}(t - mT_s)\right)}{\dfrac{\pi}{2T_s}(t - mT_s)} \cos\left(2\pi F_c(t - mT_s)\right) \tag{7.37}$$

where again the sampling rate is twice the IF bandwidth B, that is $T_s' = 1/2B$.

7.2.4 Sampling of Bandpass Signals — Half-Integer Positioning

Half-integer positioning refers to the special case for which the bandwidth of the desired signal is centered at an integer multiple of the bandwidth, that is:

$$F_c = lB \qquad l \text{ is a positive integer} \tag{7.38}$$

Figure 7.6 shows an example of half-integer band positioning for $l = 4$.

For this case, we present a special case of interest which results in a simplified digital quadrature demodulator. Assume that the sampling rate $F_s = 4B$. Sampling (7.26) at the rate $T_s = 1/4B$ and substituting $F_c = lB$, we obtain:

$$x_a(nT_s) = I(nT_s)\cos(2\pi lBnT_s) - Q(nT_s)\sin(2\pi lBnT_s)$$
$$x_a(nT_s) = I(nT_s)\cos\left(\frac{\pi l}{2}n\right) - Q(nT_s)\sin\left(\frac{\pi l}{2}n\right) \tag{7.39}$$

Note for the case where l is even, the product ln is always even whether n is odd or even. For this scenario, let $l = 2k$, then the cosine and sine argument in (7.39) is πkn and hence (7.39) reduces to:

$$x_a(nT_s) = (-1)^{kn}I(nT_s) \tag{7.40}$$

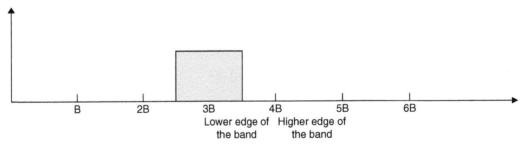

Figure 7.6 Example of half-integer band positioning for $l = 4$

This is true since $\sin(\pi kn) = 0$ for all values of k and n, and therefore the information on the quadrature component $Q(.)$ could never be recovered. Hence, we conclude that l even is not a viable choice for this sampling rate. Next, we consider the case where l is odd. In this case, the relationship in (7.39) can be expressed as:

$$x_a(nT_s) = I(nT_s)\cos(\pi) - Q(nT_s)\sin(2\pi lBnT_s)$$

$$x_a(nT_s) = \begin{cases} (-1)^{n/2}I(nT_s) & n \text{ is even} \\ (-1)^{\frac{l+n}{2}} Q(nT_s) & n \text{ is odd} \end{cases} \tag{7.41}$$

The reconstruction of bandpass signal due to half-integer positioning is developed in a very similar fashion to that of the integer positioning case.

7.2.5 Nyquist Zones

Nyquist zones subdivide the spectrum into regions spaced uniformly at intervals of $F_s/2$. Each Nyquist zone contains a copy of the spectrum of the desired signal or a mirror image of it. The odd Nyquist zones contain exact replicas of the signal's spectrum—that is, if the original signal is centered at F_c ($F_c = 0$ is the lowpass signal case)—the exact spectral replica will appear at $F_c + kF_s$ for $k = 0, 1, 2, 3 \dots$. The zone corresponding to $k = 0$ is known as the first Nyquist zone, whereas the third and fifth Nyquist zones correspond to $k = 1$ and $k = 2$, respectively.

Similarly, mirrored replicas of the signal's spectrum occur in even numbered Nyquist zones, that is the spectra are centered at $kF_s - F_c$ for $k = 1, 2, 3 \dots$. The second Nyquist zone corresponds to $k = 1$ whereas the fourth and sixth Nyquist zones correspond to $k = 2$ and $k = 3$, respectively. An example depicting odd and even Nyquist zones is given in Figure 7.7.

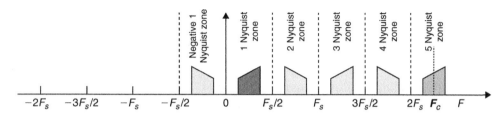

Figure 7.7 Spectral images of desired signal depicted in various Nyquist zones

A relationship between the signal's center frequency F_c at RF or IF can be expressed as:

$$F = \begin{cases} rem(F_c, F_s) & \left\lfloor \dfrac{F_c}{F_s/2} \right\rfloor \text{ is even implying that image is exact replica} \\[2em] F_s - rem(F_c, F_s) & \left\lfloor \dfrac{F_c}{F_s/2} \right\rfloor \text{ is odd implying that image is mirrored replica} \end{cases} \quad (7.42)$$

where *rem* is the remainder after division, $\lfloor \cdot \rfloor$ denotes the floor function, which rounds a number towards zero, and F is the center frequency of the image in the first Nyquist zone.

7.2.6 Sampling of Bandpass Signals — General Case

Bandpass sampling can be utilized to downconvert a signal from RF or IF to a bandpass signal at a lower IF. The bandpass signal is repeated at integer multiples of the sampling frequency. Therefore choosing the proper spectral replica of the original bandpass signal allows for downconversion. Ideally, as mentioned earlier, a bandpass signal has no frequency components below a certain frequency F_L or above a certain frequency F_U, that is, the signal or channel bandwidth is strictly defined by the bandwidth $B = F_U - F_L$.

Let the center frequency of the signal or channel F_c be defined as $F_c = F_U - B/2 = F_L + B/2$, then the bandpass sampling frequency for normal or exact spectral placement is given as [1]–[3]:

$$\frac{2F_c + B}{2n + 1} \leq F_s \leq \frac{F_c - B/2}{n} \quad 0 \leq n \leq \left\lfloor \frac{F_c - B/2}{2B} \right\rfloor \quad \text{normal spectral placement} \quad (7.43)$$

In a similar vein, for the inverted or mirrored-replica spectral placement of the signal, the sampling frequency is given as:

$$\frac{F_c + B/2}{n} \leq F_s \leq \frac{2F_c - B}{2n - 1} \quad 1 \leq n \leq \left\lfloor \frac{F_c + B/2}{2B} \right\rfloor \quad \text{inverted spectral placement} \quad (7.44)$$

The minimum sampling rates expressed in (7.43) and (7.44) do not take into account any instabilities or imperfections of the carrier or sampling frequencies [1],[4], and [5]. Therefore, when choosing a sampling frequency, the designer must take into account these imperfections to avoid any serious SNR degradation due to aliasing of the signal.

Example 7-1: Sampling of Fixed WiMAX Signals

The relationship between the sampling factor and the channel bandwidth for fixed WiMAX is shown in Table 7.1. Given this relationship, the signal versus channel bandwidth for the various signaling schemes is expressed in Table 7.2. Choose an IF center frequency of 200 MHz. Based on the boundaries defined in (7.43) and (7.44), compute the sampling rate boundaries for maximum n.

The analyses for this example are summarized in Table 7.3 and Table 7.4. It would be instructive for the reader to repeat the example for an IF frequency of 175 MHz and note what happens to the sampling rate in each case and why.

Table 7.1 Sampling factor versus nominal signal bandwidth for fixed WiMAX

Sampling Factor	Nominal Channel Bandwidth
8/7	Signal BWs that are multiples of 1.75 MHz
28/25	Signal BW that are multiples of 1.25, 1.5, 2, or 2.75
8/7	For all others not specified

Table 7.2 Signal versus channel bandwidth for fixed WiMAX

Signal Bandwidth (MHz)	Channel Bandwidth (MHz)
20	22.4
10	11.2
8.75	10
7	8
5	5.6

Table 7.3 Sampling rate boundaries for the 22.4 and 11.2 MHz channel BW case

Parameter	Value	Value
Bandwidth (MHz)	22.40	11.20
IF frequency (MHz)	200.00	200.00
f_h (MHz)	211.20	205.60
f_l (MHz)	188.80	194.40
maximum n	4.00	8.00
High sampling boundaries based on *n* max (MHz)	47.20	24.30
Low sampling boundaries based on *n* max (MHz)	46.93	24.19

Table 7.4 Sampling rate boundaries for the 10, 8, and 5.6 MHz channel BW case

Parameter	Value	Value	Value
Bandwidth (MHz)	10.00	8.00	5.60
IF frequency (MHz)	200.00	200.00	200.00
f_h (MHz)	205.00	204.00	202.80
f_l (MHz)	195.00	196.00	197.20
maximum n	9.00	12.00	17.00
High sampling boundaries based on n max (MHz)	21.67	16.33	11.60
Low sampling boundaries based on n max (MHz)	21.58	16.32	11.59

7.3 The AGC Algorithm

The purpose of the automatic gain control (AGC) algorithm is to regulate the received signal strength at the input of the ADCs such that the required signal SNR for proper decoding is met. For example, if the received signal strength is weak at the antenna, the AGC algorithm boosts the receiver gain stages in order to minimize the noise and bring the signal level to an acceptable SNR. Likewise, if the received signal strength is strong, the AGC algorithm attenuates the receiver gain stages in order to avoid signal clipping and nonlinear degradations that would otherwise deteriorate the signal SNR. In receivers that employ modern digital modulation techniques, the AGC corrects for long term fading effects due to shadowing. The short term fast fades, especially those denoted as frequency selective fades, are corrected for in the digital equalizer, be it a RAKE receiver, a decision feedback equalizer (DFE), or any other form of equalization designed to deal with this type of fading. After equalization, any remaining symbols in error are corrected, or attempted to be corrected, in the forward error correction block using a variety of appropriate coding schemes. It is important to note that the AGC must not correct for fast fades especially within a data slot, or a block of symbols within frame. Performing amplitude gain control within a coherent block of data could serve to adversely affect the equalizer or forward error correction.

The AGC loop mostly controls various analog gain and attenuation blocks at various points in the receive chain. For example, in a superheterodyne receiver, depicted in

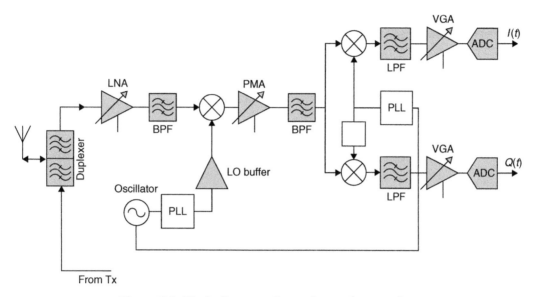

Figure 7.8 Block diagram of superheterodyne receiver

Figure 7.8, the AGC loop could switch the LNA gain from low-setting to high setting and vice versa, perform a similar function concerning the post mixer amplifier (PMA), and vary the gain of the voltage gain amplifier (VGA) to maintain a certain signal level relative to the noise in the receiver thus maintaining an acceptable SNR. The object of this discussion is not how to control the receiver line-up, but rather how to design the digital loop that estimates the input signal power and adjusts the gain accordingly to maintain a satisfactory SNR. The following analysis is applicable to most common receiver line-up architectures.

7.3.1 The Loop Detector

The first order AGC loop is depicted in Figure 7.9. The analog in-phase and quadrature input signals $I(t)$ and $Q(t)$ undergo amplifications or attenuation by the in-phase and quadrature VGAs as well as the LNA and PMA gain stages. At the output of the ADCs the discrete in-phase and quadrature signals are then squared and added to generate the instantaneous signal power:

$$r^2(n) = I^2(n) + Q^2(n) = I^2(t) + Q^2(t)\big|_{t=nT_s} \tag{7.45}$$

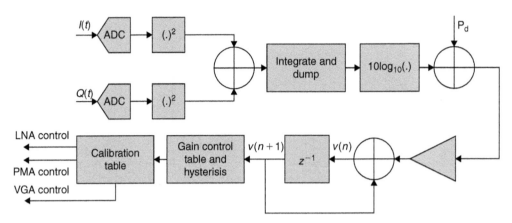

Figure 7.9 The AGC loop

The instantaneous output power provided in (7.43) is obtained at the output of the detector. Most practical detectors, however, implement an approximation to (7.45) or its square root in order to simplify the digital ASIC implementation. The most common approximation takes the form:

$$r(n) \approx \alpha \max \left\{ I(n), Q(n) \right\} + \beta \min \left\{ I(n), Q(n) \right\} \qquad (7.46)$$

The most commonly used approximation of the envelope based on (7.46) is:

$$r(n) \approx \frac{15}{16} \left[\max \left\{ I(n), Q(n) \right\} + \frac{1}{2} \min \left\{ I(n), Q(n) \right\} \right] \qquad (7.47)$$

Other approximation examples include:

$$r(n) \approx \max \left\{ I(n), Q(n) \right\} + \frac{1}{4} \min \left\{ I(n), Q(n) \right\} \qquad (7.48)$$

and:

$$r(n) \approx \max \left\{ I(n), Q(n) \right\} + \frac{3}{8} \min \left\{ I(n), Q(n) \right\} \qquad (7.49)$$

In order to evaluate the performance of the various detectors of the form presented in (7.46), it is instructive to compare their performance with the exact detector presented in (7.45). To do so, let us compare the amplitude of (7.46) to the unity vector at the output of the exact

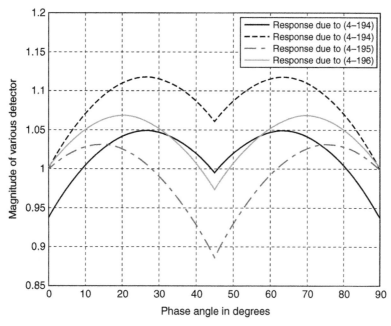

Figure 7.10 Performance of various AGC detectors based on (7.46)

detector, resulting from $I(n) = \cos(\theta)$ and $Q(n) = \sin(\theta)$. That is, compare the magnitude of (7.46) to 1 for various phase angles θ. The result of these comparisons is depicted in Figure 7.10. However, regardless of the detector used, its performance largely depends on the input data. For random white noise type input data, the performance of the AGC with an estimated detector approaches the performance of an AGC with an exact detector at the cost of reducing the attack and decay time resulting from reducing the gain value at the input of the loop filter.

In the ensuing analysis, only the true instantaneous power estimate based on (7.45) will be used. The output signals to the ADCs must not undergo any digital channel filtering before detecting the instantaneous output power in order to provide a true estimate of the signal power. This is essential, since a channel filter attenuates the interference or blocker and prevents the AGC from reacting properly to it. Certain AGC designs, however, rely on two instantaneous power estimates: one before the digital channel filters, which approximates the signal plus blocker power, and one estimate after the digital channel filters, which estimates the desired signal power without the effect of interference and blocking. In this discussion, we will limit our analysis to the case where the signal plus interference instantaneous power is used to perform gain control.

7.3.2 Loop Analysis

The integrate-and-dump function in Figure 7.9 serves to estimate the mean of the received signal power:

$$E\{r^2(n)\} = \frac{1}{N}\sum_{n=0}^{N-1}\left[I^2(n) + Q^2(n)\right] \tag{7.50}$$

Assume that the VGAs and gain stages can be modeled as linear functions of the form $10^{(\cdot)/20}$; that is, the gain stages are operating in a linear fashion without compression and the VGA gains can be varied in a monotonically increasing or decreasing trend. Then the output of the loop filter after the delay element in Figure 7.9 can be represented in state space form:

$$\nu(n+1) = \nu(n) + \mu\left[P_d - 10\log_{10}\left(10^{\nu(n)/10}\,r^2(n)\right)\right] \tag{7.51}$$

where $v(n)$ is the state at the output of the loop filter, and the gain stage $10^{\nu(n)/20}$ is an approximation of the AGC's various gain stages operating in the linear region. P_d is the desired set power of the AGC. Its value is mainly dictated by the desired SNR of the received signal. In most cases, P_d is defined in terms of the back-off necessary from ADC full scale in order to ensure no clipping in the desired signal. In some cases, however, some clipping is allowed. This occurs when the signal PAR is large enough and statistically its peak occurs less than 5% of the time.

The expectation operator serves as a lowpass filter to remove the effect of zero-mean white noise. Further note that the gain is approximated as $10^{\nu(n)/20}$ and not as $10^{\nu(n)/10}$ since this is voltage gain and not power gain. The relation in (7.51) can be further expressed as

$$\begin{aligned}\nu(n+1) &= \nu(n) + \mu\left[P_d - \nu(n) - 10\log_{10}(\hat{r}^2(n))\right]\\ &= (1-\mu)\nu(n) + \mu\left[P_d - 10\log_{10}(\hat{r}^2(n))\right]\end{aligned} \tag{7.52}$$

where $\hat{r}^2(n) = E\{r^2(n)\}$ is the mean squared input power. The mean squared input power can be estimated via an integrate-and-dump function; that is:

$$\hat{r}^2(n) = E\{r^2(n)\} \approx \frac{1}{M}\sum_{m=0}^{M-1}\left\{I^2(n-m) + Q^2(n-m)\right\} \tag{7.53}$$

where the squared values of $I(n)$ and $Q(n)$ are summed over M-samples and then presented or *dumped* to the loop where the mean squared error value is then compared with desired input signal level P_d. Note that the integrate and dump as a function is a lowpass filter; that is:

$$ID(z) = 1 + z^{-1} + z^{-2} + \ldots + z^{-M+1}$$
$$= \frac{1 - z^{-M}}{1 - z^{-1}} \qquad (7.54)$$

For a large M, the integrate-and-dump filter approximates the mean of the signal. By way of an example, compare the integrate-and-dump filter for $M = 30$ to the lowpass function $1/1 - z^{-1}$ which represents the true estimate of the mean. The results are depicted in Figure 7.11 and Figure 7.12. Note that in reality an exact mean estimator is not used if the AGC is not run in continuous mode. An integrate-and-dump version is used instead in order to *flush* the filter for various AGC updates. The length of the integration in terms of the number of samples M is typically programmable.

In the event where the in-phase and the quadrature signals resemble white Gaussian noise, the envelope $r^2(n)$ possesses a Rayleigh distribution. The distribution becomes Rician in the presence of a dominant narrowband signal.

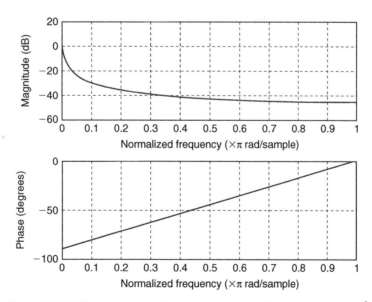

Figure 7.11 Filter response due to exact mean detector $1/1 - z^{-1}$

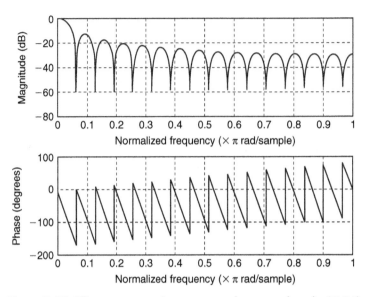

Figure 7.12 Filter response due to mean detector given in (7.54)

7.3.3 Steady State and Stability Analysis

During steady state, the mean of the error e_n is zero and $v(n + 1) = v(n) = v$. The relation in (7.52) becomes:

$$v = (1 - \mu)v + \mu\left[P_d - 10\log_{10}(\hat{r}^2(n))\right] \tag{7.55}$$

Solving for v in (7.55), we obtain the equilibrium point v as:

$$v = P_d - 10\log_{10}(\hat{r}^2(n)) \tag{7.56}$$

The equilibrium point in (7.56) is pivotal in computing the instantaneous dynamic range of the system.

Let $\chi(n)$ be the perturbation around the equilibrium point during steady state, then:

$$v + \chi(n + 1) = (1 - \mu)\left[v + \chi(n)\right] + \mu\left[P_d - 10\log_{10}(\hat{r}^2(n))\right] \tag{7.57}$$

Simplifying the relation in (7.57) we obtain:

$$\chi(n + 1) = (1 - \mu)\chi(n) - \mu v + \mu\left[P_d - 10\log_{10}(\hat{r}^2(n))\right] \tag{7.58}$$

Substituting (7.56) into (7.58) we obtain:

$$\chi(n + 1) = (1 - \mu)\chi(n) - \mu\left[P_d - 10\log_{10}\left(\hat{r}^2(n)\right)\right] + \mu\left[P_d - 10\log_{10}\left(\hat{r}^2(n)\right)\right]$$
$$= (1 - \mu)\chi(n) \tag{7.59}$$

The relation in (7.59) can be further expressed as:

$$\chi(n + 1) = (1 - \mu)^n\chi(0) \tag{7.60}$$

In order for the AGC loop to be stable, (7.60) must converge to zero as $n \to \infty$, thus imposing the relation:

$$\left|1 - \mu\right| < 1 \Rightarrow \begin{cases} \mu > 0 \\ \mu < 2 \end{cases} \tag{7.61}$$

In the above discussion, it is assumed that the AGC has the attack and decay times both dictated by the loop filter gain. Therefore, the linear AGC loop with a single pole loop filter is stable if and only if (7.61) is satisfied.

By way of an example, consider an input signal to the AGC consisting of a single sinusoid. The set of point of the AGC compared with the signal input power requires the AGC to adjust the input power by 65 dB. The output of the loop filter versus the number

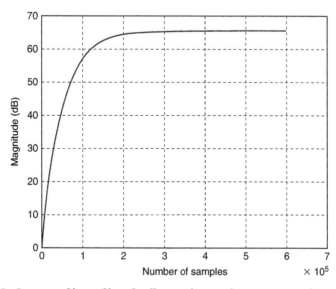

Figure 7.13 Output of loop filter feeding various gain stages as a function of time

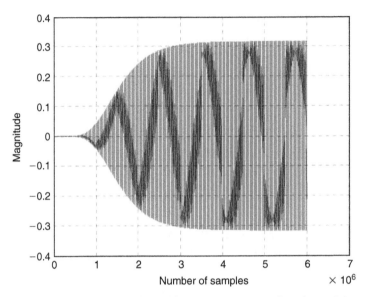

Figure 7.14 Input of in-phase signal to ADC as a function of time

of samples for a slowly but continuously adjusting AGC is shown in Figure 7.13. The in-phase input signal to the ADC as a function of the number of samples as its gain is being adjusted by the AGC is shown in Figure 7.14.

7.3.4 Peak to Average Power Ratio

7.3.4.1 PAR of a Single Tone Signal

Define the peak to average power ratio (PAR) of a given signal $r(t)$ as the ratio of the peak power of $r(t)$ to its average power:

$$PAR = 10 \log_{10} \left(\frac{P_{peak}}{P_{avg}} \right) \tag{7.62}$$

For a single tone $r(t) = A \cos(2\pi F_c t)$, the PAR can be estimated as:

$$PAR = 10 \log_{10} \left(\frac{P_{peak}}{P_{avg}} \right) = 10 \log_{10} \left(\frac{\max\left\{ |r(t)|^2 \right\}}{F_c \int\limits_{0}^{1/F_c} r^2(t)dt} \right) \tag{7.63}$$

The peak power of $r(t)$ can be simply found as:

$$P_{peak} = \max\left\{\left|A^2 \cos^2(2\pi F_c t)\right|\right\} = A^2 \tag{7.64}$$

This is true obviously for the case where $t = 2n\pi$, $n = 0,1, \ldots$ On the other hand, the average power of $r(t)$ is found as:

$$
\begin{aligned}
P_{avg} &= F_c \int_0^{1/F_c} A^2 \cos^2(2\pi F_c) dt \\
&= F_c \int_0^{1/F_c} A^2 \left[\frac{1}{2} + \frac{1}{2}\cos(4\pi F_c t)\right] dt \\
&= \frac{F_c}{2}\left\{\frac{A^2}{F_c} + \underbrace{\int_0^{1/F_c} A^2 \cos(4\pi F_c t) dt}_{=0}\right\} = \frac{A^2}{2}
\end{aligned} \tag{7.65}
$$

The PAR of $r(t)$ can then be estimated:

$$PAR = 10\log_{10}\left(\frac{P_{peak}}{P_s}\right) = 10\log_{10}\left(\frac{A^2}{A^2/2}\right) = 3dB \tag{7.66}$$

7.3.4.2 PAR of a Multi-Tone Signal

Next we extend the results obtained for a single tone in the previous section to a signal that possesses multiple tones. To simplify the analysis, consider the two-tone signal case first. Let the tones be harmonically related, that is $F_2 = MF_1$. This is certainly the case of OFDM signals where the various subcarriers are separated by a fixed frequency offset. Consider:

$$r(t) = A\cos(2\pi F_1 t) + B\cos(2\pi F_2 t) \tag{7.67}$$

The peak power of $r(t)$ can be found by first looking at the square of (7.67):

$$
\begin{aligned}
r^2(t) &= \left[A\cos(2\pi F_1 t) + B\cos(2\pi F_2 t)\right]^2 \\
&= \left[A^2 \cos^2(2\pi F_1 t) + B^2 \cos^2(2\pi F_2 t) + 2AB\cos(2\pi F_1 t)\cos(2\pi F_2 t)\right] \\
&= \frac{A^2}{2} + \frac{B^2}{2} + \frac{A^2}{2}\cos(4\pi F_1 t) + \frac{B^2}{2}\cos(4\pi F_2 t) \\
&\quad + AB\cos(2\pi(F_1 - F_2)t) + AB\cos((F_1 + F_2)t)
\end{aligned} \tag{7.68}
$$

The peak power of $r(t)$ can be found by taking the derivative of (7.68) and setting it equal to zero, that is:

$$\frac{d}{dt} r^2(t) = 0$$

$$\frac{d}{dt} r^2(t) = -2\pi A^2 F_1 \sin(4\pi F_1 t) - 2\pi B^2 F_2 \sin(4\pi F_2 t) - 2\pi AB(F_1 - F_2)$$

$$\sin(2\pi(F_1 - F_2)t) - 2\pi AB(F_1 + F_2)\sin(2\pi(F_1 + F_2)t) = 0 \qquad (7.69)$$

The relationship in (7.69) occurs for $t = 2n\pi$, $n = 0, 1, \ldots$, that is:

$$P_{peak} = \max\{r^2(t)\} = r^2(0) = (A + B)^2 \qquad (7.70)$$

The absolute value in (7.70) was dropped since $r(t)$ in this case is a real valued continuous function. Next, the average power of $r(t)$ is found as:

$$P_{avg} = MF_1 \int_0^{1/MF_1} \left[A\cos(2\pi F_1 t) + B\cos(2\pi MF_1 t) \right]^2 dt$$

$$= \frac{A^2}{2} + \frac{B^2}{2} + \frac{A^2 MF_1}{2} \underbrace{\int_0^{1/MF_1} \cos(2\pi F_1 t)dt}_{0} + \frac{B^2 MF_1}{2} \underbrace{\int_0^{1/MF_1} \cos(2\pi MF_1 t)}_{0}$$

$$+ AB \underbrace{\int_0^{MF_1} \cos(2\pi(F_1 - F_2)t)dt}_{0} + \underbrace{\int_0^{MF_1} AB\cos(2\pi(F_1 + F_2)t)dt}_{0}$$

$$= \frac{A^2}{2} + \frac{B^2}{2} \qquad (7.71)$$

The PAR is then the ratio:

$$PAR = 10\log_{10}\left(\frac{P_{peak}}{P_s}\right) = 10\log_{10}\left(\frac{(A+B)^2}{\frac{1}{2}(A^2 + B^2)}\right) = 10\log_{10}\left(2 + \frac{4AB}{A^2 + B^2}\right) \qquad (7.72)$$

At this point it is instructive to look at two different cases. The first is when the two tones are of equal power, that is $A = B$. The relationship in (7.72) then becomes:

$$PAR = 10 \log_{10} \left(\frac{P_{peak}}{P_s} \right) = 10 \log_{10} \left(2 + \frac{4AB}{A^2 + B^2} \right)$$

$$= 10 \log_{10} \left(2 + \frac{4A^2}{2A^2} \right) = 10 \log (4) \approx 6dB \qquad (7.73)$$

In (7.73), the result can be generalized to say that if the signal is made up of N tones that are of equal amplitude and that their phases are coherent in the sense that they add up constructively once over a certain period, then the PAR of this N-tone signal can be estimated as:

$$PAR = 20 \log_{10} (N) \qquad (7.74)$$

Note that, in reality, given an OFDM signal, for example, which is made up of N tones, the result presented in (7.74) is most unlikely to occur in the statistical sense due to the random nature of the data.

Next, we consider the case in which one tone is amplitude dominant over the other tone, say $A \gg B$. In this case, the relationship in (7.72) can be approximated as:

$$PAR = 10 \log_{10} \left(\frac{P_{peak}}{P_s} \right) = 10 \log_{10} \left. \left| \frac{(A + B)^2}{\frac{1}{2}(A^2 + B^2)} \right| \right|_{A+B \approx A \text{ since } A \gg B}$$

$$= 10 \log_{10} (2) \approx 3dB \qquad (7.75)$$

The result in (7.75) implies that the PAR value is mostly influenced by the amplitude of the tones. If one of the tones is dominant, for example, then the resulting PAR is that of a single tone. In reality, in a multi-tone signal, the results are somewhere in between.

7.3.4.3 PAR of an IF-Signal
In this section, we discuss the signal PAR sampled at IF. It is imperative to note that in the ensuing analysis it is assumed that the IF frequency is much larger than the signal bandwidth. In certain waveforms such as OFDM UWB, the analysis may not apply.

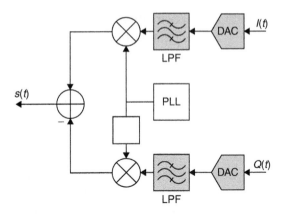

Figure 7.15 Quadrature signal modulator

Consider the IF signal $s(t)$ which is a byproduct of the in-phase and quadrature baseband signals $I(t)$ and $Q(t)$ as shown in Figure 7.15. To simplify the analysis, we ignore the effect of the lowpass filters digital or analog on the signal. The IF signal is given as:

$$s(t) = I(t)\cos(\omega t) - Q(t)\sin(\omega t) \tag{7.76}$$

The expression in (7.76) can be further expressed as the real part of $I(t) + jQ(t)$ modulated by the frequency ω, that is:

$$s(t) = \mathrm{Re}\left\{\left[I(t) + jQ(t)\right]e^{j\omega t}\right\} \tag{7.77}$$

The quadrature components of $s(t)$ can be written in terms of the envelope $r(t)$ and the phase $\theta(t)$, namely:

$$s(t) = r(t)\cos(\omega t + \theta(t)) \tag{7.78}$$

where the in-phase signal is:

$$I(t) = r(t)\cos(\theta(t)) \tag{7.79}$$

and the quadrature signal:

$$Q(t) = r(t)\sin(\theta(t)) \qquad (7.80)$$

The relationships in (7.79) and (7.80) simply imply:

$$I^2(t) + Q^2(t) = r^2(t)\left\{\cos^2(\theta(t)) + \sin^2(\theta(t))\right\} = r^2(t) \qquad (7.81)$$

or simply:

$$r(t) = \sqrt{I^2(t) + Q^2(t)} \qquad (7.82)$$

Next denote the PAR of the baseband signal as the peak of power of the envelope signal $r(t)$ divided by the average power of $r(t)$ say:

$$PAR(r) = 10\log_{10}\left(\frac{P_{peak,r}}{P_{avg,r}}\right) = 10\log_{10}\left(\frac{\max|r(t)|^2}{E\left\{|r(t)|^2\right\}}\right) \qquad (7.83)$$

where again the operator $E\{.\}$ is the expected value operator and $r(t)$ is real-valued. Next, we compare the PAR of the envelope signal at baseband $r(t)$ to the PAR of the IF signal $s(t)$. The PAR of $s(t)$ is:

$$PAR(s) = 10\log_{10}\left(\frac{\max|s(t)|^2}{E\left\{|s(t)|^2\right\}}\right) = 10\log_{10}\left(\frac{\max|r(t)\cos(\omega t + \theta(t))|^2}{E\left\{|r(t)\cos(\omega t + \theta(t))|^2\right\}}\right) \qquad (7.84)$$

The numerator of (7.84) is simply $P_{peak,r}$ since the maximum of the cosine function is 1. Recall that the assumption again is that the signal bandwidth is much smaller than the IF frequency, and therefore the signal can be considered to be constant over many periods of the modulating cosine function. Next, consider the denominator of (7.84). The average power of the signal $s(t)$ is[1]:

$$E\left\{|r(t)\cos(\omega t + \theta(t))|^2\right\} = \frac{1}{MT} \int_{t_0}^{\substack{t_0+MT \\ M\gg1}} s^2(t)\,dt$$

$$= \frac{1}{MT} \int_{t_0}^{\substack{t_0+MT \\ M\gg1}} r^2(t)\cos^2(\omega t + \theta(t))\,dt$$

$$= \frac{1}{2MT}\left\{ \int_{t_0}^{\substack{t_0+MT \\ M\gg1}} r^2(t)\,dt + \int_{t_0}^{\substack{t_0+MT \\ M\gg1}} r^2(t)\cos(2\omega t + 2\theta(t))\,dt \right\}$$

$$= \frac{P_{avg,r}}{2} + \frac{1}{2MT}\left\{ \int_{t_0}^{\substack{t_0+MT \\ M\gg1}} r^2(t)\cos(2\omega t + 2\theta(t))\,dt \right\} \qquad (7.85)$$

Assume that the signal $r(t)$ is slowly varying over MT periods of the modulating cosine function and hence it can be taken outside the integral in (7.85). Then:

$$\int_{t_0}^{\substack{t_0+MT \\ M\gg1}} r^2(t)\cos(2\omega t + 2\theta(t))\,dt \approx \hat{r} \int_{t_0}^{\substack{t_0+MT \\ M\gg1}} \cos(2\omega t + 2\theta(t))\,dt = 0 \qquad (7.86)$$

where $\hat{r}$ is a constant. Hence, the PAR of the IF signal in (7.84) can be finally defined as:

$$PAR(s) = 10\log_{10}\left(\frac{2P_{peak,r}}{P_{avg,r}}\right) = PAR(r) + 3\text{ dB} \qquad (7.87)$$

Hence it is obvious from (7.87) that the PAR of the baseband envelope signal is 3 dB less than that of the IF signal.

Another interesting facet of this discussion is to relate the PAR of the individual in-phase and quadrature components to the PAR of the envelope signal $r(t)$. If we consider $I(t)$

---

[1]Recall that $\cos^2 a = (1 + \cos 2a)/2$.

and $Q(t)$ to be zero mean statistically independent Gaussian random variables, or more realistically if the desired in-phase and quadrature signals resemble Gaussian independent random variables, with zero mean and σ^2 variance, then the square of the envelope $r^2(t)$ is chi-square distributed and consequently $r(t)$ is Rayleigh distributed with a pdf of:

$$P_r(r(t)) = \frac{r(t)}{\sigma^2} e^{-r^2(t)/2\sigma^2}, \ r(t) \geq 0 \tag{7.88}$$

The moments $r(t)$ are related to the gamma function as:

$$E\{r^n(t)\} = (2\sigma^2)^{n/2} \Gamma\left(1 + \frac{n}{2}\right) \tag{7.89}$$

where:

$$\Gamma(v) = \int_0^\infty x^{v-1} e^{-x} dx, \ v > 0$$

$$\Gamma\left(1 + \frac{v}{2}\right) = \frac{1 \times 3 \times 5 \times \dots \times (2v - 1)}{2^v} \sqrt{\pi}$$

$$\Gamma(v) = (v - 1)!, \ v \in \mathbb{N} \text{ and } v > 0$$

$$\Gamma\left(\frac{3}{2}\right) = \frac{\sqrt{\pi}}{2}, \Gamma(2) = 1 \tag{7.90}$$

This implies that the mean of the envelope $r(t)$ is:

$$E\{r(t)\} = \sqrt{\frac{\pi}{2}} \sigma \tag{7.91}$$

which is obviously nonzero compared to the zero mean of the in-phase and quadrature signal components. The variance of the envelope on the other hand is given as:

$$\sigma_r^2 = \left(2 - \frac{\pi}{2}\right) \sigma^2 \tag{7.92}$$

The power of the envelope signal can be obtained via (7.89) as:

$$E\{r^2(t)\} = 2\sigma^2 \Gamma(2) = 2\sigma^2 \tag{7.93}$$

Next, let us compare the PAR of say the in-phase signal to that of the envelope signal. Let the in-phase signal PAR be given as:

$$PAR(I) = 10 \log_{10} \left(\frac{\max |I(t)|^2}{E\{|I(t)|^2\}} \right) = 10 \log_{10} \left(\frac{M}{\sigma^2} \right) \tag{7.94}$$

and let the PAR of the envelope signal be:

$$PAR(r) = 10 \log_{10} \left(\frac{\max |r(t)|^2}{E\{|r(t)|^2\}} \right) = 10 \log_{10} \left(\frac{\max \left| \sqrt{I^2(t) + Q^2(t)} \right|^2}{2\sigma^2} \right) \tag{7.95}$$

Let $\max \left| \sqrt{I^2(t) + Q^2(t)} \right|^2 = M + \delta, \delta > 0$; then in order for the PAR of the envelope to be greater than the PAR of the in-phase component, it implies that:

$$\frac{M + \delta}{2\sigma^2} > \frac{M}{\sigma^2} \Rightarrow \delta > M \tag{7.96}$$

which is not likely for two Gaussian variables to have both maxima occur at the same instant. In fact, in an experiment of 2000 computations of the PAR of the in-phase, quadrature, and envelope signals comprised of 10,000 samples each, it was found that the relationship in (7.96) was true only a handful of times as shown in Figure 7.16. The in-phase and quadrature signals, in this case, are white Gaussian distributed with approximately zero-mean and unity variance. Figure 7.16 depicts the difference of $PAR(r)$ and $PAR(I)$, and $PAR(r)$ and $PAR(Q)$. Statistically, it can be observed that $PAR(I)$ and $PAR(Q)$ can be 3-dB or higher than $PAR(r)$. This is important to note since the AGC algorithm adjusts the gain line-up of the receiver based on the desired power of the envelope signal. And hence, the back-off from full scale takes into account the PAR of the envelope signal as seen previously and not the PARs of the in-phase and quadrature signals. However, in reality, that is not sufficient and care must be taken as to not clip the signal at the input of the converters in a non-IF sampling receiver. That is, the PAR of the in-phase and quadrature signals with respect to ADC full scale must be taken into account when setting the desired signal power in AGC algorithm.

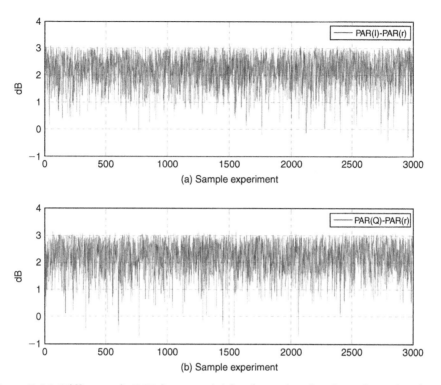

Figure 7.16 Difference in PAR between (a) in-phase signal and envelope signal and (b) quadrature signal

7.4 Appendix

7.4.1 Derivation of Analog Reconstruction Formula for Half Integer Positioning

In order to further simplify (7.36), recall that the RF or IF frequency can be expressed in terms of the bandwidth B as $F_c = (l - 1/2)B$, and that:

$$
\begin{aligned}
\cos\left(2\pi F_c t - 4\pi m F_c T_s\right) &= \cos\left(2\pi F_c t\right)\cos\left(4\pi m F_c T_s\right) + \sin E(2\pi F_c t)\sin\left(4\pi m F_c T_s\right) \\
&= \cos\left(2\pi F_c t\right)\cos\left(2\pi m\left(l - \frac{1}{2}\right)T_s\right) \\
&\quad + \sin\left(2\pi F_c t\right)\sin\left(2\pi m\left(l - \frac{1}{2}\right)T_s\right) \\
&= \cos\left(2\pi F_c t\right)\cos\left(-\pi m\right) = (-1)^m\cos\left(2\pi F_c t\right)
\end{aligned}
\tag{7.97}
$$

Furthermore, consider the relation:

$$\cos\left(2\pi F_c t - 4\pi m F_c T_s + 2\pi F_c T_s\right) = \cos\left(2\pi F_c t\right)\underbrace{\cos\left[2\pi\left(l - \frac{1}{2}\right)m - \pi\left(l - \frac{1}{2}\right)\right]}_{=0} +$$

$$\sin\left(2\pi F_c t\right)\sin\left[2\pi\left(l - \frac{1}{2}\right)m - \pi\left(l - \frac{1}{2}\right)\right]$$

$$= \sin\left(2\pi F_c t\right)\sin\left[2\pi\left(l - \frac{1}{2}\right)m - \pi\left(l - \frac{1}{2}\right)\right]$$

$$= \cos\left((m + l)\pi\right)\sin\left(2\pi F_c t\right) = (-1)^{m+l}\sin\left(2\pi F_c t\right)$$

$$(7.98)$$

Substituting the results shown in (7.97) and (7.98) into (7.36) leads directly into the relationship established in (7.37).

References

[1] Vaughn R, et al. The theory of bandpass sampling. IEEE Trans. on Signal Processing, September 1991;39(9).
[2] Liu J, Zhou X, Peng Y. Spectral arrangements and other topics in first order bandpass sampling theory. IEEE Trans. on Signal Processing, June 2001;49(6):1260–3.
[3] Choe M, Kang H, Kim K. Tolerable range of uniform bandpass sampling for software defined radio. 5th International Symposium on Wireless Personal Multimedia Communications, Volume 2, Issue 27, 30 Oct. 2002, Page(s): 840–842.
[4] Qi R, Coakly FP, Evans BG. Practical consideration for bandpass sampling. IEEE Trans. on Electronics Letters, Sept 1996;32(20):1861–2.
[5] Proakis J. Digital Communications. 2nd Edition. New York, NY: McGraw-Hill; 1989.

CHAPTER 8

Nyquist-Rate Data Conversion

The continually growing intricacy of multistandard and multimode radios in terms of desired flexibility, varied number of services, frequency bands, and complex modulation schemes is pushing the limit of common wireless transceivers. The principle aim of SDR technology is to enable such varied designs to be implemented on a single radio. A key approach to realizing this goal is by enabling further processing of the waveform, such as filtering and signal conditioning, in the digital domain, which otherwise was done in the analog domain. This can be accomplished by moving the analog/digital boundary closer to the antenna. That is not to say that the analog will be any easier, but rather that some of its complex functionality will be moved to the digital domain. The essential technology that will enable such advancement in radio design will have to come first and foremost through innovations in data conversion. The specifications of the ADC and DAC vary widely for IF sampling architecture versus, say, super-heterodyne or direct conversion.

This chapter is divided into three major sections. Section 8.1 delves into the fundamental requirements of Nyquist data converters. Topics such as quantization, oversampling, and performance are discussed in great detail. Section 8.2 addresses the various common architectures of Nyquist data converters. Popular Nyquist converters such as the FLASH converter, the pipelined and subranging converter, the folding converter, the integrating dual slope converter, and the successive approximation converter are studied. Section 8.3 contains an appendix on Gray codes.

8.1 Nyquist Converters

A Nyquist ADC, for example, assigns a digital output value based on an analog input value at a conversion or sampling rate that meets or exceeds the Nyquist rate. Such converters have a uniform quantization error and are based on FLASH architecture, pipeline or parallel architecture, and successive approximation architecture.

8.1.1 Uniform Quantization

In most practical communication systems, quantization is assumed to be a memoryless nonlinear process with many-to-one analog to discrete signal mapping. In reality, however, the quantization noise is made up of a large number of spurious signals. Since most of the spurious signals occur at frequencies well beyond the Nyquist frequency, they are typically and unfairly ignored by most analysts. However, the sampling nature of a DAC, for example, can cause each spur to be aliased back to some frequency between DC and Nyquist. In the ensuing analysis, the amplitude of the quantization noise is assumed to be uniformly distributed. This assumption is only valid when the input signal is not clipped and consequently the SQNR expression derived is only an approximation. Clipping will be treated separately in the next section. Despite this fact, however, the uniform distribution of the quantization noise is still not completely valid. By way of an example, a sinusoidal waveform is near its peaks and valleys a higher percentage of the time compared with its zero crossings, thus affecting mean square quantization error. In other words, the assumptions in the uniform quantization model are:

- The probability density function of the quantization error is uniformly distributed.

- The quantization error is stationary.

- The error signal samples are uncorrelated to the input signal samples.

- The error signal samples are uncorrelated to each other.

The amplitude of the analog signal $x_a(nT_s)$ sampled at $t = nT_s$ is mapped to a certain discrete signal $x_k(nT_s)$ belonging to a finite set of values spanning the range of the ADC, that is:

$$x_k(nT_s) \leq x_a(nT_s) \leq x_{k+1}(nT_s) \text{ for } k = 1, \ldots, K \qquad (8.1)$$

A quantization process is said to be uniform if the step size between any two given consecutive discrete levels is the same. More precisely, the step size or the resolution of the ADC δ is defined with respect to the quantization steps as:

$$\delta = x_{k+1}(nT_s) - x_k(nT_s) \text{ for all } k = 1, \ldots, K - 1 \qquad (8.2)$$

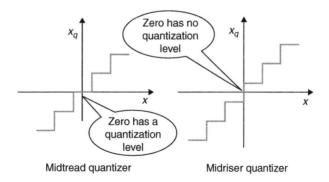

Figure 8.1 Examples of midtread and midriser quantization

A quantizer is said to be a midtread quantizer if the DC value is assigned a quantization value. On the other hand, a quantizer is said to be a midriser quantizer if the DC value is not assigned a quantization value and in essence the resulting signal is AC coupled. These two types of quantization are depicted in Figure 8.1.

Define the quantization error as the difference between the analog signal at time $t = nT_s$ and the quantized value of $x_a(nT_s)$ as:

$$e_k(nT_s) = x_a(nT_s) - x_k(nT_s) \qquad (8.3)$$

The error itself is bounded by half the step size, that is:

$$-\frac{\delta}{2} < e_k(nT_s) = x_a(nT_s) - x_k(nT_s) < \frac{\delta}{2} \qquad (8.4)$$

This is true when the analog signal level lies within the range of the ADC. However, when the absolute amplitude of the analog signal exceeds the peak voltage[1] $V_{x,FS}/2$ of the ADC, the quantization error could become larger than half the quantization step and signal clipping occurs.

The step size or resolution of the ADC can be further defined in terms of the ADC dynamic range and the number of effective bits b. Recall that K quantization levels can

---

[1] V_{FS} is the peak-to-peak voltage or full scale range of the ADC.

be represented by b binary bits if and only if $2^b \geq K$. For the purposes of this analysis, assume that $2^b = K$, then the resolution of the ADC can be defined as:

$$\delta = \frac{V_{x,Fs}}{2^b} \tag{8.5}$$

where $b > 0$. The relationship in (8.5) implies that finer resolution can be achieved for a given ADC dynamic range by increasing the number of bits representing the discrete signal.

Excluding the case where the input signal is outside the range of the ADC, we assume in the ensuing analysis that the input signal is zero mean white Gaussian process. The resulting quantization error can then be assumed to be stationary, additive, and uniformly distributed in the interval $(-\delta/2, \delta/2)$. Furthermore, the error samples are assumed to be uncorrelated, that is:

$$E\{e_k(nT_s)e_k(nT_s + T_s)\} = E\{e_k(nT_s)\}E\{e_k(nT_s + T_s)\} \tag{8.6}$$

In order for this analysis to hold true, the input signal is assumed to span the entire dynamic range (full scale) of the ADC. Define the signal-to-quantization-noise ratio (SQNR) as the ratio of input signal power to the quantization noise power:

$$SQNR = 10 \log_{10}\left(\frac{P_x}{P_e}\right) \tag{8.7}$$

The probability distribution function (PDF) of the error signal is depicted in Figure 8.2. Assuming the error signal to be zero mean, then quantization noise power is given as:

$$P_e = E\{e_k^2(t)\} - \underbrace{E^2\{e_k(t)\}}_{=0} = \sigma_e^2 \tag{8.8}$$

where σ_e^2 is the variance of the quantization error signal.

The variance of the error can be estimated from the PDF as:

$$P_e = \sigma_e^2 = E\{e_k^2(t)\} = \int_{-\delta/2}^{\delta/2} e_k^2(t)P(e_k(t))de(t) \tag{8.9}$$

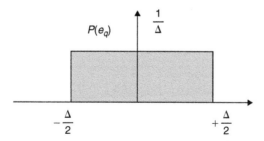

Figure 8.2 PDF of quantization error signal

where $P(e_k(t)) = 1/\delta$ is the PDF of the error signal. Evaluating the integral in (8.9), we obtain:

$$P_e = \sigma_e^2 = \int_{-\delta/2}^{\delta/2} e_k^2(t)P(e_k(t))de(t) = \frac{1}{\delta}\int_{-\delta/2}^{\delta/2} e_k^2(t)de(t) = \frac{\delta^2}{12} \tag{8.10}$$

The power spectral density of the noise is then given as:

$$S_e(f) = \frac{P_e}{F_s} \tag{8.11}$$

over the entire band $[-F_s/2, F_s/2]$. Note that, when discussing the power spectral density at baseband in either the in-phase or quadrature channels—that is, the noise interval is now $[0, F_s/2]$—the power spectral density in (8.11) is then:

$$S_{e,baseband}(f) = \frac{2P_e}{F_s} \tag{8.12}$$

This is an important distinction to make when dealing with bandpass sampling.

The SQNR can then be expressed as:

$$SQNR = 10\log_{10}\left(\frac{P_x}{P_e}\right) = 10\log_{10}\left(\frac{12\sigma_x^2}{\delta^2}\right) \tag{8.13}$$

where σ_x^2 is the variance of the zero-mean input signal. Substituting (8.5) into (8.13), we obtain:

$$SQNR = 10 \log_{10} \left(\frac{12 \times 2^{2b} \times \sigma_x^2}{V_{x,FS}^2} \right) = 10.8 + 6.0206b + 10 \log_{10} \left(\frac{\sigma_x^2}{V_{x,FS}^2} \right) \quad (8.14)$$

Recall that the peak to average ratio (PAR) in dB or crest factor was defined as:

$$PAR = 10 \log_{10} \left(\frac{P_{peak}}{P_s} \right) = 10 \log_{10} \left(\frac{V_{x,FS}^2}{4\sigma_x^2} \right) = 10 \log_{10} \left(\frac{1}{\kappa^2} \right) \quad (8.15)$$

where κ is known as the waveform loading factor, defined as the ratio of the input signal RMS voltage to the full scale voltage of the ADC, and $P_{peak} = (V_{x,FS}/2).^2$ In modulation schemes that employ I and Q analog downconversion, the PAR discussed above becomes only relevant to a single ADC and is not to be confused with the PAR typically associated with the signal envelope at baseband or the PAR of the signal envelope modulated at RF, as previously mentioned in Chapter 7.

At this point it is important to note that the SQNR expressed in (8.14) is applicable only when the input signal to the ADC occupies the full scale of the converter. If the maximum input signal is less than half the ADC full scale, say X_{max}, then the relationship for the SQNR would have to be modified to reflect the range of the input signal. The relation in (8.14) is then modified such that:

$$SQNR = 10 \log_{10} \left(\frac{12 \times 2^{2b} \times \sigma_x^2}{X_{max}^2} \right) = 10.8 + 6.0206b + 10 \log_{10} \left(\frac{\sigma_x^2}{X_{max}^2} \right) \quad (8.16)$$

The other extreme of underloading the ADC occurs when the input signal to the ADC exceeds the full scale range of the converter; then clipping is said to occur. Clipping and its adverse effects will be discussed in Section 8.1.4. In communication systems,

---

[2]In most textbooks, the bandwidth is the lowpass signal bandwidth, which is typically defined as half the IF signal bandwidth. Therefore, the oversampling ratio is divided by two times the lowpass bandwidth, which is equivalent to the IF bandwidth.

an automatic gain control (AGC) loop is employed to keep the input signal to the ADC within the desired range, which typically results in the highest SNR.

8.1.2 Effect of Quantization on ADC Output Signal Spectrum

The analysis provided herein is a precise investigation of the spectral content of the output signal of a Nyquist ideal or nonideal ADC resulting from a single tone input. Let the input tone $x_a(t) = A \cos(2\pi t/T)$ sampled at less than the Nyquist rate. The output, similar to the input, is also a periodic signal with period T, thus implying that it can be represented by Fourier series expansion. Harmonics occurring at frequencies greater than the Nyquist frequency are aliased back between DC and $F_s/2$. The Fourier series will in turn provide the magnitude of the various harmonics created by the ADC's quantization process. The frequency of these harmonics that lie in band can be obtained via the relation:

$$F_n = |nF - mF_s| < F_s/2 \text{ for a given integer } m \tag{8.17}$$

where $F_n = 1/T$ and F_n is the nth harmonic. The Fourier series expansion of the quantized periodic signal $y(t)$ can be expressed as:

$$y(t) = a_0 + \sum_{l=1}^{\infty} \{a_l \cos(2\pi Flt) + b_l \sin(2\pi Flt)\} \tag{8.18}$$

where:

$$a_0 = \frac{1}{T} \int_0^T y(t)$$

$$a_l = \frac{2}{T} \int_0^T y(t) \cos(2\pi Flt) dt$$

$$b_l = \frac{2}{T} \int_0^T y(t) \sin(2\pi Flt) dt \tag{8.19}$$

However, due to the fact that $y(t)$ is even symmetric, the $\{b_l\}$ coefficients in (8.19) are zero, and thus the quantized signal can be represented by the a_0 and $\{a_l\}$ coefficients only.

Next, define a set of measured threshold levels that correspond to a set of input amplitudes associated with the transition from one output word to another, say ϑ_k

for $k = 0,1,\ldots, M$. For a midtread quantizer with b bits, the number of thresholds is $M + 1 = 2^b$, one of which is equal to zero. On the other hand, a midriser has $M + 1 = 2^b + 1$ thresholds. Traditionally, however, there are only $M - 1$ practical thresholds where by convention $\vartheta_0 = -\infty$ and $\vartheta_M = \infty$. Define $\hat{x}_k$ as the input amplitude situated midway between two consecutive threshold levels; then the output of the ADC as referred to the input can be expressed as:

$$y(t) = \sum_{k=Q_{min}-1}^{Q_{max}} \hat{x}_k \Im(t,t_k,t_{k+1}) \quad \text{for } 0 < t < T/2 \tag{8.20}$$

where $y(t) = y(T - t)$ for $T/2 \le t < T$

$$\Im(t,t_k,t_{k+1}) = \begin{cases} 1 & \text{if } t_k \le t < t_{k+1} \\ 0 & \text{otherwise} \end{cases} \tag{8.21}$$

is the indicator function, Q_{min} is the smallest integer such that $-A < \vartheta_{Q_{min}}$, and Q_{max} is the largest integer such that $\vartheta_{Q_{max}} < A$. The variables t_k, and consequently t_{k+1}, in (8.20) correspond to the time that the sinusoid crosses the kth threshold and are given by the relation

$$t_k = T \cos^{-1}\left(-\frac{\vartheta_k}{2\pi A}\right) \quad \text{for } Q_{min} \le k \le Q_{max} \tag{8.22}$$

Substituting the result in (8.20) in a_l in (8.19), we obtain:

$$\begin{aligned} a_l &= \frac{4}{T} \int_0^{T/2} \sum_{k=Q_{min}-1}^{Q_{max}} \hat{x}_k \Im(t,t_k,t_{k+1}) \cos(2\pi lT/T)dt \\ &= \frac{4}{T} \sum_{k=Q_{min}-1}^{Q_{max}} \hat{x}_k \int_{t_k}^{t_{k+1}} \cos(2\pi lT/T)dt \\ &= \frac{2}{\pi l} \sum_{k=Q_{min}-1}^{Q_{max}} \hat{x}_k \left\{ \sin\left(\frac{2\pi l}{T}t_{k+1}\right) - \sin\left(\frac{2\pi l}{T}t_k\right) \right\} \end{aligned} \tag{8.23}$$

The term that corresponds to the *nth* harmonic in (8.23) is $a_l \cos(2\pi lt/T)$ and consequently the average power in the *nth* harmonic is $P_l = a_l^2/2$; compared to the average power in the fundamental, $P_1 = a_1^2/2$ can be best expressed as the ratio in dBc as:

$$P_{harmonic,l} = 10 \log_{10} \left(\frac{a_l^2}{a_1^2} \right) \tag{8.24}$$

Using this analysis, the interested researcher may derive an expression for a precise total harmonic distortion due to a specific number of tones.

8.1.3 Effect of Oversampling on Uniform Quantization

In uniform quantization, the bandwidth of the quantization noise is from DC to $F_s/2$. Increasing the sampling frequency does not increase the quantization power, but rather spreads the quantization noise up to $F_s/2$. The noise power within the bandwidth of the signal, however, gets scaled by the oversampling ratio according to the relation:

$$OSR = \frac{F_s}{B} \tag{8.25}$$

where it is important to remind the reader that B is the IF bandwidth[2]. This process is depicted in Figure 8.3.

Given the noise power within the channel bandwidth σ_e^2/OSR, the SQNR becomes:

$$
\begin{aligned}
SQNR &= 10 \log_{10} \left(\frac{\sigma_x^2}{\sigma_e^2} OSR \right) = 10 \log_{10} \left(\frac{12 \times 2^{2b} \times \sigma_x^2}{V_{x,FS}^2} \frac{F_s}{B} \right) \\
&= 10 \log_{10} \left(\frac{3\kappa^2 2^{2b} F_s}{B} \right) \\
&= 4.7712 + 6.02b + 10 \log_{10} \left(\frac{F_s}{B} \right) - PAR \\
&= 4.7712 + 6.02b + 10 \log_{10}(OSR) - PAR
\end{aligned}
\tag{8.26}
$$

The oversampling term in (8.26) is often referred to as processing gain. The improvement in SQNR due to OSR occurs only after the signal has been filtered via a channel filter. That is, in order to obtain the improvement in SQNR due to oversampling, the signal at

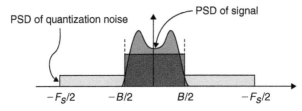

Figure 8.3 Spreading of quantization noise due to oversampling reduces the noise power within the desired channel bandwidth

the output of the ADC has to be filtered and its bandwidth reduced to the occupied signal bandwidth. Finally, the quantization noise power of an oversampled signal can be found by integrating (8.11) over the entire bandwidth:

$$\int_{-B/2}^{B/2} S_e(f)df = \frac{BP_e}{F_s} = \frac{1}{OSR} P_e \tag{8.27}$$

Example 8-1: ADC Requirements for IF Sampling Receiver

Consider a WiMAX IF sampling receiver operating at an IF frequency of 175 MHz. After analysis of this particular receiver, it was found that the required ADC SQNR is 35 dB with a maximum channel equivalent bandwidth of 26 MHz (for the 20-MHz occupied signal bandwidth case). Assume that the PAR is 14 dB. What is the required effective number of bits for the ADC? Assume that any reasonable clock rate can be generated and that the modem designers expect normal spectral placing. Realistically, a filter with such narrow band at this IF frequency is very difficult to build. Furthermore, the effect of blockers, spurs, and other degradations may force the SQNR of the ADC to be higher than 35 dB. For the sake of comparison, repeat the example for a channel bandwidth (noise equivalent bandwidth) of 70 MHz. Compare the results.

The first step is to estimate the minimum sampling rate required at 175 MHz. First let's determine the range of n:

$$\frac{2F_c + B}{2n + 1} \leq F_s \leq \frac{F_c - B/2}{n} \quad 0 \leq n \leq \left\lfloor \frac{F_c - B/2}{2B} \right\rfloor \quad \text{normal spectral placement}$$

$$0 \leq n \leq \left\lfloor \frac{F_c - B/2}{2B} \right\rfloor \Rightarrow 0 \leq n \leq \left\lfloor \frac{175 - 26/2}{2 \times 26} \right\rfloor = 3 \tag{8.28}$$

Choose the sampling rate based on the maximum number of n:

$$\frac{2F_c + B}{2n + 1} \le F_s \le \frac{F_c - B/2}{n} \Rightarrow 53.71\,\text{MHz} \le F_s \le 54\,\text{MHz} \tag{8.29}$$

Choose 54 MHz as the sampling rate. Next, consider the SQNR relationship:

$$SQNR = 4.7712 + 6.02b + 10\log_{10}\left(\frac{F_s}{B}\right) - PAR$$

$$b = ceil\left\{\frac{SQNR - 4.7712 - 10\log_{10}\left(\frac{F_s}{B}\right) + PAR}{6.02}\right\}$$

$$= ceil\left\{\frac{35 - 4.7712 - 10\log_{10}\left(\frac{54}{26}\right) - 14}{6.02}\right\} = 7bits \tag{8.30}$$

The results are summarized in Table 8.1.

Table 8.1 Sampling rate boundaries for the 22.4 and 11.2 MHz channel BW case

Parameter	Value
Upper band (MHz)	213.00
Lower band (MHz)	187.00
Bandwidth (MHz)	26.00
IF frequency (MHz)	175.00
f_h (MHz)	188.00
f_l (MHz)	162.00
maximum n	3.00
High sampling boundaries based on max n (MHz)	54.00
Low sampling boundaries based on max n (MHz)	53.71
F_s (MHz)	54.00
PAR (MHz)	14.00
OSR	2.08
SQNR (dB)	35.00
Number of bits	7.00

8.1.4 Performance and Requirements of the Nyquist Converter

8.1.4.1 ADC Overloading and Signal Clipping

So far we have assumed that the received signal is within the full scale or range of the ADC. In this case, the distortion was characterized as quantization noise which is white and uniformly distributed. In practical systems, another type of distortion is often encountered known as signal clipping. Clipping occurs when the input signal to the ADC exceeds its range.

In the following analysis, assume that the input signal $x(t)$ is Gaussian distributed with zero mean and variance σ_x^2. The PDF of the input signal is given as

$$p(x(t)) = \frac{1}{\sigma_x \sqrt{2\pi}} e^{-\frac{x^2}{2\sigma_x^2}} \tag{8.31}$$

The total power of the clipped signal is:

$$P_{clip} = 2 \int_{V_{FS}/2}^{\infty} \left(x(t) - \frac{V_{x,FS}}{2} \right)^2 p(x(t)) dx(t) = 2\sqrt{\frac{2}{\pi}} \frac{\sigma_x^2}{\mu^3} e^{-\mu^2/2} \tag{8.32}$$

where[3] $\mu = V_{x,FS}/(2\sigma_x)$. Recall that for a zero-mean Gaussian process, the signal power is σ_x^2 and hence the signal to clipping ratio SCR is given as [1],[2]:

$$SCR = \sqrt{\frac{\pi}{8}} \mu^3 e^{\mu^2/2} \tag{8.33}$$

The total signal-to-noise ratio due to clipping and quantization in the linear scale is then given as:

$$\frac{1}{SNR_{Total}} = \frac{1}{SCR} + \frac{1}{SQNR}$$

$$SNR_{Total} = \frac{SCR \times SQNR}{SCR + SQNR} \tag{8.34}$$

[3]In this analysis, we assume that the signal clips at $V_{x,FS}/2$. This implies that the maximum quantization step occurs at half full scale.

Note that in practical communication systems, some clipping is usually allowed, especially waveforms with high PAPR. The reason is that a waveform with such a high crest factor tends to rarely reach its peak power and most of the instantaneous power occurs around the average power, as would be statistically expected. This implies that, in order to obtain optimal converter performance, it is more desirable to allow the received signal to better utilize the ADC dynamic range at the expense of some clipping. In other words, the best SNR performance is obtained by allowing the ADC to provide the best statistical representation of the desired signal.

8.1.4.2 *Performance Degradation Due to Aperture Jitter*
ADCs are not only limited by the number of effective bits and quantization noise, but also by sampling time uncertainty caused by the ADC's clock jitter. In this section we discuss the effect of clock jitter on the received signal under uniform sampling conditions.

Uniform sampling implies that the samples generated at the output of the ADC are equidistantly spaced in time at a rate equal to the inverse of the sampling frequency F_s. However, this regularity in sampling can be disturbed if a slight uncertainty is applied to the position of the samples. This ambiguity in the sampling process is known as jitter, which in itself can be modeled as a random variable. The ultimate effect of jitter is degradation to the signal SNR. To illustrate this phenomenon, consider the single-tone signal:

$$x(t) = A \sin(2\pi F t) \tag{8.35}$$

Sampling (8.35) at F_s we obtain:

$$x(n) \cong x_a(t = nT_s)\big|_{T_s = 1/F_s} = A \sin\left(2\pi \frac{F}{F_s} n\right) = A \sin(2\pi f n) \tag{8.36}$$

where $f = F/F_s$ is the normalized frequency. The sampled signal in (8.36) is obtained under ideal conditions. Due to jitter, however, an uncertainty in the sample position occurs and hence the sampled signal becomes:

$$x(n) \approx x_a(nT_s) + \Delta t(nT_s) \frac{\delta}{\delta t} x_a(t)\big|_{t = nT_s} \tag{8.37}$$

for small time jitter $\Delta t(t)$ at time t, as illustrated in Figure 8.4. Note that $\Delta t(nT_s)$ denotes the time jitter at the sampling instant nT_s. In general, for Nyquist converters, the error

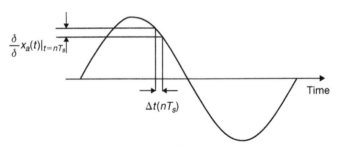

Figure 8.4 Time jitter and its corresponding phase jitter for a single-tone sampled signal

signal is assumed to be spectrally white. In response to a full swing harmonic input signal at the maximum input frequency, the RMS time jitter $\Delta t_{rms}(nT_s)$ can be related to the resolution of the ADC as [3],[4]

$$\Delta t_{rms}(nT_s) = \frac{1}{2\pi B 2^b}\sqrt{\frac{2}{3}}OSR \tag{8.38}$$

where, as previously defined, b is the number of bits in the ADC and OSR is the oversampling ratio. The relation in (8.38) is based on the assumption that the error signal due to sampling jitter is white and serves only as a good approximation.

For the single-tone case, the derivative term in (8.37) is:

$$\frac{\delta}{\delta}x_a(t)\Big|_{t=nT_s} \approx 2\pi A f \cos(2\pi f n) \tag{8.39}$$

Assuming that the time jitter is zero mean white process, then the signal error power can be related to the jitter power as:

$$\sigma_x^2\Big|_{\sigma_x^2=E\{\delta x^2\}} = 2\pi^2 A^2 f^2 \,\sigma_t^2\Big|_{\sigma_t^2=E\{\delta t^2\}} \tag{8.40}$$

From (8.40) the signal to jitter power ratio for a single tone can then be computed:

$$SNR_{jitter} = 10\log_{10}\left(\frac{A^2/2}{2\pi^2 A^2 f^2 \sigma_t^2}\right) = 10\log_{10}\left(\frac{1}{4\pi^2 f^2 \sigma_t^2}\right) \tag{8.41}$$

Therefore the jitter SNR is related to the sampling clock SNR as:

$$SNR_{jitter} = 10 \log_{10} \left(\frac{1}{\sigma_t^2} \right) + 10 \log_{10} \left(\frac{F_s^2}{4\pi^2 F^2} \right)$$

$$= SNR_{clk} + 10 \log_{10} \left(\frac{F_s^2}{4\pi^2 F^2} \right) \tag{8.42}$$

From (8.42), it can be seen that the shape of the phase noise spectrum of the sampling clock is overlaid on top of the sampled data.

The relationship in (8.40) is widely used to express integrated phase noise power σ_x^2 in terms of jitter power σ_t^2 and shows that both measures are byproducts of the same phenomenon. Furthermore, (8.40) also implies that for a given time jitter, a tone with higher frequency will have higher phase error.

The analysis above can be easily extended to a multitone scenario. If the input signal is made up of N tones of equal amplitudes A, then the SNR due to jitter in this case is:

$$SNR_{jitter} = 10 \log_{10} \left(\frac{1}{4\pi^2 \sigma_t^2 \sum\limits_{n=0}^{N-1} f_n^2} \right) \tag{8.43}$$

where f_n corresponds to the nth-tone frequency. The expression in (8.43) assumes that all tones are of equal power but at different frequencies. This case is mostly helpful when evaluating ADCs and a closed form expression due to jitter is required.

Next, a more general case is discussed following the steps provided in [5]. Assume that the input to the ADC is a zero-mean signal with a flat uniformly distributed spectrum. The spectrum is bounded by an upper frequency limit F_U and a lower frequency limit F_L as shown in Figure 8.5.

Recall that Parseval's theorem states that the power of a given signal in the time domain is equal to the power of the signal in the frequency domain; that is, the power in either domain is conserved:

$$\int\limits_{-\infty}^{+\infty} |x_a(t)|^2 dt = \int\limits_{-\infty}^{+\infty} |X_a(F)|^2 \, dF \tag{8.44}$$

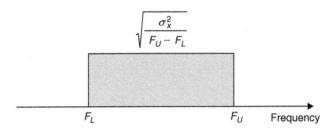

Figure 8.5 Power spectral density of input ADC signal expressed in watts/Hz

Furthermore, the Fourier transform of the derivative of $x_a(t)$ is:

$$\Im\left\{\frac{\partial}{\partial t}x_a(t)\right\} = j2\pi F\Im\left\{x_a(t)\right\} \tag{8.45}$$

From (8.44) and (8.45) the power in the derivative of $x_a(t)$ is:

$$E\left\{\left|\frac{\partial}{\partial t}x_a(t)\right|^2\right\} = \int_{-\infty}^{\infty}\left|j2\pi Fx_a(t)\right|^2 dt = 4\pi^2\int_{-\infty}^{\infty}\left|FX_a(F)\right|^2 dF \tag{8.46}$$

Given the uniform spectrum of the input signal, depicted in Figure 8.5, the power relation in (8.46) becomes:

$$E\left\{\left|\frac{\partial}{\partial t}x_a(t)\right|^2\right\} = 4\pi^2\int_{-\infty}^{\infty}\left|FX_a(F)\right|^2 dF$$

$$= 4\pi^2\int_{F_L}^{F_U}\frac{\sigma_x^2 F^2}{F_U - F_L}dF$$

$$= \frac{4}{3}\pi^2\sigma_x^2(F_U - F_L)^2 \tag{8.47}$$

In an IF sampling receiver, let the center frequency of the signal in the proper Nyquist zone be F_c, and the bandwidth of the signal centered at F_c be B; then we can express the upper frequency bound as $F_U = F_c + B/2$ and the lower frequency bound as $F_L = F_c - B/2$, and the SNR due to clock jitter is:

$$SNR_{jitter} = \frac{E\left\{x_a^2(t)\right\}}{E\left\{\Delta t^2(nT_s)\left(\frac{\partial}{\partial t}x_a^2(nT_s)\right)^2\right\}}$$

$$= \frac{\sigma_x^2}{\frac{4}{3}\pi^2\sigma_t^2\sigma_x^2\left(F_c^2 + \frac{B^2}{12}\right)} = \frac{1}{\frac{4}{3}\pi^2\sigma_t^2\left(F_c^2 + \frac{B^2}{12}\right)} \qquad (8.48)$$

Note that for a single-tone scenario—that is, for $B = 0$—the relationship in (8.48) reduces to that of the single-tone case. Furthermore, if the center frequency of the signal F_c is much larger than the signal bandwidth, then the SNR due to jitter can be treated as a single carrier case. However, if that is not the case, then the relationship in (8.48) must be used.

The analysis so far has not taken the aliasing effect into account. That is, so far it has been assumed that the desired signal is properly sampled such that with added bandwidth due to jitter, no aliasing signal is folding back onto the desired sampled signal. In the event where the signal bandwidth is close to the edge of the Nyquist zone, then the noise caused by jitter, if spectrally significant, can alias back in-band, causing further degradation of the SNR. Therefore, the higher the Nyquist zone of the sampled signal the better the phase noise requirements of the sampling clock have to be. This phenomenon in itself is a limitation of IF sampling receivers.

8.1.4.3 Performance Degradation Due to Aperture Time

An ideal sample and hold (S/H) circuit samples a waveform instantaneously. A practical S/H, however, requires an aperture turnoff time τ. Aperture turn off time is the time during which the current flowing from the current sources decreases linearly (see [6]). An aperture timing error of τ seconds results in a timing error of $\Delta t = \tau/2$ seconds. The resulting SNR can then be determined as:

$$SNR_{Aperture\_time} = 10\log_{10}\left(\frac{F_s}{B(\pi\tau F)^2}\right) \qquad (8.49)$$

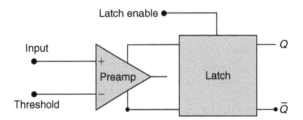

Figure 8.6 Conceptual diagram of comparator architecture

Often, aperture jitter and aperture time are measured simultaneously in the laboratory resulting in an effective aperture uncertainty given as:

$$\sigma_{total} = \sqrt{\sigma_t^2 + (\tau/2)^2} \tag{8.50}$$

Like quantization noise, the S/H aperture uncertainty produces nonlinear distortion in the form of spurs rather than random noise.

8.1.4.4 Conversion Errors

Conversion errors occur during the process of, say, digitizing an analog input to a digital word that is greatly inaccurate. Conversion errors are due to the inability of the comparators to make a proper decision within the allocated time.[4] Depending on the converter architecture, making an erroneous decision can inaccurately affect subsequent decisions. A conceptual block diagram of a comparator is shown in Figure 8.6. The comparator "indecision" takes place when the difference between the analog input and the nearest threshold is not sufficient to drive the comparator into saturation, a condition known as metastability. Metastability is illustrated in Figure 8.7. Errors resulting from metastability are often referred to as "rabbits" or "sparkle codes." Sparkle codes in FLASH converters can be reduced by employing Gray encoding in between the thermometer logic and the binary codes.

Given the differential input voltage at the time of latching to the comparator $\Delta V_{input}(t)$, the output voltage $V_{output}(t)$ can be approximated as

$$V_{output}(t) = G\Delta V_{input}(t)e^{t/\tau} \tag{8.51}$$

---

[4]In FLASH architectures, for instance, this time is the elapsed time after the comparator output is latched.

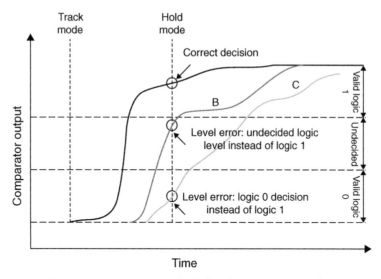

Figure 8.7 Comparator logic level output versus time

where G is the gain of the preamplifier, τ is the regeneration time constant of the latch, and t is the elapsed time after the comparator output has latched. For small values of $\Delta V_{input}(t)$, it takes a longer time for the output to reach a valid logic level. Figure 8.7 illustrates three different, though exaggerated, scenarios. During track mode, waveform A reaches a valid logic level of "1" starting from a logic level of "0" well before the hold mode is asserted. Waveform B, on the other hand, is short of the voltage level needed to assert a logic 1 and falls into an undecided logic level, thus causing an output error. Waveform C is most extreme in the sense that when the hold mode is asserted, the output voltage is still at logic level 0 and the comparator output is definitely in error. In high resolution FLASH converters, depicted in Figure 8.8, a trade-off is typically made between the converter speed and its bit resolution. The faster the converter speed, the smaller the time difference between track mode and hold mode. This is true since the differential input voltage becomes smaller and smaller as the resolution significance of the respective bits becomes less and less. This in essence requires longer track time in order to allow the output voltage of the preamplifier to reach the appropriate level, thus implying slower conversion speed.

Given a parallel/series architecture, let n be the number of bits converted in the first cycle; then a logic error resulting from one of the comparators being in a metastable state results

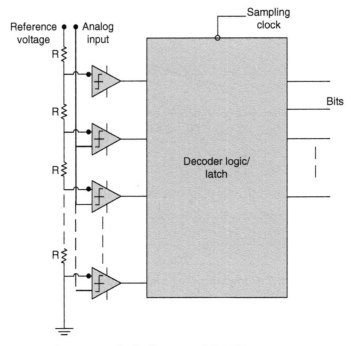

Figure 8.8 Block diagram of FLASH converter

in the output of the converter being in error, especially when subsequent LSB decisions depend on the result of the bits in the first cycle. The resulting converter error, for a parallel architecture, is assumed to be uniformly distributed and its mean square value for a b-bit converter is expressed in the following relation:

$$E\left\{e^2(t)\right\} = \frac{2^{2b}\delta^2}{12} \tag{8.52}$$

where δ is the step size of the quantizer. The uniform error distribution is true since the output of the converter is not correlated with the input signal and resembles a quantization error in nature. For a parallel/series architecture, let p be the probability of an MSB conversion error; then the mean square error due to MSB conversion is given as the conditional probability:

$$E\left\{e^2(t)/\text{MSB conversion error}\right\} = \frac{2^{2b}\delta^2}{12}p \tag{8.53}$$

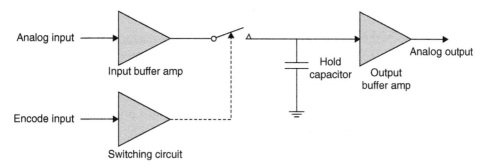

Figure 8.9 Sample and hold amplifier structure

For the remaining LSBs $b - n$, the conditional mean square error is:

$$E\left\{e^2(t)/\text{LSB conversion error}\right\} = \frac{2^{2(b-n)}\delta^2}{12}q \qquad (8.54)$$

where q is the LSB conversion error. The resulting total noise due to conversion error can then be expressed as:

$$E\left\{e^2(t)\right\} = \frac{2^{2(b-n)}\delta^2}{12}(p - 2^{-2n}q) \qquad (8.55)$$

Finally, the relationship in (8.55) can be used to derive the signal-to-conversion-error ratio:

$$SNR_{\text{conversion error}} = \frac{E\left\{s^2(t)\right\}}{E\left\{e^2(t)\right\}} OSR = \frac{3\kappa^2}{(p + 2^{-n}q)} OSR \qquad (8.56)$$

where κ was previously defined as the loading factor of the input signal. Note that the difficulty in obtaining reasonable/practical probabilities for p and q before a converter is designed makes it hard for the system analyst to obtain sound SNR estimates for level errors. Nonetheless, the thought process in obtaining such a figure projects certain insights when choosing converter architecture.

8.1.4.5 Performance Degradation Due to Droop
Droop is simply defined as the change of the amplitude in the S/H output voltage during the hold mode. This drop in voltage, or droop, is mainly caused by the discharge of the hold capacitor resulting from leakage in the S/H switch as well as the input current of the output buffer amplifier, as depicted in Figure 8.9.

Droop, if reasonably small, can be modeled by altering the threshold levels in the way in which droop affects them. For series/parallel converter architectures, threshold levels affected by droop possess an effective level error d. Given the number of bits in the converter b, there exist $2^b - 1$ total thresholds in the converter. Furthermore, assume that the total number of bits converted in the first cycle (or per cycle) is n; then there exist $2^n - 1$ droop-affected thresholds. Consequently, the mean square level error resulting from droop is given via the relation:

$$\Delta^2 = \frac{2^n - 1}{2^b - 1} d^2 \tag{8.57}$$

Note that for FLASH architecture the mean square level error is d^2. Similarly, a mean square level expression for successive approximation architecture can be found as:

$$\Delta^2 = \frac{2^n - 1}{2^b - 1} d^2 \sum_{l=1}^{L-1} 2^{n(l-1)} \left(\frac{L - l}{L - 1} \right)^2 \tag{8.58}$$

where L is total number of conversion cycles per conversion. The droop in this case is the total droop occurring between the time of the first and last conversion cycles. The resultant degradation can then be expressed in terms of SNR as:

$$SNR_{droop} = 10 \log_{10} \left(\frac{2^{2b} \kappa^2}{4\Delta^2} OSR \right) \tag{8.59}$$

where κ is the loading factor defined earlier as applicable to the waveform only,[5] and OSR is the oversampling ratio.

8.1.4.6 Performance Degradation Due to DC Offset

In most practical systems, DC offsets resulting from the analog front end as well as the converter itself are calibrated out. A DC offset compensation algorithm is typically used to remove the effect of small DC offsets at the output of the converter. However, the effect of DC offset is more detrimental on quadrature baseband sampling

---

[5]Waveform only in this context means signal without additive noise. Certain analysts include the effect of noise when computing the loading (or crest) factor of a received signal.

schemes than it is on IF sampling or low IF sampling receivers. In IF sampling receivers, the impact is mainly felt on the dynamic range of the ADC, consequently causing clipping, for example, to start at lower signal levels. This directly impacts the SQNR performance of the ADC. In analog quadrature sampling, however, DC offsets can be particularly troublesome to discrete phase lock loops and automatic frequency control algorithms.

8.1.4.7 Differential and Integral Nonlinearity

Differential nonlinearity (DNL) of a converter is a static performance parameter defined as the maximum deviation of the converter step width and the ideal value of 1 LSB between two consecutive codes. For an ideal ADC, for instance, the DNL corresponds to 0 LSB, thus implying that each analog step is equivalent to 1 LSB. A DNL specification of 1 LSB or less ensures that the converter's transfer function is monotonic with no missing codes.

To better understand DNL, recall that for an ADC for example, the analog input value to the converter $x_a(t,i)$ is converted to an M-bit digital value via the relation:

$$
\begin{aligned}
x_a(t,i)\big|_{t=nT_s} &\cong x(nT_s) + e_q(nT_s) \\
&= x_{ref} \sum_{m=0}^{M-1} b_m(nT_s)2^m + e_q(nT_s), \quad \text{for } b_m(.) = 0 \text{ or } 1
\end{aligned}
\tag{8.60}
$$

where x_{ref} represents a reference voltage, current, or charge, b_0 (.) is the LSB of the converter, b_{M-1} (.) is the most significant bit (MSB), and e_q (.) is the quantization error. The variable i associated with the analog input $x_a(t,i)$ indicates the quantization signal level at the output of the converter associated with $x_a(t,i)$. Given that the signal amplitude varies between zero and full scale voltage $V_{x,Fs}$, then the ideal step representing the LSB of the converter (or resolution) is as defined in (8.5) or $\delta = V_{x,Fs}/2^b$.

Define DNL as the relationship between two analog input values resulting in adjacent quantization signal levels at the output of the ADC as:

$$
DNL(i) = \frac{x_a(t,i) - x_a(t,i-1) - \delta}{\delta}
\tag{8.61}
$$

The DNL phenomenon is illustrated in Figure 8.10.

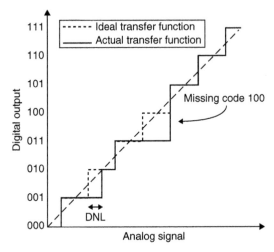

Figure 8.10 Transfer function of a 3-bit ADC

INL, on the other hand, is defined as the integral of DNL errors, thus implying that good INL means good DNL. Therefore, for a total number of L-codes the INL is defined as:

$$INL = \sum_{i=0}^{L-1} DNL(i) \tag{8.62}$$

INL, illustrated in Figure 8.11, compares an ideal ADC transfer function to an ADC with a transfer function that exhibits nonlinear behavior.

8.1.4.8 Second and Third Order Nonlinearity
Consider the nonlinearity model defined in Chapter 3:

$$y(t) = \sum_{n=0}^{\infty} \beta_n x^n(t) = \beta_0 + \beta_1 x(t) + \beta_2 x^2(t) + \beta_3 x^3(t) + \cdots \tag{8.63}$$

where $y(t)$ is the output of a nonlinear device, in this case an ADC or a DAC, excited by a bounded input $x(t)$. The results developed in Chapter 5 are applicable to data converters

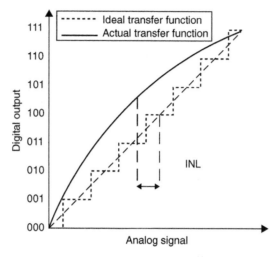

Figure 8.11 Transfer function of ideal 3-bit ADC compared to an actual ADC transfer function exhibiting INL error

and will be repeated here in short for completeness. For a single-tone input $x(t) = \alpha \cos(2\pi F t)$, the signal-to-second-harmonic ratio can be computed as:

$$SNR_{\text{second harmonic}} = 10 \log_{10} \left(\frac{\beta_1^2}{\beta_2^2} \frac{4}{\alpha^2} \right) \tag{8.64}$$

Similarly, the signal-to-third-harmonic ratio can also be computed as:

$$SNR_{\text{third harmonic}} = 10 \log_{10} \left(\frac{\beta_1^2}{\beta_3^2} \frac{16}{\alpha^4} \right) \tag{8.65}$$

Note that, although second and third order nonlinearity analysis alone is not sufficient to characterize the nonlinear behavior of a converter, one can use these analysis tools to develop an understanding of certain key parameters of ADCs and DACs.

8.1.4.9 Converter Thermal Noise
In quadrature sampling, the dominant sources of noise in the ADCs and DACs are digital in nature. However, in IF sampling transceivers, analog noise can play a key role

in determining the performance of the system. For example, the noise figure parameter of an ADC in IF transceivers is more dominant since it is closer to the front when compared, say, to superheterodyne or a direct conversion transceiver employing baseband sampling.

The sources of analog noise are, for example, the wideband amplifier, the sample and hold block in both ADCs and DACs, as well as the buffer amplifier in the DAC. Let the noise power spectral density referenced to the input of the converter be given as $N_0 = KTF$ where K is Boltzmann's constant given as 1.38×10^{-23} joules/kelvin, and T is the temperature in kelvins.[6] In practice, the wideband is followed by a lowpass filter whose noise equivalent bandwidth is B_e, then the in-band portion of the thermal noise is defined as

$$N_T = N_0 B_e \frac{F_s}{B} \tag{8.66}$$

where B is the IF sampling bandwidth.

8.1.4.10 Antialiasing Filter Requirements

The antialiasing filter requirements are partially dictated by the data converter and the architecture of the transceiver. For instance, it is desirable that the aliasing signal components after filtering cause minimal degradation to the desired signal after analog-to-digital conversion. To maximize the signal dynamic range, the distortion power due to aliasing can, for instance, be specified to less than the minimum distortion level present in 1/2 LSB of dynamic range. That is, the power in 1/2 LSB of dynamic range can be expressed as:

$$P_{1/2LSB} = 10 \log_{10} \left(\frac{1}{2^{2(b+1)}} \right) \text{dBc} \tag{8.67}$$

The relation expressed in (8.67) is referenced from the converter's full scale dynamic range. This requirement is not always easily met, and it is up to the system analyst to choose the amount of converter dynamic range that will be occupied by aliasing components. Therefore, once the overall architecture is chosen, careful analysis of the

---

[6]$T = 290$ kelvins is considered to be room temperature.

required instantaneous ADC dynamic range and blocking requirements will reveal the amount of filtering required. Another important factor to consider is the radio architecture. For example, direct conversion architecture requires two quadrature converters: that is, one converter for the in-phase component and one for the quadrature component. Compare this to IF sampling architecture, for example, which requires only one converter operating at IF frequency. The former, however, can operate at a lower sampling rate, thus imposing less linearity restrictions on the converter and generally less power consumption than the latter. Furthermore, IF sampling architecture tends to impose higher dynamic range requirements on the converter.[7] From Example 8.2, given an allowable degradation of a certain number of LSBs, it is obvious that the higher the dynamic range of the converter, the higher the required stop-band attenuation of the filter. On the other hand, IF sampling architecture is superior to quadrature sampling in terms of DC offset and I/Q amplitude and phase imbalance degradation. However, as mentioned earlier, a trade-off between filter stop-band rejection, converter resolution, and transceiver architecture is always in order for any given design in order to determine the optimal solution in terms of performance, cost, and power consumption. A comparison between sampling strategies and antialiasing filter complexity for various transceiver architectures is summarized in Table 8.2.

Table 8.2 Sampling strategy and antialiasing filter complexity in various transceiver architectures

Architecture	Sampling Approach
Direct conversion	Quadrature baseband sampling, moderate antialiasing requirements
Super-heterodyne with analog baseband	Quadrature baseband sampling, moderate to simple antialiasing requirements
Super-heterodyne with IF sampling	IF sampling with hard antialiasing filtering requirement
Low IF	IF sampling with moderate antialiasing filtering requirement

[7]For instance, IF sampling architecture requires 3-dB higher instantaneous PAR than quadrature ADC architecture.

Example 8-2: ADC Requirements for IF Sampling Receiver

Consider a UMTS direct conversion receiver targeting LTE (long term evolution) applications. The maximum anticipated occupied signal bandwidth for LTE is 20 MHz at RF. Compute the antialiasing filter requirement such that the distortion due to aliasing components is less than or equal to 1/2 LSB. Assume a direct conversion receiver, and that the ADC for either the in-phase or quadrature channel has 7 bits. Repeat the analysis for 8-, 9-, and 10-bit ADCs.

Recall that the occupied signal bandwidth at baseband is 10 MHz after downconversion. Assume that the sampling rate is 40 MHz. How does the requirement of having less than 1/2 LSB of filter rejection affect the complexity of the antialiasing filter? Note that, realistically, LTE requires 10-bit ADCs to meet the 64-QAM modulation requirement. For the 8-bit case, and taking into account practical filter design considerations, is the 1/2-LSB power attenuation a reasonable requirement?

Applying the relationship presented in (8.67) in order for the distortion power not to exceed the power present in 1/2 LSB quantization level, the required filter rejection for the 7-bit ADC is:

$$P_{1/2LSB} = 10\log_{10}\left(\frac{1}{2^{2(b+1)}}\right)$$

$$= 10\log_{10}\left(\frac{1}{2^{2(7+1)}}\right) = -48.16 \text{ dBc} \qquad (8.68)$$

The analysis for 10, 11, and 12 bit ADC is summarized in Table 8.3.

Simulation results showing various filter responses are given in Figure 8.12.

Table 8.3 Required filter attenuation versus number of ADC bits

Number of Bits	Required Filter Attenuation
7	−48.16
8	−54.19
9	−60.21
10	−66.23

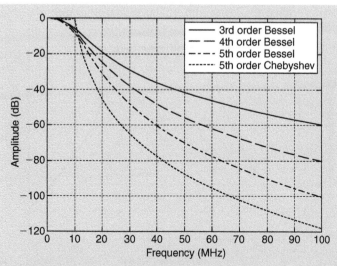

Figure 8.12 Magnitude response of third, fourth, and fifth order Bessel filters along with a Chebyshev fifth order filter

Table 8.4 Analog baseband filter type versus integrated power

Filter Type	Integrated Power in (dBc)
3rd order Bessel	−40.8017
4th order Bessel	−67.8056
5th order Bessel	−94.1226
5th order Chebyshev	−128.9729

The integrated power between the Nyquist rate and 100 MHz for various analog filters is shown in Table 8.4. It is obvious from Table 8.4 that the third order Bessel filter does not provide enough rejection to reach the equivalent of 1/2 LSB or less degradation. On the other hand, the fourth order Bessel filter provides enough rejection to meet the 1/2 LSB or less degradation requirement. The fifth order Bessel and the fifth order Chebyshev provide yet more rejection; however, they may be an overkill given the proposed ADC's resolution.

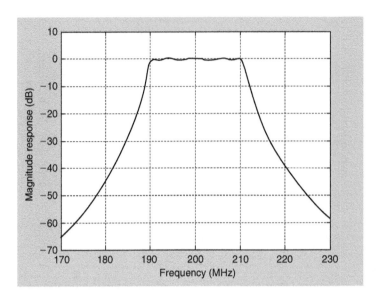

Figure 8.13 Magnitude of fifth order bandpass Chebyshev type 1 filter

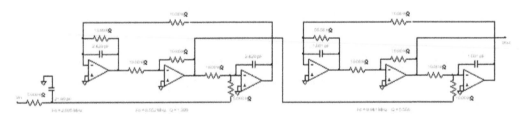

Figure 8.14 Active circuit of lowpass Chebyshev type 1 fifth order filter

In light of Example 8-2, let's compare the complexity of the analog filter preceding the ADC in an IF sampling receiver to that preceding the ADC in a direct conversion zero IF transceiver. The aim of this analysis is to compare the complexity of the filter in the former architecture to the combined complexity needed to precede the in-phase and quadrature ADC in the later architecture. Again, consider a 20-MHz LTE channel. The analog filter at baseband has a passband of 10 MHz, whereas the bandpass filter has passband of 20 MHz centered at an IF frequency of say 200 MHz. The frequency response of the bandpass fifth order Chebyshev filter is depicted in Figure 8.13.

The circuit implementations of both lowpass and bandpass filters using active devices are depicted in Figure 8.14 and Figure 8.15, respectively. Just by comparing the circuits

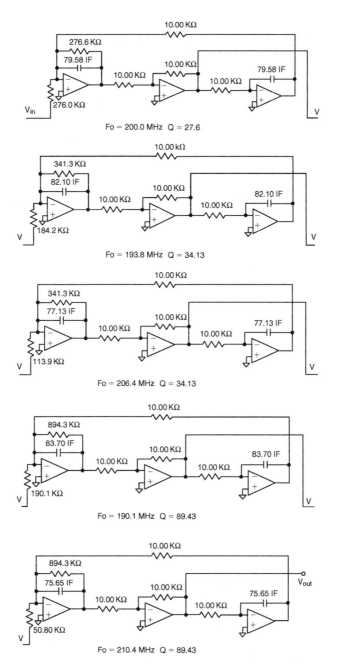

Figure 8.15 Active circuit of bandpass Chebyshev type 1 fifth order filter

and looking at component count in both filters, it is obvious that the bandpass filter is more complex in nature than both analog lowpass filters combined. Furthermore, the nonlinearity and noise figure requirements are higher for the bandpass filter than the lowpass filters. The noise figure has to be better since the filter is higher up in the chain and hence affects overall noise figure of the system more than the lowpass analog filter. Furthermore, due to the higher instantaneous PAR requirement, the spur-free dynamic range due to IIP2 and IIP3 has to also be higher than that of the lowpass analog filters. Therefore, it can be easily concluded that the IF receiver would have to levy more in its performance trade-off on the ADC than on the bandpass filter. Hence, the system designer must rethink the whole analog and digital line-up when designing an IF sampling receiver.

8.2 Overview of Nyquist Sampling Converter Architectures

In this section, the various Nyquist sampling converters are briefly discussed. The discussion focuses on the advantages and disadvantages of common converter architectures and highlights relevant distinctive features. Thus far, the two common types of Nyquist converters used in SDR designs are based on parallel and pipelined architectures.

8.2.1 The FLASH Architecture

The FLASH or parallel converter, as it is often called, is the fastest type of ADC which can be used as a standalone, low-resolution converter or as a fundamental building block of other converters. One such example of the latter is the multibit delta-sigma ($\Delta\Sigma$) modulator. State-of-the-art FLASH converters used for wireless applications today can use up to 6-bit resolution and have sampling speeds over 1 Gps. One such application for this type of converter is MBOA-UWB.

An N-bit FLASH ADC relies on 2^N resistors and $2^N - 1$ comparators arranged in a totem structure as depicted in Figure 8.8. The reference voltage of each comparator, tapped from the resistor-chain, is 1 LSB higher than the one below it in the totem structure. For a given input voltage $x_a(t)$, some of the comparators below a certain point in the totem will have their reference voltage lower than the input voltage and a logic 1 output is produced. The comparators above that same point will have their reference voltage above the input voltage and a logic 0 output is produced. The outputs of the various comparator stages, also known as thermometer logic, are further decoded to produce the desired N-bit binary ADC output.

The input voltage $x_a(t)$ is applied to the input of the totem and received by all the comparators at once. Each comparator has an inherent delay and the output of the thermometer is deferred by only one comparator delay. The decoder logic and latches have a relatively small delay. This in essence contributes to the speed of the FLASH converter in transforming an analog input signal to its equivalent binary output signal. In theory, given a perfectly dynamically matched set of comparators, the intrinsic delay in each comparator acts as a sample and hold circuit and hence a specific sample and hold circuit is not necessary, as is the case with other converters that require such a circuit.

FLASH converters suffer from various degradations and nonidealities such as INL, DNL, kickback noise, sparkles in the thermometer logic, metastability, nonlinear input capacitance, slew-dependent sampling instant of comparators, and degradation due to clock jitter. We have discussed most of these issues in detail earlier in the chapter. FLASH ADCs also suffer degradation in performance due to input capacitance. This degradation serves to limit the allowable signal bandwidth and introduces certain harmonics in the sampled signal output.

8.2.2 Pipelined and Subranging Architecture

The pipelined ADC is one of the most common converter architectures used in signal processing applications. State-of-the-art pipelined architecture has a resolution of up to 12 or 13 bits with speeds exceeding 100 Msps. Pipelined converters with calibration circuitry can provide up to 16 bits of resolution at the expense of reduced conversion speed.

The pipelined ADC overcomes the design complexities of the FLASH converter at the expense of conversion speed. However, pipelined ADCs can be designed to achieve higher resolution at reduced cost. The hardware cost of a FLASH ADC tends to be exponential with respect to resolution, whereas that of a pipelined ADC tends to be linear as stages are added to increase the converter's bit width.

A general pipelined ADC, described in Figure 8.16, is made up of multiple stages cascaded in series driven by nonoverlapping clocks. The first stage is comprised of a sample and hold, an ADC, typically a FLASH, and a DAC, as shown in Figure 8.17. The output of the sample and hold in the first stage is digitized by the first ADC to produce b_1 bits. These bits serve as the MSBs of the converter and are crucial to the accuracy of the converter. The output of the ADC is consequently converted back to analog using b_1-bit DAC. The DAC output is then subtracted from the output of the sample and hold

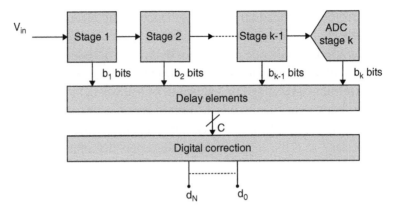

Figure 8.16 The pipelined ADC converter

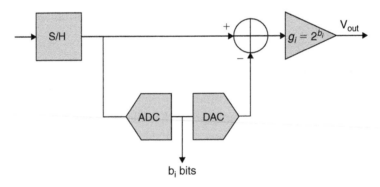

Figure 8.17 A typical pipeline stage

circuit, resulting in a smaller signal to be digitized. Note that at this stage the required range of the second-stage ADC is smaller than that of the first stage, and hence the name *subranging* ADC. The output signal is then amplified and fed to the second stage, where the process is repeated.

In the interstage amplifier, possibly designed in switched-capacitor circuits, g_i must be designed to settle within 1/2 clock period, which may prove to be difficult and power consuming. Furthermore, the gain accuracy of the interstage amplifier is crucial, especially in the early stages, to achieve the desired resolution.

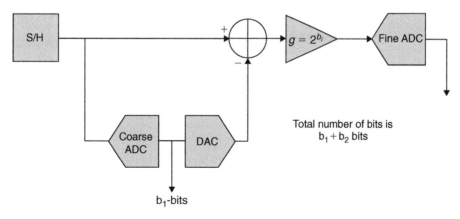

Figure 8.18 A subranging two-step ADC

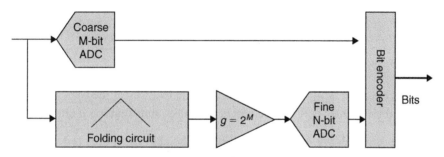

Figure 8.19 A folding ADC

A subranging ADC, on the other hand, is similar to a pipelined ADC and typically uses a pipelined stage with a *coarse* ADC followed by a *fine* ADC, as shown in Figure 8.18 for two-step architecture. Subranging ADCs are medium resolution ADCs providing more than 10 bits digital output with reasonable power consumption. The two-step ADC samples at speeds up to 350 Msps.

8.2.3 Folding Architecture

Folding ADCs are moderate-resolution high-speed data converters. The architecture itself can be thought of as a blend of FLASH and two-step ADC concepts. A typical folding ADC uses fewer comparators than a FLASH converter and hence consumes less power and less area. The folding ADC involves a coarse ADC to generate the MSBs, a folding circuit, and a fine ADC to generate the LSBs. A basic folding ADC, shown in Figure 8.19, is based

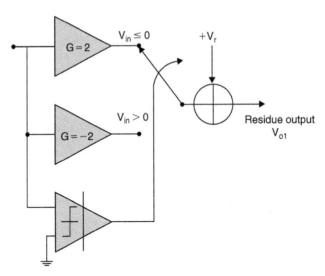

Figure 8.20 Conceptual Gray-code folding circuit

on a parallel architecture while borrowing some of the features of the two-step architecture without the need for a DAC or a sample and hold circuit, thus making it simpler to design.

The basic premise of folding ADCs is as follows. The analog input is fed into a coarse ADC in order to obtain the MSBs in a similar manner to that shown earlier in the two step architecture. The same input signal is fed to a folding circuit. The residue—that is, the output of the folding circuit—is in turn fed to a fine ADC after normalization to obtain the LSBs. The MSBs and LSBs are fed to an encoder to obtain the digital output. The folding circuit serves to reduce the number of comparators used in folding ADCs as opposed to FLASH ADCs.

Next we discuss a basic folding circuit depicted in Figure 8.20, as presented in [7]. The output of the comparator is normalized and fed to the ADC. Assume that the input signal ranges from $-V_r$ to $+V_r$; then for $V_{in} = -V_r$, the residue output, say $V_{o1} = 2V_{in} + V_r = -2V_r + V_r = -V_r$. On the other hand, for an input signal $V_{in} = V_r$, the residue output signal becomes $V_{o1} = -2V_{in} + V_r = -2V_r + V_r = -V_r$. Next assume that the input signal is equal to zero, then the output signal becomes $V_{o1} = -2V_{in} + V_r = -2 \times 0 + V_r = V_r$. The resulting transfer function of the folding circuit is shown in Figure 8.21.

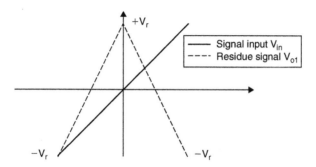

Figure 8.21 Basic folding circuit response

Example 8-3: A trivial folding ADC

The purpose of this example is to simply illustrate the concept behind a folding ADC and not necessarily to provide a depiction of an actual device. Assume we are given a three-bit folding ADC such that the MSB is obtained via a slicer (serving in this case as the coarse ADC comprised of a single-bit ADC) and the last two LSBs are obtained via a fine ADC driven by a folding circuit. Assume that the input to the ADC system is limited between $-V_r$ and $+V_r$, where V_r is the reference voltage of the ADC. Enumerate the various outputs of the folding circuit in response to various input voltage levels.

In a three-bit ADC, the input voltage level varies between $-V_r$ and $+V_r$ with converter resolution obtained from (8.5) as $\delta = V_{FS}/2^b = 2V_r/2^3 = V_r/4$. The input-output relationship for the folding circuit in response to various input thresholds and as a function of binary Gray code is summarized in Table 8.5. Obtaining a Gray code from a binary code, and vice versa, is summarized in Section 8.3. The input-output voltage relationship is specified in the previous section.

Table 8.5 Input-output relationship of folding circuit with relation to Gray code for a two-bit fine converter

Gray Code	Input Voltage to Folding Circuit	Output Voltage to Folding Circuit	Corresponding 2-Bit Gray Code
000	$-V_r$	$-V_r$	00
001	$-3/4V_r$	$-1/2V_r$	01
011	$-1/2V_r$	0	11
010	$-1/4V_r$	$1/2V_r$	10
110	$1/4V_r$	$1/2V_r$	10
111	$1/2V_r$	0	11
101	$3/4V_r$	$-1/2V_r$	01
100	V_r	$-V_r$	00

It is interesting to note from Example 8-3 that, as the input signal swings over the entire converter full scale, that is between $-V_r$ and $+V_r$, the output of the folding circuit, the residue signal, swings over the entire full scale twice. That is, as the input signal varies between $-V_r$ and $+V_r$, the residue signal varies from $-V_r$ to $+V_r$, and then back to $-V_r$. This implies that if the input signal frequency is bounded by the Nyquist frequency F_N, then the frequency of the residue signal is twice F_N. Depending on the desired resolution of the fine ADC, the folding circuit depicted in Figure 8.20 is cascaded multiple times. This causes the residue signal output to fold multiple times in response to the input signal swinging from $-V_r$ to $+V_r$ as shown in Figure 8.22 for single, double, and quadruple

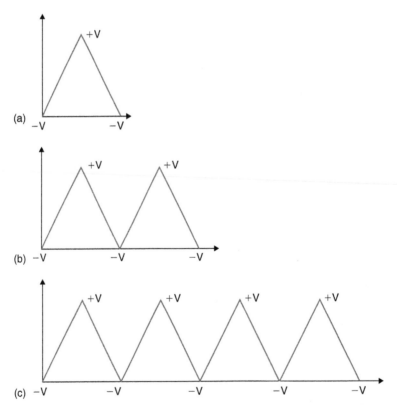

Figure 8.22 Folding circuit response: (a) single folding, (b) double folding, and (c) quadruple folding

folding. This increased resolution comes at the cost of the folding circuit operating at frequencies significantly higher than the Nyquist frequency.

The folding circuit suffers from serious nonlinearity in the folding process. This can be best described as smoothing of the response around the folding corners of the residue signal, as compared with the ideal residue signal as depicted in Figure 8.23. This deviation from the ideal response transpires as a DNL error in the digital output of the ADC. This effect becomes much more pronounced at higher frequencies and thus limits the resolution of this type of converter.

8.2.4 Integrating Dual-Slope Architecture

The integrating multislope ADC is a slow high-resolution converter not typically suited for SDR applications, as will be demonstrated. In this section, only the dual-slope ADC will be discussed (see Figure 8.24). The working operation of the converter can be explained simply as follows: the *unknown* input signal V_{in} is applied to the integrator op-amp circuit for a *known* integration interval, say T_s. This interval is determined by the counter allowing for the integrating capacitor C to accumulate its voltage to a value equivalent to the average of the input voltage, provided that T_s is sufficient time. After T_s seconds, the counter is reset and the switch closes towards V_{REF}, which is a *known* negative voltage reference. This in essence *de-integrates* the capacitor and drives its voltage to zero. The time T_{int} it takes for the capacitor to discharge, which is tracked by

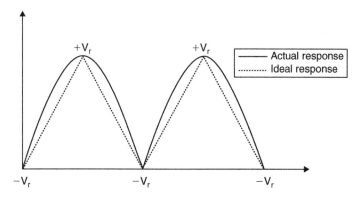

Figure 8.23 Effect of nonlinearity on folding circuit response (solid line) compared with ideal response (dashed line)

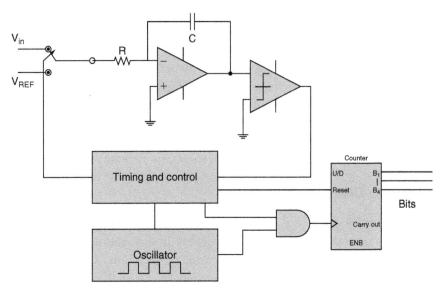

Figure 8.24 Conceptual dual-slope integrating ADC

the counter, is proportional to the output voltage and in essence is a digital representation of it.

This architecture provides for high resolution conversion at the expense of conversion time. Errors resulting from bias current and offset voltages of the integrating amplifier and the comparator can be cancelled digitally. This architecture, despite its high resolution, is not suitable for IF receivers.

8.2.5 Successive Approximation Architecture

The successive approximation ADC is not particularly attractive for SDR, especially for IF-sampling receivers. However, the architecture will be briefly described for the sake of completeness. Successive approximation, depicted in Figure 8.25, is an iterative approach. The major blocks of this converter are the sample and hold, the comparator, a successive approximation register (SAR), and a DAC. At the start of each conversion, the SAR bits are all set to zero and the output of the DAC is also zero. Starting with the MSB, the bit is changed to logic 1 and the output of the DAC is compared with the input signal. If the input signal voltage is larger than that of the DAC output voltage, the bit

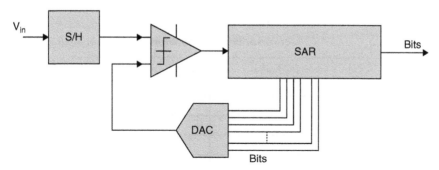

Figure 8.25 Successive-approximation ADC

stays at logic 1, otherwise it is set to logic 0. Once this step is accomplished, then the second MSB is set to logic 1, and then again the output voltage of the DAC is compared to the voltage of the sample and hold. If the output of the sample and hold voltage is larger than that of the DAC output voltage, the bit is set to logic 1, otherwise it resets to logic 0. The process continues until all the bits in the SAR are either set or reset. Note that the higher the resolution of the converter, the longer it takes to perform a single conversion. The conversion time grows exponentially as a function of the converter's resolution. In this architecture, the sample and hold circuits and DAC are critical for its performance.

8.3 Appendix

8.3.1 Gray Codes

Sparkle codes at the output of the thermometer logic and metastability can be mitigated in FLASH converters by using Gray encoding. The Gray code[8] is then converted to a binary code if necessary. A binary reflected Gray code $g_0 g_1 \ldots g_{n-1} g_n$ can be generated from a binary code $d_0 d_1 \ldots d_{n-1} d_n$ using the following simple procedure. First, generate the Gray-code LSB g_n by examining the digit d_{n-1}. If $d_{n-1} = 1$, then g_n is the inverted binary value of d_n, otherwise $g_n = d_n$. To generate g_{n-1}, examine the binary digit d_{n-2}. Again, if $d_{n-2} = 1$, then g_{n-1} equals the inverse of d_n, otherwise let $g_n = d_n$. Repeat the above procedure until you reach g_0, which is equal to d_0.

---

[8]Named after Frank Gray of Bell Labs [8].

Table 8.6 Three-bit binary code and its
corresponding Gray code

Binary Code (d_0,d_1,d_2)	Gray Code (g_0, g_1, g_2)
000	000
001	001
010	011
011	010
100	110
101	111
110	101
111	100

On the other hand, given a Gray code number $g_0g_1...g_{n-1}g_n$, the equivalent binary number $d_0d_1...d_{n-1}d_n$ can be generated following the procedure presented herein. Given the Gray code nth digit, the following sum can be computed:

$$S_n = \sum_{l=0}^{n-1} (g_l \bmod 2) \qquad (8.69)$$

If S_n is logic 1, then replace g_n by its binary inverse. If S_n is a logic 0, then leave g_n unchanged. Repeat the procedure outlined in (8.70) for all the bits in the code word to obtain the equivalent binary number. Table 8.6 presents a three-bit binary code and its corresponding Gray code generated according to the procedure outlined above.

The generation of the Gray code from the output of the thermometer logic for the three bit case can be designed via a simple logic circuit comprised of basic logic gates and latches. Assuming that the output of the thermometer logic is $c_1,c_2,c_3,c_4,c_5,c_6,c_7$ then the Gray code output is given according to the relation:

$$\begin{aligned}
g_0 &= c_1\bar{c}_3 + c_5\bar{c}_7 \\
g_1 &= c_2\bar{c}_6 \\
g_2 &= c_4
\end{aligned} \qquad (8.70)$$

The Gray code, if necessary, can then be translated to binary code via another logic circuit.

References

[1] Mestdagh D, Spruyt P, Brian B. Effect of amplitude clipping in DMT-ADSL Transceivers. IEEE Trans. on Electronics Letters, July 1993;29(15):1354–5.

[2] Al-Dahir N, Cioffi J. On the uniform ADC bit precision and clip level computation for a Gaussian signal. IEEE Trans. on Signal Processing, February 1996;44(2):434–8.

[3] Bang-Sup Song. High speed pipelined ADC. Tutorial at European Solid-State Circuits Conference Sept. 2002.

[4] Arkesteijn V, Klumperink E, Nauta B. ADC clock jitter requirements for software radio receivers. IEEE Conf. on Vehicular Technology, 2004;3:1983–5.

[5] Smith P. Little known characteristics of phase noise. Analog Devices Application Note AN-741, Norwood, Massachusetts, www.analog.com.

[6] Rezavi B. Principles of Data Conversion System Design. New York, NY: John Wiley and Sons; 1995.

[7] Kester W. ADC Architecture VI: Folding ADCs. Analog Devices Application Note, Norwood, Massachusetts, www.analog.com.

[8] Gray F. *Pulse code communication*, March 17, 1953 (filed Nov. 1947) US Patent 2,632,058.

CHAPTER 9

$\Delta\Sigma$ *Modulators for Data Conversion*

In wireless communication, $\Delta\Sigma$ modulators are widely used as data converters to construct oversampled ADCs and DACs and they play an integral part in modern frequency synthesis. $\Delta\Sigma$ modulators operate at higher conversion or sampling rates than would be required by the Nyquist criterion. For this reason these types of data converters are known as oversampling converters. Despite their popularity, $\Delta\Sigma$ converters have yet to be adopted in SDR radios that employ IF sampling architecture. The obvious reasons are the required dynamic range and increased sampling rate required to support a multistandard multimode operation at IF. However, in wireless architectures that rely on quadrature demodulation such as direct conversion, low-IF, and super-heterodyne, $\Delta\Sigma$ converters have presented viable low-power, low-cost data conversion solutions.

This chapter is divided into six major sections. Section 9.1 presents an overview of the signal processing operation of $\Delta\Sigma$ converters. Section 9.2 compares the performance of continuous-time versus discrete-time converters. The SQNR performance for first and higher order loops is given in Section 9.3. Section 9.4 discusses bandpass of $\Delta\Sigma$ converters. Section 9.5 discusses common converter architectures such as modulators with multibit quantizers and the MASH architecture. The chapter concludes with a small section on the nonidealities of $\Delta\Sigma$ converters.

9.1 The Concept of $\Delta\Sigma$ Modulation

In lowpass $\Delta\Sigma$ ADC, the analog input signal is converted to a digital signal using a coarse quantizer [1]. The frequency content of the digital signal approximates well

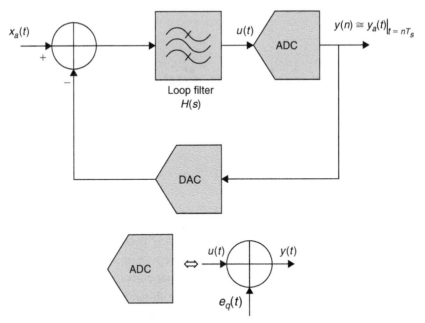

Figure 9.1 Conceptual basic $\Delta\Sigma$ data converter

the input signal in a narrow frequency range. The quantization noise of the converter increases with frequency due to the highpass shaping function imposed on it by the loop filter in the closed loop form shown in Figure 9.1.

The basic components of a continuous $\Delta\Sigma$-modulator as depicted in Figure 9.1 are: the loop filter with transfer function $H(s)$, the quantizer or ADC clocked at the sampling rate F_s, and a DAC placed in the feedback portion of the loop. To understand the basic function of a $\Delta\Sigma$ converter, consider the difference between the analog input signal $x_a(t)$ and the output signal of the quantizer $y(n)$ at the output of the DAC. This signal is shaped by the loop filter $H(s)$ resulting in a new filtered signal, say $u(t)$, which in turn constitutes the input to the quantizer. The output of the quantizer is the signal $y(n)$, which is the sampled version of $y(t)$ sampled at the instance nT_s. The quantizer itself, as discussed earlier, is a nonlinear element. In its simplest form, the quantizer can be a single-bit ADC, sometimes referred to as a slicer. On the other hand, the quantizer can be comprised of a high-speed multibit ADC.

For simplicity of analysis, assume that the combined quantizer-DAC response is linear and that it can be represented as a simple summation as shown in Figure 9.1. Then in the Laplace domain, the signal $u(t)$ can be expressed as[1]

$$U(s) = H_a(s)[X_a(s) - Y(s)] \qquad (9.1)$$

The quantization noise $e_q(t)$ is additive and uniformly distributed. The output signal $y(t)$ can be expressed in terms of the quantization noise and input signal to the quantizer as

$$Y(s) = U(s) + E_q(s) \qquad (9.2)$$

Substituting (9.1) into (9.2) and rearranging we obtain

$$Y(s) = \frac{H_a(s)}{1 + H_a(s)} X_a(s) + \frac{1}{1 + H_a(s)} E_q(s) \qquad (9.3)$$

The output signal $y(t)$ is the sum of the input signal $x_a(t)$ filtered by the signal transfer function:

$$STF(s) = \frac{H_a(s)}{1 + H_a(s)} \qquad (9.4)$$

and the quantization noise $e_q(t)$ filtered by the noise transfer function:

$$NTF(s) = \frac{1}{1 + H_a(s)} \qquad (9.5)$$

The impact of these filters on the signal and noise will be illustrated in an example.

---

[1]MASH modulator stands for *multi-stage noise shaping* modulator.

Example 9-1: Role of STF and NTF on Input Signal and Quantization Noise

Assume that the loop filter of the $\Delta\Sigma$ modulator depicted in Figure 9.1 is a simple integrator, that is, $H(s) = 1/s$. This integrator is a simple lowpass filter. Derive STF(s) and NTF(s). Given the magnitude response of STF(s) and NTF(s), what can you deduce about their characteristics?

The *STF(s)* and *NTF(s)* can be derived simply as

$$STF(s) = \frac{H_a(s)}{1 + H_a(s)} = \frac{1}{s + 1}$$

$$NTF(s) = \frac{1}{1 + H_a(s)} = \frac{s}{s + 1} \tag{9.6}$$

The magnitude responses of *STF(s)* and *NTF(s)* are given as

$$|STF(\Omega)|^2 = \frac{1}{1 + \Omega^2}$$

$$|NTF(\Omega)|^2 = \frac{\Omega^2}{1 + \Omega^2} \tag{9.7}$$

The magnitude response of (9.7) at DC is given as

$$\lim_{\Omega \to 0} |STF(\Omega)|^2 = \lim_{\Omega \to 0} \left(\frac{1}{1 + \Omega^2} \right) = 1$$

$$\lim_{\Omega \to 0} |NTF(\Omega)|^2 = \lim_{\Omega \to 0} \left(\frac{\Omega^2}{1 + \Omega^2} \right) = 0 \tag{9.8}$$

Likewise, evaluating the magnitude response of (9.7) at infinity gives

$$\lim_{\Omega \to \infty} |STF(\Omega)|^2 = \lim_{\Omega \to 0} \left(\frac{1}{1 + \Omega^2} \right) = 0$$

$$\lim_{\Omega \to \infty} |NTF(\Omega)|^2 = \lim_{\Omega \to 0} \left(\frac{\Omega^2}{1 + \Omega^2} \right) = 1 \tag{9.9}$$

The relations (9.8) and (9.9) imply that *STF(s)* is a lowpass filter acting on the input signal, whereas *NTF(s)* is a highpass filter acting on the quantization noise. The consequence of the latter tends to minimize the effect of the quantization noise on the signal by *shaping* it away from the signal. Compared to a Nyquist ADC, the quantization noise in $\Delta\Sigma$ modulators is not uniformly distributed but rather follows a highpass distribution as dictated by *NTF(s)*.

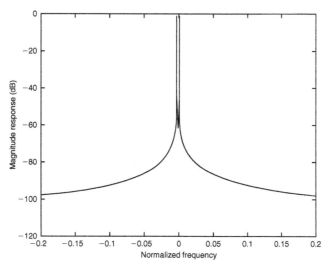

Figure 9.2 Input signal consisting of single sinusoid to a first order ΔΣ modulator

An immediate observation from Example 9-1 is that the higher the sampling rate of the system, the less impact the quantization noise has on the desired signal. Thus, oversampling in ΔΣ modulators tends to improve the SQNR. This improvement will be discussed in more detail later on. However, to summarize, it is desired that the signal and noise transfer functions minimize the effect of the quantization noise inside the signal band without degrading the desired signal. This is done by allowing the noise transfer function to significantly reduce its gain inside the desired signal bandwidth, thus lessening the degradation due to quantization noise on the signal. Conversely, it is desired that the signal transfer function act to protect the signal by *filtering* the quantization noise outside the desired signal band. This in essence implies that an ideal *NTF(s)* acts in an inverse manner to *STF(s)*.

In order to further illustrate the process of shaping the quantization noise, consider an input signal consisting of a single sinusoid as depicted in Figure 9.2. The output of the quantizer in response to the sinusoid of a first order ΔΣ modulator is depicted in Figure 9.3. A first order ΔΣ modulator employs a loop filter with a simple integrator. Higher order loops will employ loop filters that have a corresponding number of integrators. Note that the quantization noise is *shaped* away from the sampled tone. The input signal is oversampled by a factor of 8. Note that increasing the loop from a first to a third order improves the SQNR, as is evident when comparing Figure 9.3 with Figure 9.4.

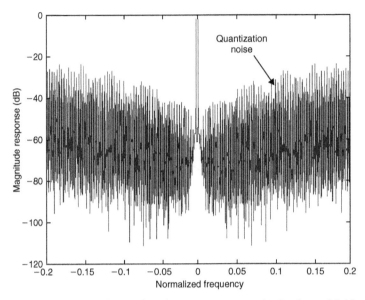

**Figure 9.3 Output of quantizer in response to a single sinusoidal input
of a first order $\Delta\Sigma$ modulator**

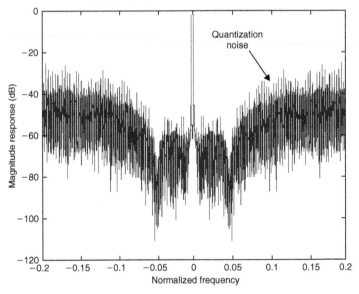

**Figure 9.4 Output of quantizer in response to a single sinusoidal input
of a third order $\Delta\Sigma$ modulator**

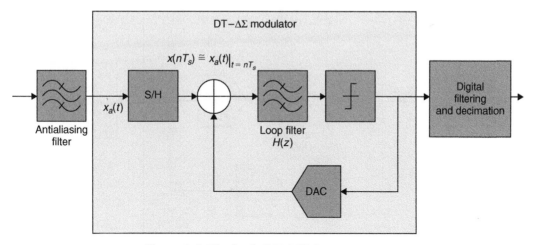

Figure 9.5 The basic DT $\Delta\Sigma$ data converter

The oversampling process and the order of the loop, and their impact on SQNR, will be discussed later on.

9.2 Comparison between Continuous-Time and Discrete-Time $\Delta\Sigma$ Modulation

9.2.1 A Brief Comparison between CT and DT $\Delta\Sigma$ Modulators

$\Delta\Sigma$ modulators can be designed either in continuous-time (CT) or in discrete-time (DT). In the previous section, for example, the modulator was assumed to be a CT $\Delta\Sigma$-modulator. The distinction between DT and CT modulators will be explained herein. A DT $\Delta\Sigma$ modulator implies that the analog input signal to the converter is first presented to a sample and hold circuit after filtering via an antialias filter, as shown Figure 9.5. The loop filter $H(z)$ is implemented using a switched capacitor or a switched current circuit. DT $\Delta\Sigma$ modulators designed in monolithic switched capacitor circuits, for example, possess a high degree of linearity in comparison to CT $\Delta\Sigma$ modulators whose various blocks are designed using analog integrators. These analog blocks are implemented using RC filters employing op-amps, operational transconductance amplifiers (OTA), or $_{gm}C$-filters [1].

Unlike CT $\Delta\Sigma$ modulators, where the sampling operation takes place inside the loop as depicted in Figure 9.6, the input to the DT $\Delta\Sigma$ modulator is already sampled and

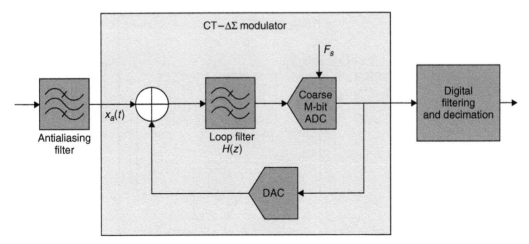

Figure 9.6 The basic CT ΔΣ data converter

presented to the input of the DT loop in discrete form. This in itself is a disadvantage since any errors due to the sampling process, including sampling clock jitter, are added to the input signal and filtered via *STF*(.) and do not undergo any shaping. This, however, is not the case in CT ΔΣ modulators since the sampling circuit, a single or multi-bit ADC, is inside the loop. This implies that the anomalies due to the sampling process are shaped by the loop's *NTF*(.), as is the case with quantization noise, and hence tend to deteriorate the desired signal less when compared to a DT ΔΣ modulator. Another advantage of having the sampling process take place inside the loop is the inherent antialiasing function performed by the loop filter itself thus levying some of the filtering requirement off of the antialiasing filter. However, having the sampling process take place inside the loop is not without its problems. CT ΔΣ modulators require the ADC circuit to settle very fast[2] in order to provide the feedback circuit with the accurate quantization level needed to generate the feedback signal. The quantizer in a DT ΔΣ modulator, on the other hand, is required to settle within half a sampling period.

Another important difference between CT and DT ΔΣ data converters is in the realization of the loop filter. In DT implementations, switched capacitor circuits are

---

[2]This is the metastability problem discussed earlier and typically associated with FLASH converters.

limited by the OTA bandwidth thus limiting the maximum allowable clock rate. Given the same technology, a CT ΔΣ data converter can be clocked at a higher rate than a comparable DT ΔΣ data converter. Having said that, however, the integrator gain, which depends on an *RC* or *gmC* product in a CT ΔΣ loop filter, is prone to process variation and thus tends to significantly affect the loop filter's transfer function. This, it turns out, is not a major concern for integrators designed using switched capacitor circuits for DT ΔΣ modulators since only the relative mismatch between capacitors, which tends to be small, affects the overall transfer function of the loop filter [2].

Finally, it is important to point out that DT ΔΣ modulators, due to their discrete-time nature, lend themselves to be easily studied using digital signal processing techniques. The linear loop can be easily modeled in the *z*-domain. Over the years, a wealth of mathematical and software tools has evolved to enable the analysis and simulation of DT ΔΣ modulators. Today, a system architect designing a CT ΔΣ modulator starts his or her analysis by specifying the modulator requirements in the discrete-time domain and then using discrete-to-continuous time mapping obtains the CT ΔΣ modulator's requirements. This design process will be discussed in the next section.

9.2.2 Equivalence between CT and DT ΔΣ Modulators

In order to further analyze the performance of a CT ΔΣ modulator, it is most insightful to model its behavior as a DT ΔΣ modulator. The reason, again as stated above, is that in the past, the majority of research in this area focused on discrete-time implementations, and consequently a large body of publications and tools has been developed to analyze DT ΔΣ modulators. Therefore, most engineers who wish to design a CT ΔΣ modulator start by analyzing a DT ΔΣ modulator and then convert the design to continuous time.

In order to ensure the equivalence between the two loops, it is required that the DT ΔΣ modulator have an identical impulse response to its corresponding CT ΔΣ modulator at every sampling instance taken at the sampling rate T_s at the input to the quantizer. This in turn implies that the discrete output sequences of both loops must be identical in response to all input signals. To further illustrate this point consider the CT ΔΣ modulator depicted in Figure 9.7. The ADC itself can be modeled as a sampler followed by a quantizer.

Figure 9.7(b) shows that the continuous-time open loop response from the output of the quantizer to its input *acts like a discrete time filter*. This is true since the output of the

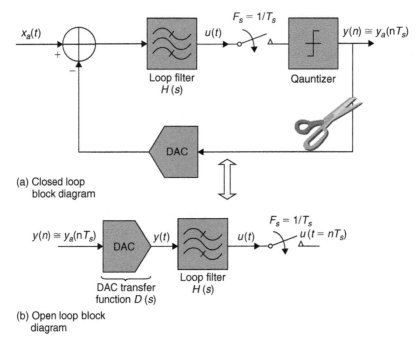

Figure 9.7 CT $\Delta\Sigma$ data converter block diagram: (a) closed loop and (b) open loop

quantizer, which is the input to the open loop, is discrete and its output, which is the input to the quantizer, is also discrete. This is true despite the fact that the loop filter and the DAC are both operating in continuous time. Therefore, the continuous open loop transfer function has an equivalent discrete time transfer function $H(z)$. This identical behavior implies that the time domain response of both continuous-time and discrete-time transfer functions can be expressed as

$$Z^{-1}\{H(z)\} \cong L^{-1}\{H_a(s)D_a(s)\}\Big|_{t=nT_s} \forall t \tag{9.10}$$

where $D_a(s)$ is the DAC's linear impulse response and $Z^{-1}\{.\}$ and $L^{-1}\{.\}$ designate the inverse Z-transform and inverse Laplace-transform operators. This process in essence ensures that the response of the $NTF(s)$ of the continuous-time loop will produce identical results when sampled at T_s to the $NTF(z)$ of the discrete-time loop.

From this discussion, we infer that the impulse response $h(n)$ of the discrete-time loop filter must be the sampled version of the convolution between the continuous-time loop filter $h_a(t)$ and the DAC linear response $d_a(t)$ at the sampling instance

$$h(n) \cong d_a(t) * h_a(t) = \left. \int_{-\infty}^{\infty} d_a(\tau)h_a(t - \tau)d\tau \right|_{t=nT_s} \tag{9.11}$$

This mapping is none other than the impulse invariance mapping used in traditional digital IIR filter design. Assuming that the analog impulse response can be expressed in parallel fraction expansion form [3]:

$$H_a(s) = \sum_{k=1}^{N} \frac{\alpha_k}{s - p_k}$$
$$h_a(t) = \sum_{k=1}^{N} \alpha_\kappa e^{p_k t} \quad \text{for } t \geq 0 \tag{9.12}$$

where $\{\alpha_k(t)\}$ are the coefficients of the analog loop filter $h_a(t)$ and $\{p_k(t)\}$are its poles. Consequently, it can be shown that the transfer function of the equivalent discrete-time loop filter can be expressed as

$$H(z) = \sum_{k=1}^{N} \frac{\alpha_k}{1 - e^{p_k T_s} z^{-1}}, \quad \text{where } z_k = e^{p_k T_s} \quad \text{for } k = 1, 2, ..., N \tag{9.13}$$

Although the poles of the digital filters are mapped via the relationship $z_k = e^{p_k T_s}$, the zeros in the analog and digital domain do not conform to the same relationship. Furthermore, the discrete-time model provides a satisfactory model of the continuous time loop only when the sampling rate is sufficiently high.

9.2.3 Feedback DAC Linear Models

The results concerning the mapping of the loop filter from the continuous-time domain to the discrete-time domain discussed above apply to the DAC's transfer function mapping as well. In this section, we will discuss two popular DAC models, namely the non-return-to-zero (NRZ) DAC and the return-to-zero (RZ) DAC. Both of these DACs have temporal transfer functions that are made up of rectangular pulses and tend to be performance

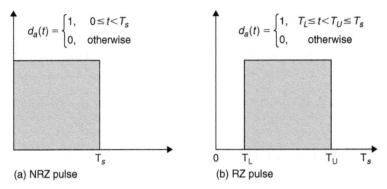

Figure 9.8 DAC pulses: (a) NRZ pulse and (b) RZ pulse

limited by the sampling clock jitter which directly degrades the input signal. More exotic DACs with nonrectangular transfer functions exist that are more tolerant of sampling clock jitter at the expense of circuit complexity. Furthermore, a DAC transfer function can be chosen along with an appropriate continuous-time loop filter to enable various transformations between CT and DT designs, thus enabling a CT $\Delta\Sigma$ modulator to match the behavior of the intended DT $\Delta\Sigma$ modulator.

The output of an NRZ DAC remains constant over the entire sampling period, as shown in Figure 9.8. The time domain response of an NRZ DAC is given as

$$d_{a,NRZ}(t) = \begin{cases} 1 & 0 \le t < T_s \\ 0 & \text{otherwise} \end{cases} \tag{9.14}$$

In the Laplace domain, the relation in (9.14) can be expressed as

$$D_{a,NRZ}(s) = \frac{1 - e^{-sT_s}}{s} \tag{9.15}$$

This function is also known as zero-order hold. The output of a generalized RZ DAC, on the other hand, can be expressed in the time-domain as

$$d_{a,RZ}(t) = \begin{cases} 1 & T_L \le t < T_U \le T_s \\ 0 & \text{otherwise} \end{cases} \tag{9.16}$$

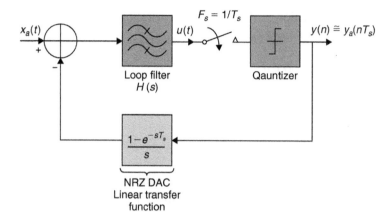

Figure 9.9 Linear model of CT ΔΣ modulator with NRZ DAC in the feedback path

The two most common values for T_L and T_U are the following two pairs:

- $T_L = 0$ and $T_U = T_s/2$, in this case the DAC is often referred to as RZ DAC.

- $T_L = T_s/2$ and $T_U = T_s$; in this case the DAC is referred to as half delay RZ DAC.

In the Laplace domain, the relation in (9.16) is given as

$$D_{a,RZ}(s) = \frac{e^{-sT_L}(1 - e^{-sT_U})}{s} \tag{9.17}$$

In lowpass modulator design, RZ DACs tend to be less susceptible to nonlinearities due to the fact that the area under a practical pulse has dependencies on the level of the preceding and following pulses [4],[5].

9.2.3.1 CT ΔΣ Modulator with NRZ DAC in the Feedback

At this point, let us re-examine the equivalence between the CT and DT ΔΣ modulators for the case where the feedback DAC has a rectangular pulse. The linear model employing an NRZ DAC is depicted in Figure 9.9. Applying the relationship in (9.10) to the model shown in Figure 9.9, the discrete-time transfer function can be related to the continuous-time transfer function as

$$h(t = nT_s) = Z^{-1}\{H(z)\} = L^{-1}\left\{\frac{1 - e^{sT_s}}{s}H_a(s)\right\}\bigg|_{t=nT_s} \tag{9.18}$$

In the time domain, the relation in (9.18) is the convolution:

$$h(t = nT_s) = h_a(t)*d_a(t)\Big|_{t=nT_s} = \int_{-\infty}^{\infty} h_a(t - \tau)d_a(\tau)d\tau\Big|_{t=nT_s}$$

$$= \int_0^{T_s} h_a(t - \tau)d\tau\Big|_{t=nT_s} \tag{9.19}$$

Since $d_a(t)$ is a pulse of a given width T_s, the relationship in (9.19) is referred to as the pulse response of the continuous-time loop. Given that the loop filter assumes the mathematical form given in (9.12):

$$\int_0^{T_s} h_a(t - \tau)d\tau\Big|_{t=nT_s} = \int_0^{T_s} \sum_{k=1}^{N} \alpha_k e^{P_k(t-\tau)}\, d\tau\Big|_{t=nT_s} \quad \text{for } t > T_s \tag{9.20}$$

The relation in (9.20) can be further manipulated by bringing the integral inside the summation as

$$h(t = nT_s) \cong \int_0^{T_s} \sum_{k=1}^{N} \alpha_k e^{P_k(t-\tau)}\, d\tau\Big|_{t=nT_s} = \sum_{k=1}^{N} \alpha_k e^{P_k t} \int_0^{T_s} e^{-P_k \tau}\, d\tau\Big|_{t=nT_s} \quad \text{for } t \geq T_s$$

$$= \sum_{k=1}^{N} \frac{\alpha_k e^{P_k t}}{-P_k}(e^{-P_k T_s} - 1)\Big|_{t=nT_s} \quad \text{for } t \geq T_s$$

$$= \sum_{k=1}^{N} \frac{\alpha_k e^{P_k n T_s}}{-P_k}(e^{-P_k T_s} - 1) \quad \text{for } t \geq T_s \tag{9.21}$$

The integrals in (9.20) and (9.21) are expressed for $t > T_s$. However, it is also imperative to examine the transfer function for $t < T_s$, that is

$$\int_0^{0 \leq t < T_s} h_a(t - \tau)d\tau\Big|_{t=nT_s} = \int_0^{0 \leq t < T_s} \sum_{k=1}^{N} \alpha_k e^{P_k(t-\tau)}\, d\tau\Big|_{t=nT_s} \quad \text{for } 0 \leq t < T_s \tag{9.22}$$

Further manipulation of the integral and summation in (9.22) results in a relation similar to the one expressed in (9.21):

$$
h(0) \cong \int_0^t \sum_{k=1}^N \alpha_k e^{p_k(t-\tau)} d\tau \Bigg|_{\substack{t=nT_s \\ t<T_s \Rightarrow n=0}} = \sum_{k=1}^N \alpha_k e^{p_k t} \int_0^t e^{-p_k \tau} d\tau \Bigg|_{\substack{t=nT_s \\ n=0}} \quad \text{for } 0 \le t < T_s
$$

$$
= \sum_{k=1}^N \frac{\alpha_k e^{p_k t}}{-p_k}(e^{-p_k t} - 1) \Bigg|_{\substack{t=nT_s \\ n=0}} \quad \text{for } 0 \le t < T_s
$$

$$
= \sum_{k=1}^N \frac{\alpha_k e^{p_k n T_s}}{-p_k}(e^{-p_k n T_s} - 1) \Bigg|_{n=0} = 0 \text{ for } 0 \le t < T_s
$$

$$(9.23)$$

The discrete-time response in (9.23) is only valid for $n = 0$, since the upper limit for the integral is t such that $0 \le t < T_s$. Therefore, given that $t = nT_s$ and $t < T_s$, this implies that $n = 0$ in (9.23), resulting in $h(0) = 0$! This implies that the first sample in the loop pulse response is zero, further implying a delay that exists in the numerator of the pulse-invariant transfer function [6]. Therefore, the equivalent discrete-time loop filter in (9.23) can be obtained by using the definition of the Z-transform

$$
H(z) = \sum_{n=-\infty}^{\infty} h(n)z^{-n} = \sum_{n=1}^{\infty} \left\{ \sum_{k=1}^N \frac{\alpha_k e^{p_k n T_s}}{-p_k}(e^{-p_k T_s} - 1) \right\} z^{-n}
$$

$$
= \sum_{k=1}^N \left\{ \frac{\alpha_k}{-p_k}(e^{-p_k T_s} - 1) \sum_{n=1}^{\infty} (e^{p_k T_s} z^{-1})^n \right\}
$$

$$
= \sum_{k=1}^N \frac{\alpha_k}{-p_k} \frac{(1 - e^{p_k T_s})z^{-1}}{1 - e^{p_k T_s} z^{-1}} \qquad (9.24)
$$

Note, however, that the relationships described in (9.21) and (9.23) imply that the loop response is continuous at T_s.

Example 9-2: Equivalence between First Order CT and DT $\Delta\Sigma$ Modulators

The purpose of this example is not in its practical value but rather to illustrate the theory presented thus far. Consider the continuous-time loop filter transfer function of the CT loop $\Delta\Sigma$ modulator depicted in Figure 9.9 as $H_a(s) = 1/(s + 0.4)$. For the sake of simplicity, it is assumed that $T_s = 1$. It is further implied from Figure 9.9 that the DAC response is NRZ. Find the equivalent discrete-time loop filter.

The analog loop filter is given as

$$H_a(s) = \sum_{k=1}^{N} \frac{\alpha_k}{s - p_k} = \frac{1}{s + 0.4} \tag{9.25}$$

This type of filter is known as a lossy integrator and is depicted in Figure 9.10. The transfer function of this integrator is given in the Laplace domain as

$$H_a(s) = \frac{V_o(s)}{V_{i1}(s) - V_{i2}(s)} = \frac{g_1}{sC + g_2} \tag{9.26}$$

The transfer function of this filter is depicted in Figure 9.11.

Using the relationship in (9.24), the equivalent discrete time loop filter is given as

$$H(z) = \sum_{k=1}^{N} \frac{\alpha_k}{-p_k} \frac{(1 - e^{p_k T_s})z^{-1}}{1 - e^{p_k T_s} z^{-1}} = \frac{1}{0.4} \frac{(1 - e^{0.4})z^{-1}}{1 - e^{0.4} z^{-1}}$$

$$= \frac{-1.23 z^{-1}}{1 - 1.49 z^{-1}} \tag{9.27}$$

Note that the transfer function of the equivalent DT $\Delta\Sigma$ modulator, depicted in Figure 9.12, has retained its lowpass characteristics:

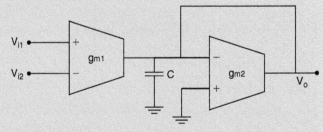

Figure 9.10 OTA realization of a lossy integrator (also known as a $_{gm}$-C filter)

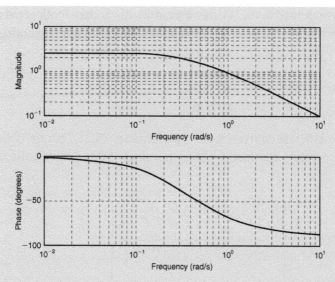

Figure 9.11 Magnitude and phase response of continuous-time loop filter

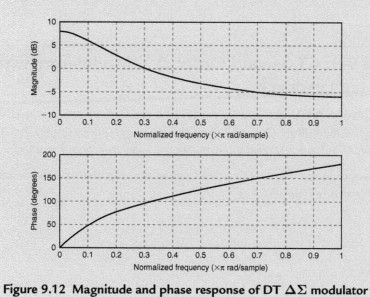

Figure 9.12 Magnitude and phase response of DT ΔΣ modulator

The relationship presented in (9.24), however, does not address collocated poles. For example, in second order lowpass $\Delta\Sigma$ modulators we encounter two collocated poles of the form

$$\frac{\beta_k}{(1 - e^{p_k T_s} z^{-1})^2} \tag{9.28}$$

where β_k is the coefficient equivalent to α_k presented in (9.21), but for collocated double poles. The corresponding continuous-time equivalent of this double collocated pole expression is

$$\left\{ \frac{(1 - e^{-p_k T_s} - p_k T_s)\dfrac{s}{T_s} + p_k^2}{(s - p_k)^2 (1 - e^{-p_k T_s})^2} \right\} \beta_k \tag{9.29}$$

which in turn shows that there exist two collocated poles in the continuous-time domain. Transformations that address more than two collocated poles are appropriately discussed in [7].

Example 9-3: CT Second Order $\Delta\Sigma$ Modulators with Distributed Feedback

Consider the second order DT $\Delta\Sigma$ modulator with distributed feedback topology depicted in Figure 9.13. Assume that the feedback DAC is NRZ and that the lowpass filter inside the loop I(z) is the first order integrator $z^{-1}/(1 - z^{-1})$. What is the equivalent CT-$\Delta\Sigma$ modulator loop filter?

Applying the open loop procedure described earlier, the loop filter transfer function is obtained by analyzing Figure 9.14 as

$$H(z) = -\kappa_2 I(z) - \kappa_1 \kappa_2 I^2(z)$$

$$= -\kappa_2 \frac{z^{-1}}{1 - z^{-1}} - \kappa_1 \kappa_2 \left(\frac{z^{-1}}{1 - z^{-1}} \right)^2 \tag{9.30}$$

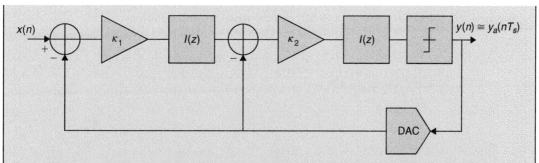

Figure 9.13 Second order DT ΔΣ modulator with distributed feedback topology

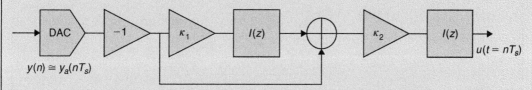

Figure 9.14 Open loop of second order DT ΔΣ modulator

For the first expression of (9.27), the equivalent Laplace domain is given by simply manipulating

$$X_a(F) = \begin{cases} \dfrac{1}{F_s} X\left(\dfrac{F}{F_s}\right) & -\dfrac{F_s}{2} < F < \dfrac{F_s}{2} \\ 0 & \text{otherwise} \end{cases} \tag{9.31}$$

That is,

$$I(z) = \frac{z^{-1}}{1 - z^{-1}} \Rightarrow I(s) = \left.\frac{1/(T_U - T_L)}{T_s}\frac{1}{s}\right|_{\substack{NRZ \Rightarrow T_U = 1 \\ T_L = 0}} = \frac{1}{T_s s} \tag{9.32}$$

Similarly, the second expression in (9.30) can be derived from (9.29) as

$$I^2(z) = \left(\frac{z^{-1}}{1-z^{-1}}\right)^2 \Rightarrow \left\{ \begin{array}{l} I^2(s) = \left. \dfrac{\dfrac{T_U+T_L-2}{2T_s(T_U-T_L)}s + \dfrac{1}{T_s^2(T_U-T_L)}}{s^2} \right|_{\substack{NRZ \Rightarrow T_U=1 \\ T_L=0}} \\[40pt] I^2(s) = \dfrac{-T_s s + 1}{T_s^2 s^2} \end{array} \right. \tag{9.33}$$

The loop transfer function in continuous-time can then be expressed as

$$\begin{aligned} H(s) &= -\kappa_2 I(s) - \kappa_1\kappa_2 I^2(s) \\ &= -\kappa_2\frac{1}{T_s s} + \frac{\kappa_1\kappa_2 T_s s}{T_s^2 s^2} - \frac{\kappa_1\kappa_2}{T_s^2 s^2} \\ &= \frac{-\kappa_2 + \kappa_1\kappa_2}{T_s s} - \frac{-\kappa_1\kappa_2}{T_s^2 s^2} \\ &= \frac{g_1}{s} + \frac{g_2}{s^2} \end{aligned} \tag{9.34}$$

The second order CT $\Delta\Sigma$ modulator is depicted in Figure 9.15.

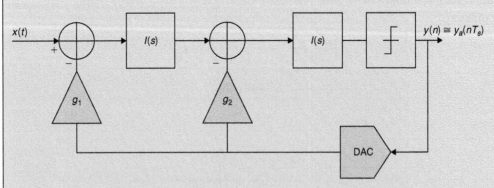

Figure 9.15 Second order CT $\Delta\Sigma$ modulator with distributed feedback topology

9.2.3.2 CT ΔΣ Modulator with RZ DAC in the Feedback

In order to obtain an equivalent expression to (9.24) for an RZ DAC or an RZ DAC with half delay, we need to replace the zero-order hold expression in (9.15) with:

$$D_{a,RZ}(s) = \frac{1 - e^{-sT_s/2}}{s} \quad \text{for a simple RZ} \quad \text{where } T_L = 0 \text{ and } T_U = T_s/2$$

$$D_{a,RZ}(s) = e^{-sT_s/2} \frac{(1 - e^{-sT_s/2})}{s} \quad \text{for a half-delay RZ} \quad \text{where } T_L = T_s/2 \quad \text{and} \quad T_U = T_s$$

$$(9.35)$$

according to (9.17). This in turn implies that the integration limits in (9.19) are changed from $[0, T_s]$ for the NRZ case to $[0, T_s/2]$ for the RZ case and $[T_s/2, T_s]$ for the half delay RZ case. This implies that the discrete-time equivalent loop filter for both cases is given as

$$H(z) = \sum_{k=1}^{N} \frac{\alpha_k e^{p_k T_s/2}}{-p_k} \frac{(1 - e^{p_k T_s/2})z^{-1}}{1 - e^{p_k T_s} z^{-1}} \quad \text{for simple RZ DAC case}$$

$$H(z) = \sum_{k=1}^{N} \frac{\alpha_k}{-p_k} \frac{(1 - e^{p_k T_s/2})z^{-1}}{1 - e^{p_k T_s} z^{-1}} \quad \text{for half delay RZ DAC case} \qquad (9.36)$$

In a similar vein to the development of collocated poles for the NRZ case, expressions for the RZ and half-delay RZ transformation can also be found. In this section, we will present those pertinent to two collocated poles which unavoidably arise from second order lowpass ΔΣ modulators. For the simple RZ DAC, the pulse transformation is

$$\left\{ \frac{\left[1 - e^{-p_k T_s/2} - p_k T_s \left(1 - \frac{1}{2} e^{-p_k T_s/2} \right) \right] \frac{s}{T_s} + p_k^2 \left(1 - \frac{1}{2} e^{-p_k T_s/2} \right)}{(s - p_k)^2 (1 - e^{-p_k T_s/2})^2} \right\} \beta_k e^{-p_k T_s} \qquad (9.37)$$

Similarly, an expression for the pulse transformation for the half-delay RZ DAC is given as

$$\left\{ \frac{(1 - 0.5 p_k T_s - e^{-p_k T_s/2}) \frac{s}{T_s} + 0.5 p_k^2}{(s - p_k)^2 (1 - e^{-p_k T_s/2})^2} \right\} \beta_k e^{-p_k T_s} \qquad (9.38)$$

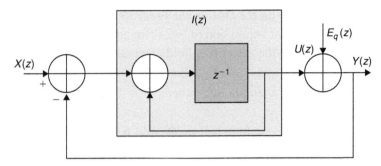

Figure 9.16 The linear model of the first order $\Delta\Sigma$ modulator

Similar pulse transformation expressions for triple and higher numbers of collocated poles can also be derived and applied in order to obtain the equivalent continuous-time loop filter expression.

9.3 SQNR Performance of $\Delta\Sigma$ Modulators

9.3.1 The First Order Loop

The analysis for the linear loop will be done in the z-domain. In the linear model, the quantization error is shown simply as additive and signal independent. This simplification of the model could lead to serious flaws in the analysis. However, the advantage of the linear model is its simplicity in capturing the essence of noise shaping in this type of converters. Let $X(z)$ be the input signal to the first order $\Delta\Sigma$ modulator. The loop filter is an integrator defined by the function $I(z) = z^{-1}/(1 - z^{-1})$. The output signal $Y(z)$ is the sum of the input signal to the quantizer $U(z)$ plus the quantization noise $E_q(z)$ or simply

$$Y(z) = U(z) + E_q(z) \tag{9.39}$$

In turn, the input to the quantizer is given according to Figure 9.16 as

$$U(z) = \frac{z^{-1}}{1 - z^{-1}}\{X(z) - Y(z)\} \tag{9.40}$$

Substituting (9.40) into (9.39) and solving for $Y(z)$ we obtain

$$Y(z) = z^{-1}X(z) + (1 - z^{-1})E(z)$$
$$Y(z) = STF(z)X(z) + NTF(z)E(z) \tag{9.41}$$

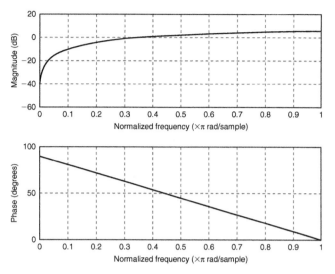

Figure 9.17 Magnitude and phase response of $NTF(z) = 1 - z^{-1}$

From (9.41), we note that the signal is merely delayed by the $STF(.)$ whereas the quantization error is filtered by the $NTF(.)$. In this case, $STF(.)$ is merely an allpass function whereas $NTF(.)$ is a highpass FIR filter, namely $NTF(z) = 1 - z^{-1}$ as depicted in Figure 9.17.

In the frequency domain, the magnitude squared response of $NTF(.)$ is found by substituting $z = e^{j2\pi f}$ into (9.41), where f is the normalized frequency defined as the ratio of the analog frequency to the sampling frequency F/F_s, and taking the magnitude square:

$$
\begin{aligned}
\left| NTF(e^{j2\pi f}) \right|_{f=F/F_s}^2 &= \left| 1 - e^{-j2\pi f} \right|^2 \\
&= \left| e^{-j2\pi f/2} (e^{j2\pi f/2} - e^{-j2\pi f/2}) \right|^2 \\
&= \left| 2je^{-j2\pi f/2} \underbrace{\frac{(e^{j2\pi f/2} - e^{-j2\pi f/2})}{2j}}_{=\sin(\pi f)} \right|^2
\end{aligned}
$$

$$
\left| NTF(e^{j2\pi F/F_s}) \right|^2 = 4\sin^2\left(\frac{\pi F}{F_s}\right) \quad \text{where } F/F_s \text{ was substituted for } f \qquad (9.42)
$$

Recall that, due to oversampling, the quantization noise is now spread between DC and the sampling frequency. The inband quantization noise *at baseband* is limited between $[0, B/2]$ where B is the RF signal bandwidth for the in-phase or quadrature channel. However, when looking at the overall degradation of the quantization noise one must take into account the degradation in both in-phase and quadrature channels or simply over the entire bandwidth B.

Finally, the quantization noise power, shaped by the $NTF(\cdot)$'s highpass characteristic, is the integral of the power spectral density $S_e(f) = P_e/F_s$ of the quantization noise multiplied by squared magnitude of $NTF(.)$

$$
P_{e,\Delta\Sigma_1} = \int_{-B/2}^{B/2} S_e(F)|NTF(F)|^2 dF = \frac{P_e}{F_s} \int_{-B/2}^{B/2} |NTF(F)|^2 \, dF
$$
$$
= 4 \frac{P_e}{F_s} \int_{-B/2}^{B/2} \sin^2\left(\frac{\pi F}{F_s}\right) dF \tag{9.43}
$$

where P_e is the quantization noise power assumed to be white. For a large OSR such that $\sin^2(\pi F/F_s) \approx (\pi F/F_s)^2$, or simply $F/F_s \ll 1$, the relationship in (9.43) becomes the approximation

$$
P_{e,\Delta\Sigma_1} \approx 4 \frac{P_e}{F_s} \int_{-B/2}^{B/2} \left(\frac{\pi F}{F_s}\right)^2 dF = \frac{4}{3} \frac{P_e}{F_s^3} \pi^2 F^3 \Big|_{F=-B/2}^{F=B/2}
$$
$$
= \frac{1}{3} \frac{P_e \pi^2}{OSR^3} \tag{9.44}
$$

where $OSR = F_s/B$ is directly substituted into (9.44).

Recall that the SQNR was defined earlier in Chapter 7 as the ratio of signal power to quantization noise power:

$$
SQNR = 10 \log_{10}\left(\frac{P_x}{P_{e,\Delta\Sigma_1}}\right) = 10 \log_{10}\left(\frac{3 OSR^3 P_x}{P_e \pi^2}\right)
$$
$$
= 10 \log_{10}\left(\frac{3}{\pi^2}\right) + 30 \log_{10}(OSR) + 10 \log_{10}\left(\frac{P_x}{P_e}\right) \tag{9.45}
$$

Substituting

$$SQNR = 10\log_{10}\left(\frac{12 \times 2^{2b} \times \sigma_x^2}{V_{x,FS}^2}\right) = 10.8 + 6.0206b + 10\log_{10}\left(\frac{\sigma_x^2}{V_{x,FS}^2}\right) \qquad (9.46)$$

into (9.45), the SQNR becomes:

$$SQNR = 10.8 + 6.0206b + 10\log_{10}\left(\frac{3}{\pi^2}\right) + 30\log_{10}(OSR) + 10\log_{10}\left(\frac{\sigma_x^2}{V_{x,FS}^2}\right) \qquad (9.47)$$

Or alternatively the SQNR is given as

$$SQNR = 4.7712 + 6.0206b + 10\log_{10}\left(\frac{3}{\pi^2}\right) + 30\log_{10}(OSR) - PAR \qquad (9.48)$$

Note that each doubling of the OSR increases the SQNR by 9.03 dB compared to 3 dB for a Nyquist ADC. Therefore, OSR has a much more profound effect on SQNR for $\Delta\Sigma$ modulators compared with Nyquist converters. This generalization will become more evident in the coming sections.

Finally, we must realize that the linear model of the first order $\Delta\Sigma$ modulator predicts that the loop is unconditionally stable. This conclusion is arrived at, however, by assuming a linear model for the quantizer! Hence a true analysis of the loop stability must take into account the nonlinearity associated with the quantizer. A DC signal input to the loop such that $x(n) = constant > 1$ for all n, or even a slowly varying signal, causes the output of the integrator, the input to the quantizer $u(n)$, to grow without bound.

On the other hand, if the input signal is rational and bounded such that $|x(n)| < 1$, the output $u(n)$ becomes periodic. The periodic sequence, known as limit cycles or idle tones, depends on the input signal. Idle tones do not cause instability in the loop and can be removed via digital filtering provided they fall outside the desired signal bandwidth. For an irrational bounded DC input $|x(n)| < 1$, the output sequence is not periodic but rather quasi-periodic.

Example 9-4: Limit Cycles in First Order $\Delta\Sigma$ Modulators

Consider the first order $\Delta\Sigma$ modulator depicted in Figure 9.18. Let the input signal be a rational constant number $x(n) = 3/4$. Assume that the slicer satisfies the following function:

$$y(n) = \begin{cases} 1 & u(n) \geq 0 \\ -1 & u(n) < 0 \end{cases} \tag{9.49}$$

Assume the initial conditions $y(0) = 1$ and $u(0) = 0$. What is the periodic output sequence of the loop? What is the average of $y(n)$? Plot the spectrum of $y(n)$.

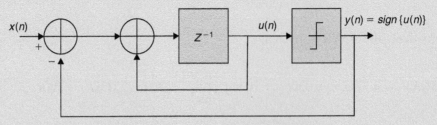

Figure 9.18 First order $\Delta\Sigma$ modulator loop

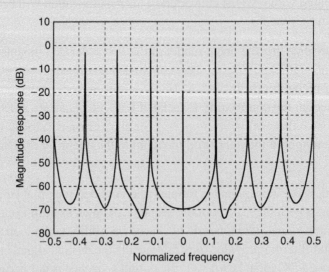

Figure 9.19 Spectrum of output sequence $y(n)$ due to DC input $x(n) = 3/4$

The difference equation governing the loop in Figure 9.18 is

$$u(n) = x(n-1) + u(n-1) - sign\{y(n-1)\}, \quad n = 1, 2... \tag{9.50}$$

The output sequence $y(n)$ is the sequence $\{...1 \quad -1 \quad 1 \quad 1 \quad 1 \quad 1 \quad 1 \quad 1 \quad 1 \quad -1...\}$. That is, it is a sequence of seven 1s followed by a -1. This periodic sequence has an average of 0.75 which is equal to the DC input signal. The output spectrum depicted in Figure 9.19 shows that all the idle tones have the same magnitudes.

Repeat this example for the constant input $x(n) = 0$.

Given a rational DC input signal, say λ/ρ, such that λ and ρ are relatively prime (i.e., $\gcd(\lambda, \rho) = 1$) and λ, ρ, then two types of spectral components will ensue: in-band tones and out of band tones:

$$f_{ib,n} = \frac{n\lambda}{2\rho} F_s, \qquad n = 1, 2... \text{ for in-band tones}$$

$$f_{ob,n} = \frac{n(\rho - \lambda)}{2\rho} F_s, \quad n = 1, 2... \text{ for out of band tones} \tag{9.51}$$

Another point worth mentioning is the periodicity of (9.50). For both odd λ and ρ the resulting sequence is periodic with period ρ. If either λ or ρ is odd and the other is even, then the ensuing sequence resulting from (9.50) is periodic with period 2ρ.

9.3.2 The Second and Higher Order Loops

The second order $\Delta\Sigma$ modulator depicted in Figure 9.20 is comprised of two integrators in the signal path. In order to reduce the delay of the desired output signal, only one integrator has a delay element in the signal path. The input to the quantizer $U(z)$ is deduced from Figure 9.20:

$$U(z) = \left(\frac{z^{-1}}{1 - z^{-1}}\right)\left(\frac{1}{1 - z^{-1}}\right)(X(z) - Y(z)) - \left(\frac{z^{-1}}{1 - z^{-1}}\right)Y(z) \tag{9.52}$$

At the output of the quantizer, the signal is related to the input signal $X(z)$ and $U(z)$:

$$Y(z) = U(z) + E(z) \tag{9.53}$$

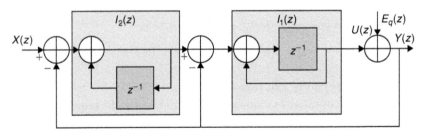

Figure 9.20 The linear model of the second order $\Delta\Sigma$ modulator

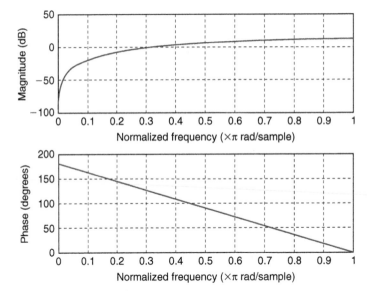

Figure 9.21 Magnitude and phase response of $NTF(z) = (1 - z^{-1})^2$

Substituting (9.52) into (9.53) and solving for $Y(z)$:

$$
\begin{aligned}
Y(z) &= z^{-1}X(z) + (1 - z^{-1})^2 \ E(z) \\
&= STF(z) \ X(z) + NTF(z) \ E(z)
\end{aligned}
\tag{9.54}
$$

Similar to the first order case, $STF(.)$ in (9.54) causes a mere delay to the input signal, whereas $NTF(.)$ is a second order highpass transfer function as depicted in Figure 9.21.

The square of the magnitude response of $NTF(.)$ in the frequency domain is derived in a similar manner to (9.42):

$$\left|NTF(e^{j2\pi f})\right|_{f=F/F_s}^2 = 16\sin^4\left(\frac{\pi F}{F_s}\right) \tag{9.55}$$

Approximating the sine function for a large OSR as

$$\sin^4\left(\frac{\pi F}{F_s}\right) \approx \left(\frac{\pi F}{F_s}\right)^4 \quad \text{for large } OSR \tag{9.56}$$

results in the quantization noise power:

$$P_{e,\Delta\Sigma_1} = \int_{-B/2}^{B/2} S_e(F)\,|NTF(F)|^2\,dF = 16\frac{P_e}{F_s}\int_{-B/2}^{B/2}\sin 4\left(\frac{\pi F}{F_s}\right)dF$$

$$\approx 16\frac{P_e}{F_s}\int_{-B/2}^{B/2}\left(\frac{\pi F}{F_s}\right)^4 dF = \frac{1}{5}\frac{P_e\pi^4}{OSR^5} \tag{9.57}$$

The SQNR for the second order loop can then be derived:

$$SQNR = 10.8 + 6.0206b + 10\log_{10}\left(\frac{5}{\pi^4}\right) + 50\log_{10}(OSR) + 10\log_{10}\left(\frac{\sigma_x^2}{V_{x,FS}^2}\right)$$

$$= 4.7712 + 6.0206b + 10\log_{10}\left(\frac{5}{\pi^4}\right) + 50\log_{10}(OSR) - PAR \tag{9.58}$$

From the relationship in (9.58), we note that for every doubling of the sampling rate we get 2.5 bits of added converter resolution. Like first order $\Delta\Sigma$ modulators, second order loops also suffer from idle tones.

Increasing the order of the loop to Nth order implies cascading N-integrators in the loop. The resulting *SQNR* for an arbitrary order loop is

$$SQNR = 10.8 + 6.0206b + 10\log_{10}\left(\frac{2N+1}{\pi^{2N}}\right)$$

$$+ (2N+1)10\log_{10}(OSR) + 10\log_{10}\left(\frac{\sigma_x^2}{V_{x,FS}^2}\right)$$

$$= 4.7712 + 6.0206b + 10\log_{10}\left(\frac{2N+1}{\pi^{2N}}\right)$$

$$+ (2N+1)10\log_{10}(OSR) - PAR \tag{9.59}$$

From (9.59), it is implied that the *SQNR* increases approximately by 6 dB per additional bit in the quantizer. Furthermore, the *SQNR* increases by $6(N+1/2)$ every time we double the OSR, resulting in $N+1/2$ bit increase in converter resolution.

9.4 Bandpass $\Delta\Sigma$ Modulators

So far, the discussion concerning $\Delta\Sigma$ modulators has centered on lowpass converters. However, as discussed earlier, most SDR implementations rely on the versatility of a bandpass data converter in order to operate with a variety of waveforms. Bandpass $\Delta\Sigma$ modulators retain most of the advantages that their corresponding lowpass counterparts have. Nevertheless, as with lowpass $\Delta\Sigma$ modulators, bandpass $\Delta\Sigma$ modulators are ideal when operating on a relatively narrowband signal when compared with the sampling frequency.

As mentioned earlier, in a digital sampling receiver, the bandpass converter, in this case a $\Delta\Sigma$-modulator, converts the analog bandpass signal into a digital bandpass signal followed by digital downconversion, say via a digital quadrature mixer, for example, as depicted in Figure 9.22. A distinguishing feature of bandpass $\Delta\Sigma$ modulators from their counterpart lowpass converters is noise shaping. Instead of shaping the noise away from DC, a bandpass $\Delta\Sigma$-modulator shapes the noise from a certain frequency away from DC, as shown in Figure 9.23.

In lowpass $\Delta\Sigma$ modulators, the basic filtering block is a lowpass filter. For a bandpass modulator, the integrators are replaced with resonators. This can be accomplished via the

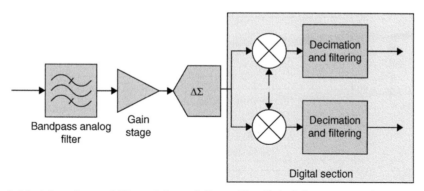

Figure 9.22 A bandpass $\Delta\Sigma$ modulator followed by digital downconversion and filtering

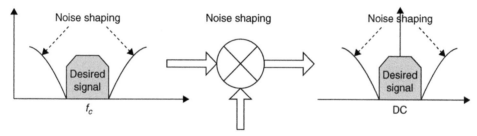

Figure 9.23 The process of bandpass sampling via a $\Delta\Sigma$ modulator and then converting the signal to baseband via a digital mixer

transformation $z^{-1} \rightarrow -z^{-2}$. For example, consider the output of a first order linear loop shown in (9.41):

$$Y(z) = z^{-1}X(z) + (1 - z^{-1})E(z) \tag{9.60}$$

The bandpass structure can be easily obtained by replacing z^{-1} by z^{-2} in (9.60)

$$Y(z) = -z^{-2}X(z) + (1 + z^{-2})E(z) \tag{9.61}$$

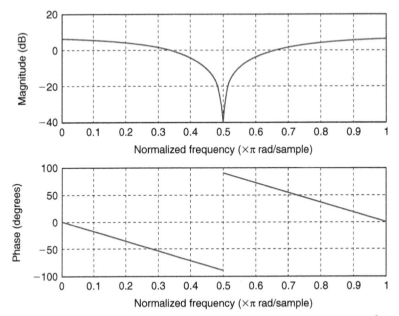

Figure 9.24 Magnitude and phase response of $NTF(z) = 1 + z^{-2}$

The NTF of (9.61), depicted in Figure 9.24, has zeros at the frequencies $F_s/4$ or $z = \pm j$ on the unit circle. The stability of the resulting bandpass modulator using this transformation is achieved by simply ensuring that the lowpass modulator is stable. On the other hand, the generalized N-path transformation $z^{-1} \rightarrow \pm z^{-N}$ can also be used. In this case, the modulator dynamics are preserved however the order of the filter and its complexity have increased for $N > 2$.

So far, the location of the bandpass frequency is a simple fraction of the sampling frequency. A more generalized transformation, one that does not preserve the lowpass loop dynamics, is given as

$$z^{-1} \rightarrow \frac{-1}{z^{-1}} \left(\frac{1 + \alpha z^{-1}}{\alpha + z^{-1}} \right) \quad \text{where } |\alpha| < 1 \tag{9.62}$$

In this case, a negative α results in the loop passband being centered closer to DC, whereas a positive α results in the passband being centered closer to $F_s/2$.

Example 9-5: Lowpass to Bandpass Transform of $\Delta\Sigma$ Modulator

Using the transform mapping provided in (9.62), design a first order $\Delta\Sigma$ modulator, such that the bandpass center frequency is at $0.234F_s$. Provide both the STF(.) and the NTF(.). Recall that the lowpass $STF(z) = z^{-1}$ and the $NTF(z^{-1}) = 1 - z^{-1}$.

First, let's derive both the STF(.) and NTF(.) expressions based on (9.62). The lowpass STF(.) is z^{-1}, or the equivalent bandpass STF(.) is

$$STF(z) = \frac{-1}{z^{-1}} \left(\frac{1 + \alpha z^{-1}}{\alpha + z^{-1}} \right) \text{ where } |\alpha| < 1 \tag{9.63}$$

Expressing STF(.) in the frequency domain, we obtain

$$STF(f) = \frac{-1}{e^{-j2\pi f}} \left(\frac{1 + \alpha e^{-j2\pi f}}{\alpha + e^{-j2\pi f}} \right) \tag{9.64}$$

where f is the normalized frequency defined as $f = F/F_s$. The magnitude of (9.64) is 1. This is not surprising since we anticipate that the STF(.) will retain its all-pass characteristics for all values of α.

Next consider the bandpass NTF(.). The transfer function is obtained by substituting (9.62) into $1 - z^{-1}$, or

$$\begin{aligned} NTF(z) &= 1 + \frac{1}{z^{-1}} \left(\frac{1 + \alpha z^{-1}}{\alpha + z^{-1}} \right) \\ &= \frac{1 + 2\alpha z^{-1} + z^{-2}}{\alpha z^{-1} + z^{-2}} \end{aligned} \tag{9.65}$$

In order to center the bandpass loop around $0.234F_s$, we must place a null (or a zero) at that frequency, that is

$$1 + 2\alpha z^{-1} + z^{-2} = 0 \Rightarrow z^{-1} = -\alpha \pm j\sqrt{\alpha^2 - 1} \tag{9.66}$$

In the frequency domain, (9.66) becomes

$$e^{-j2\pi f} \Big|_{f=0.234F_s/F_s} = e^{-j2\pi \times 0.234} = -\alpha \pm j\sqrt{\alpha^2 - 1} \tag{9.67}$$

Solving for α in (9.67) results in

$$\alpha = -\cos(2\pi \times 0.234) = -0.1004 \tag{9.68}$$

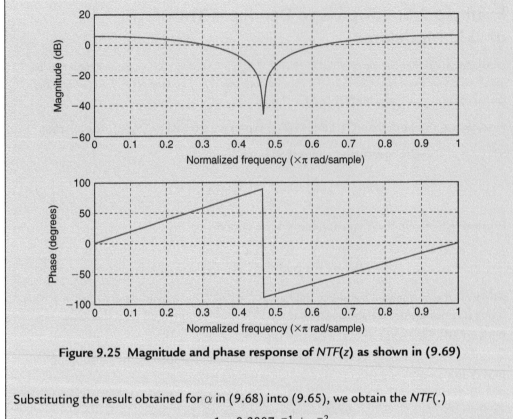

Figure 9.25 Magnitude and phase response of *NTF*(z) as shown in (9.69)

Substituting the result obtained for α in (9.68) into (9.65), we obtain the *NTF*(.)

$$NTF(z) = \frac{1 - 0.2007z^{-1} + z^{-2}}{-0.1004z^{-1} + z^{-2}} \qquad (9.69)$$

The NTF(.) obtained in (9.69) is depicted in Figure 9.25. Recall that the Nyquist frequency in the plot is normalized to 1 and consequently F_s is equal to 2.

The design methodology employed thus far is known as the lowpass prototype method. Other methods, such as the generalized filter approximation method, are out of the scope of this book. Similar to their lowpass counterparts, bandpass modulators suffer from nonidealities in the circuits, thus resulting in deterioration in the desired signal SNR. One such example: in switched-capacitor design, the values of the coefficients are set by the capacitor ratios. Any mismatch in these ratios could cause the nulls (or zeros) to shift from their desired frequency, thus increasing the quantization noise in the modulator output and degrading the signal quality.

9.5 Common Architectures of $\Delta\Sigma$ Modulators

9.5.1 $\Delta\Sigma$ Modulators with a Multibit Quantizer

There are many advantages to employing a multi-bit quantizer instead of a single bit slicer in a $\Delta\Sigma$ modulator. For one, when the signal input to the quantizer occupies full scale, the quantization noise is reduced by 6.02 dB for every added bit of resolution. Furthermore, a multibit quantizer serves to further stabilize the loop by reducing the effective gain variance of the input of the quantizer. Increasing the number of bits in the quantizer serves to further attenuate the inband quantization noise without having to increase the order of the loop.

The obvious drawback of multibit quantizers is that in CT modulators the actual quantizer is an ADC. For both CT and DT modulators, the feedback DAC is comprised of a multibit DAC. Any DAC errors are only filtered by the STF(.), which in turn may be an all-pass function in the signal band, and hence these errors tend to be additive to the signal and further degrade the performance. In order to alleviate this problem, the DAC is required to be extremely linear, which tends to be costly in terms of power consumption and circuit complexity.

9.5.2 Multi-Stage or MASH $\Delta\Sigma$ Modulators

Multi-stage modulators, also known as MASH[3] or cascaded modulators, are built by cascading two or more $\Delta\Sigma$ modulators in tandem. The modulators consist of first or higher order modulators, as shown in Figure 9.26 The input to the first stage of the modulator is the typical desired signal to be modulated. The input of every consequent stage is the quantization error of its preceding stage.

In order to illustrate the workings of a multi-stage converter, consider the output of the first converter according to Figure 9.26:

$$Y_1(z) = STF_1(z)X(z) + NTF_1(z)E_{q1}(z) \tag{9.70}$$

---

[3]MASH modulator stands for *m*ulti-st*a*ge noise *sh*aping modulator.

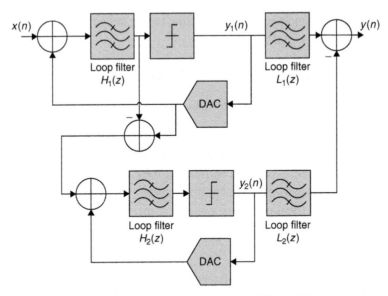

Figure 9.26 A two stage cascaded $\Delta\Sigma$ modulator

where $E_{q1}(.)$ is the quantization error due to the first loop quantizer and $STF_1(.)$ and $NTF_1(.)$ are the signal and noise transfer functions of the first loop. In a similar fashion, the output of the second loop in response to $E_{q1}(z)$ as an input is

$$Y_2(z) = STF_2(z)E_{q1}(z) + NTF_1(z)E_{q2}(z) \qquad (9.71)$$

where, similar to the first loop, $E_{q2}(.)$ is the quantization error due to the second loop quantizer and $STF_2(.)$ and $NTF_2(.)$ are the signal and noise transfer functions of the second loop. Next, consider the output of the modulator as a linear combination of $Y_1(.)$ and $Y_2(.)$:

$$Y(z) = L_1(z)Y_1(z) - L_2(z)Y_2(z) \qquad (9.72)$$

Substituting (9.70) and (9.71) into (9.72), we obtain:

$$\begin{aligned} Y(z) = L_1(z)\Big\{STF_1(z)X(z) + NTF_1(z)E_{q1}(z)\Big\} \\ -L_2(z)\Big\{STF_2(z)E_{q1}(z) + NTF_2(z)E_{q2}(z)\Big\} \end{aligned} \qquad (9.73)$$

Design $L_1(.)$ and $L_2(.)$ such that the relation:

$$L_1(z)NTF_1(z) = L_2(z)STF_2(z) \tag{9.74}$$

is satisfied. Furthermore, let

$$L_1(z) = STF_2(z) \quad \text{and} \quad L_2(z) = NTF_1(z) \tag{9.75}$$

Substituting both (9.74) and (9.75) into (9.73), we obtain

$$Y(z) = STF_1(z)STF_1(z)X(z) - NTF_1(z)NTF_2(z)E_{q2}(z) \tag{9.76}$$

The implication of the result obtained in (9.75) can be simply illustrated through an example. Consider both loops to be first order $\Delta\Sigma$ modulators; that is, let

$$\begin{aligned} STF_1(z) &= STF_1(z) = z^{-1} \\ NTF_1(z) &= NTF_2(z) = 1 - z^{-1} \end{aligned} \tag{9.77}$$

The output of the MASH modulator is then

$$Y(z) = z^{-2}X(z) - (1 - z^{-1})^2 E_{q2}(z) \tag{9.78}$$

This structure is known as 1-1 MASH. The shaped error at the output of the MASH converter $e_{q2}(.)$, which can be shown to be also white and uniformly distributed, is in response to the output of the first $\Delta\Sigma$ modulator, which is noise. It can be seen from (9.78) that, although both loops are first order, the cascaded response acts as if the loop is a second order loop. Likewise, if we let

$$\begin{aligned} STF_1(z) &= STF_2(z) = z^{-2} \\ NTF_1(z) &= NTF_2(z) = (1 - z^{-1})^{-2} \end{aligned} \tag{9.79}$$

again, according to (9.76), the output of the MASH modulator, even though it is constructed as two second order $\Delta\Sigma$ modulators in cascade, acts as a fourth order $\Delta\Sigma$ modulator:

$$Y(z) = z^{-4}X(z) - (1 - z^{-1})^4 E_{q2}(z) \tag{9.80}$$

This MASH structure is known as a 2-2 MASH. This is true given that neither modulator overloads. To generalize, an *M-N-P* MASH is a modulator made up of three cascaded $\Delta\Sigma$ modulators such that the order of the first loop is *M*, the second loop is *N*, and finally the third loop is *P*.

The difference, however, between a MASH modulator made up of a cascade of two second order $\Delta\Sigma$ modulators and a corresponding fourth order $\Delta\Sigma$ modulator is that the former, theoretically, achieves a fourth order response by simply satisfying that the two second order $\Delta\Sigma$ modulators are stable. That is, the stability analysis of the MASH structure is that of a second order modulator and not of a fourth order modulator, as would be required otherwise.

Note however, if the relationship in (9.74) is not satisfied, that is

$$L_1(z)NTF_1(z) - L_2(z)STF_2(z) = F(z), \ F(z) \neq 0 \qquad (9.81)$$

then the quantization noise due to the output of the first loop $e_{q1}(.)$ will appear at the output of the MASH converter, adding to the quantization noise of the system. The effect of this error, due to this imperfection, can be catastrophic to the performance of the data converter. The impact of $e_{q1}(.)$ leaking to the output degrades the overall SQNR. This leakage is mainly due to component mismatches that prevent (9.75) from being satisfied. It is important to note that component mismatches tend to affect single loop design much less than their cascaded counterparts. Finally, although we have restricted this analysis to two cascaded loops, the math can be extended to higher order MASH modulators.

9.6 Further Nonidealities in $\Delta\Sigma$ Modulators

In this section we briefly discuss the nonidealities in both continuous-time and discrete-time $\Delta\Sigma$ modulators. The aim is not to provide a comprehensive treatise on the subject, but rather to convey some of the design challenges associated with oversampled converters.

Most designs that rely on op-amps as building blocks for $\Delta\Sigma$ modulators can suffer from finite op-amp DC gain and finite op-amp bandwidths, resulting in nonzero settling time, among other nonidealities.

Finite op-amp DC gain, for example, causes the zeros of the NTF(.) to be shifted away from DC in the case of the lowpass converter, resulting in less attenuation of the quantization noise in the desired band. From a system design perspective, this implies that the integrators in the loop filter are no longer ideal but rather *leaky*. Leaky integrators have proven to be particularly problematic for multi-loop $\Delta\Sigma$ modulators.

Degradations ensuing from finite op-amp unity gain bandwidth result in increased quantization noise in the desired signal band. This particular type of degradation is particularly troubling for CT modulators compared with DT modulators. In order to avoid this problem, the op-amp unity gain bandwidth must be chosen several orders of magnitude higher than the sampling rate.

DAC nonidealities in CT $\Delta\Sigma$ modulators can also be a significant source of degradation to the desired signal. One such ideality is excess loop delay which can be thought of as a constant delay between the ideal DAC pulse and the DAC pulse implemented in the loop. In high-speed $\Delta\Sigma$ modulators, excess loop delay can shift the DAC pulse so that part of it overlaps the next sampling period. A further degradation due to finite slew rate causes the rising and falling edges of the DAC pulse to be mismatched, causing further degradation to the signal.

References

[1] Galton I. Delta-sigma data conversion in wireless transceivers. Invited Paper, IEEE Trans. on Microwave Theory and Techniques January 2002;50(1):302–15.

[2] Chan KT, Martin KW. Components for a GaAs delta-sigma modulator oversampled analog-to-digital converter. Proc. IEEE Int. Conf. on Circuits and Systems 1992;3:1300–303.

[3] Thurston AM, Pearce TH, Hawksford MJ. Bandpass implementation of the sigma-delta A-D conversion technique. Proc. IEE International Conf. on Analog-to-Digital and Digital-to-Analog Conversion Sept. 1991;81–6 Swansea, U.K.

[4] Signore BPD, Kerth DA, Sooch NS, Swanson EJ. A monolithic 20-b delta-sigma A/D converter. IEEE Journal on Solid-State Circuits December 1990;25(6):1311–16.

[5] Shoai O, Snelgrove WM. Design and implementation of a tunable 40 MHz-70 MHz Gm-C bandpass $\Delta\Sigma$ modulator. IEEE Trans. on Circuit and Systems II July 1997;44(7):521–30.

[6] Shoai O. Continuous-Time Delta-Sigma A/D Converters for High Speed Applications: Carlton University; 1995 Ph.D. Dissertation.

[7] Cherry JA, Snelgrove WM. Continuous-Time Delta-Sigma Modulators for High Speed A/D Conversion. Boston: Kluwer Academic Publishers; 1999.

CHAPTER 10

Multirate Digital Signal Processing

Modern communication standards such as WiMAX and UMTS-LTE rely on multiple signaling bandwidths and modulation schemes to provide the user with a variety of services. At times, these services require high data rate communication utilizing complex modulation schemes for video and file transfer applications, while at other times the user simply desires to establish a simple phone call using a voice application. This flexibility implies that the radio has to be capable of processing signals at various data rates. Multirate digital signal processing techniques provide the tools that enable the SDR to process the data signals with varying bandwidths. *Interpolation*, for example, can be used to increase the data rate of a given signal. *Decimation* is concerned with lowering the data rate of a given signal to obtain a new signal with a smaller data rate. *Filtering* and *fractional delay processing* are also powerful tools that are used by the SDR to alter the data rate of a signal.

This chapter is divided into three main sections. In Section 10.1, the basics of interpolation and decimation in the context of integer or rational fractions are presented. The underlying theory of multistage decimation and interpolation is discussed in detail. Section 10.2 delves into the details of polyphase structures, half-band filters, and cascaded-integrator comb filters. The pros and cons of each approach are discussed. Finally, Section 10.3 presents certain popular approaches to rational and irrational sampling rate conversions, namely using Lagrange interpolation and the Farrow filter.

10.1 Basics of Sample Rate Conversion

Consider the sampled sequence $\{x(n)\}$ obtained by sampling the analog signal $x_a(t)$ at the sampling rate of F_s. In general terms, sampling rate conversion is the process of

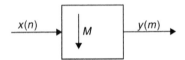

Figure 10.1 A downsampling block diagram

obtaining a new sampling sequence $\{y(m)\}$ of $x_a(t)$ directly from $\{x(n)\}$. The newly obtained samples of $\{y(m)\}$ are equivalent to the sampled values of $x_a(t)$ generated at the new sampling rate of F_s'.

In this section, we discuss decimation and interpolation by integer values. Both operations are considered fundamental to multirate signal digital signal processing.

10.1.1 Decimation by an Integer M

10.1.1.1 Single-Stage Decimation

Decimation or downsampling by an integer is the process of decreasing the sampling rate of a given sequence $\{x(n)\}$ by an arbitrary factor M. The new sequence $\{y(m)\}$ can be expressed as a function of the old sequence $x(n)$ as will be shown shortly [1]–[3]. A signal flow representation of the downsampling process is shown in Figure 10.1. The variable m denotes the variable of the new sequence sampled at the new sampling rate.

The downsampling operation can be thought of as a two-step process. First, consider the original sequence $\{x(n)\}$ depicted in Figure 10.2(a) for $M = 3$. Multiplying $\{x(n)\}$ by the sequence:

$$\varphi_M(m) = \begin{cases} 1 & \text{for } m = Mn, \ M \in \mathbb{N}^+ \\ 0 & \text{otherwise} \end{cases} \tag{10.1}$$

results in a new sequence depicted in Figure 10.2(b). Discarding the zeros as expressed in (10.1) results in a new sequence at the reduced sampling rate as shown in Figure 10.2(c). This new sequence, in theory, is made up of every Mth sample of $\{x(n)\}$.

This process can be further illustrated in the z-domain. Consider the z-transform of $\{x(n)\}$:

$$X(z) = \sum_{n=-\infty}^{\infty} x(n)z^{-n} \tag{10.2}$$

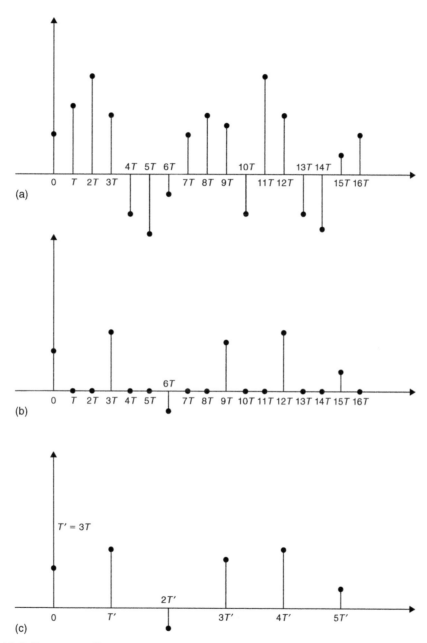

Figure 10.2 Downsampling as a two-step process: (a) original signal, (b) replacing the *M*−1 samples with zero samples, and (c) removing the zero samples to reduce the sampling rate

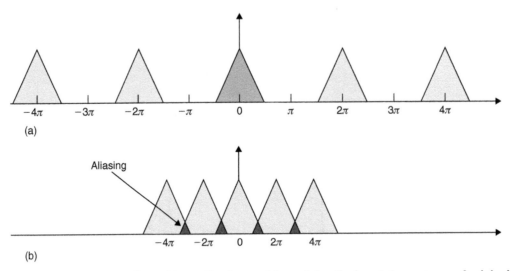

(a)

(b)

Figure 10.3 Downsampling without filtering could result in aliasing: (a) spectrum of original signal, (b) aliasing around the folding frequency causes overlap of the sampled images thus causing degradation to the original signal

The z-transform of the sequence $\{y(m)\}$, the decimated version of $\{x(n)\}$, is a series of shifted images of $X(z)$.

However, decimation without filtering can cause degradation in the signal due to aliasing. The reason is that a sampled signal repeats its spectrum every 2π radians, as shown in Figure 10.3(a). Decimation without filtering could cause the sampled images to overlap, as shown in Figure 10.3(b), depending on the signal bandwidth. To prevent aliasing from occurring, an antialiasing filter is typically used prior to downsampling.

Sufficient *linear* filtering before downsampling, as depicted in the block diagram in Figure 10.4, avoids unwanted aliasing in the downsampled signal [1]. The entire operation in Figure 10.4 can be thought of as a linear but time-variant filtering operation.

Ideally, the lowpass filter satisfies the following spectral characteristics

$$H_{ideal}(e^{j\omega}) = \begin{cases} 1, & -\dfrac{\pi}{M} \leq \omega \leq \dfrac{\pi}{M} \\ 0, & \text{otherwise} \end{cases} \qquad (10.3)$$

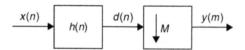

Figure 10.4 Lowpass filter followed by a downsampling operation

The spectrum of the input signal $X(e^{j\omega})$ is assumed to be nonzero in the frequency interval $-\pi \leq \omega \leq \pi$. Spectrally, the purpose of the filter is to decrease the bandwidth of the signal $x(n)$ by a factor M. The resulting sequence $d(n)$ can be expressed in terms of the convolution of $x(n)$ with the filter transfer function $h(n)$:

$$d(n) = \sum_{l=0}^{\infty} h(l)x(n-l) \tag{10.4}$$

After downsampling by M, the resulting sequence $y(m)$ is

$$y(m) = d(Mm) = \sum_{k=0}^{\infty} h(k)x(Mm-k) \tag{10.5}$$

It is important to note, as mentioned earlier, that the combination of linear filtering with downsampling results in a linear but time-varying operation.

To understand the spectral implication of downsampling on the signal, define the sequence:

$$\bar{d}(n) = \begin{cases} d(n) & n = 0, \pm M, \pm 2M, \ldots \\ 0 & \text{otherwise} \end{cases} \tag{10.6}$$

and the complex number on the unit circle as the root of 1 for a given M as

$$W_M = e^{-j2\pi/M} = \sqrt[M]{-1} \tag{10.7}$$

Furthermore, let us revisit the pulse train sequence in (10.1), which can be expressed as

$$\varphi_M(n) = \frac{1}{M} \sum_{k=0}^{M-1} e^{j\frac{2\pi}{M}kn} = \frac{1}{M} \sum_{k=0}^{M-1} W_M^{kn} \tag{10.8}$$

Multiplying the relation in (10.8) by the sequence $d(n)$ we obtain the sequence:

$$\bar{d}(n) = d(n)\varphi_M(n) \tag{10.9}$$

Comparing the relations in (10.6) and (10.9), we obtain the equality:

$$y(m) = d(Mm) = d(m)\varphi_M(m) = \bar{d}(m) \tag{10.10}$$

Again, expressing the z-transform of $y(m)$ using (10.10):

$$
\begin{aligned}
Y(z) &= \sum_{m=-\infty}^{\infty} y(m)z^{-m} \\
&= \sum_{m=-\infty}^{\infty} \bar{d}(Mm)z^{-m} \\
&= \sum_{n=-\infty}^{\infty} \bar{d}(n)z^{-n/M}
\end{aligned} \tag{10.11}
$$

The last step in (10.11) is valid since $d(n)$ is zero except when n is a multiple M, and hence the result:

$$Y(z) = \bar{D}(z^{1/M}) \tag{10.12}$$

Substituting (10.10) and (10.8) for $\varphi_M(n)$ and $\bar{d}(n)$ into (10.12), then $\bar{D}(z)$ becomes

$$
\begin{aligned}
\bar{D}(z) &= \sum_{n=-\infty}^{\infty} \bar{d}(n)z^{-n} = \sum_{n=-\infty}^{\infty} d(n)\varphi_M(n)z^{-n} \\
&= \sum_{n=-\infty}^{\infty} \left\{ d(n)\frac{1}{M}\sum_{k=0}^{M-1} W_M^{kn} \right\} z^{-n} \\
&= \frac{1}{M}\sum_{k=0}^{M-1}\sum_{n=-\infty}^{\infty} d(n)(W_M^k z^{-1})^n
\end{aligned} \tag{10.13}
$$

Note, however, that the inner summation in (10.13) is

$$\sum_{n=-\infty}^{\infty} d(n)(W_M^k z^{-1})^n \cong D(W_M^k z) \tag{10.14}$$

This simply implies that $\bar{D}(z)$ can be obtained by substituting (10.14) for the inner summation of (10.13):

$$\bar{D}(z) = \frac{1}{M}\sum_{k=0}^{M-1} D(W_M^k z) \tag{10.15}$$

Finally, comparing the results in (10.12) and (10.15), we obtain:

$$Y(z) = \frac{1}{M} \sum_{k=0}^{M-1} D(W_M^k z^{1/M})$$
(10.16)

The result obtained in (10.16) can be expressed in the frequency domain by replacing W_M by $e^{-j2\pi/M}$ and $z = e^{-j\omega}$. The resulting relation is known as the discrete-time Fourier transform (DTFT) of $Y(z)$:

$$Y(e^{j\omega}) = \frac{1}{M} \sum_{k=0}^{M-1} D\left(e^{j\frac{\omega-2\pi k}{M}}\right)$$
(10.17)

Note that $Y(e^{j\omega})$ consists of multiple spectral images replicated equidistantly at intervals of $2\pi k/M$, for $k = 0, 1$, etc.

Furthermore, realizing from (10.4) that $d(n)$ is none other than $x(n)$ filtered by $h(n)$ implies:

$$Y(e^{j\omega}) = \frac{1}{M} \sum_{k=0}^{M-1} \underbrace{H\left(e^{j\frac{\omega-2\pi k}{M}}\right) X\left(e^{j\frac{\omega-2\pi k}{M}}\right)}_{=D\left(e^{j\frac{\omega-2\pi k}{M}}\right)}$$
(10.18)

Note that the term in (10.18) for $k = 0$ represents the signal of interest:

$$Y_{desired}(e^{j\omega}) \approx \frac{1}{M} H(e^{j\omega/M}) X(e^{j\omega/M})$$
(10.19)

For an ideal lowpass filter as defined in (10.3), the desired decimated signal becomes:

$$Y_{desired}(e^{j\omega}) = \frac{1}{M} X(e^{j\omega/M}), \quad \text{for } -\pi \leq \omega \leq \pi$$
(10.20)

It is important to note that filtering is necessary only when decimation can cause the various spectral images of $X(e^{j\omega})$ to overlap.

At this point it is instructive to examine the linear time-variant implication of the decimation process. As mentioned earlier, a delay of M samples before the decimator is equivalent to a delay of one sample after the decimator, which in the z-domain equates to

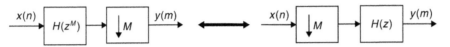

Figure 10.5 Linear time-variant property of downsampling

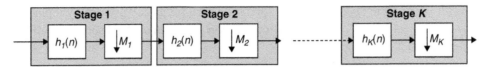

Figure 10.6 A conceptual K-stage decimator

z^{-M} before the decimator and to z^{-1} after the decimator. Hence the relationship given in (10.18) can be expressed as

$$Y(e^{j\omega}) = \frac{1}{M} \sum_{k=0}^{M-1} H\left(e^{j\frac{\omega-2\pi k}{M}}\right) X\left(e^{j\frac{\omega-2\pi k}{M}}\right) = \frac{1}{M} \left\{\sum_{k=0}^{M-1} X\left(e^{j\frac{\omega-2\pi k}{M}}\right)\right\} H(e^{j\omega}) \quad (10.21)$$

The relationship in (10.21) indicates that, for a given input sequence, the same output sequence can be obtained by placing a lowpass filter either before or after the decimator, as shown in Figure 10.5. This property is known as the Noble identity for decimation. The advantage lies in the savings gained in computational complexity by placing the filter after the downsampler rather than before. This topic will be further discussed in the coming sections.

10.1.1.2 Multistage Decimation

In certain SDR applications, the received signal waveform is oversampled by several orders of magnitude over the Nyquist rate. In this case, decimating the signal in a single stage may result in a very costly and complex filtering arrangement. In this case, it may be more efficient to implement the sampling rate conversion in multiple decimation stages as shown in Figure 10.6 [4]. The gradual decrease in sampling rate results in a simplified design of the various filter stages. The ensuing design can still meet the overall decimation and filtering requirements. The savings are accomplished by reducing the number of multiples and adds (MADs) per second.

The condition for allowing multiple stage decimation is that M, the total decimation factor, is not a prime number and that it can be written as the product:

$$M = \prod_{i=1}^{K} M_i = M_1 \times M_2 \times \cdots M_K \quad (10.22)$$

where M_1, M_2, ..., M_K are the decimation factors associated with the respective stages in a multistage decimator. This in turn implies that the original sampling rate F_s and the final sampling rate $F_{s,K}$ are related according to:

$$F_{s,K} = \frac{F_s}{M} = \frac{F_s}{\prod\limits_{i=1}^{K} M_i} \tag{10.23}$$

The relationship between two successive sampling rates can be expressed as

$$F_{s,i} = \frac{F_{s,i-1}}{M_i} \tag{10.24}$$

Before delving into the mathematical details, let's first illustrate the multistage decimation process from a spectral perspective. A spectral depiction of a three-stage decimation process showing the desired frequency response of each stage is presented in Figure 10.7. Assume that the data is given at a sampling rate $F_{s,1}$ and that the desired final sampling rate is $F_{s,3}$. The passband frequency of the desired signal is f_{pass} and the ultimate desired stopband frequency is f_{stop}. The purpose of the last stage of the decimation filter is clear and that is to limit the signal bandwidth such that $f_{stop} \leq F_{s,3}/2$. In general, this implies that, in order to prevent any degradation due to aliasing, the stopband frequency of the last stage must obey the sampling theorem, that is

$$f_{stop} \leq \frac{F_{s,K}}{2} \tag{10.25}$$

In general, the passband and stopband frequencies of the ith stage filter can then be defined according to Figure 10.8. The passband characteristics of the ith stage filter must be flat between DC and f_{pass} in order to prevent any distortion to the passband signal. On the other hand, in order to ensure that no aliasing components fall into the desired signal band, the stopband cutoff frequency of the ith stage filter must conform to the relation:

$$f_{stop,i} \leq F_{s,i} - f_{stop} \tag{10.26}$$

The result in (10.26) implies what was discussed earlier: that the stopband cutoff frequency of the last stage is $f_{stop,K} = F_{s,K} - f_{stop} = 2f_{stop} - f_{stop} = f_{stop}$.

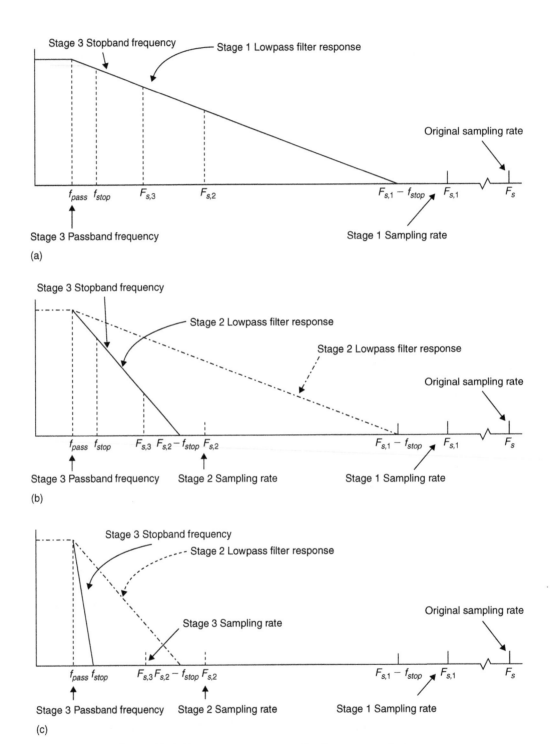

Figure 10.7 Frequency response of a three-stage decimator

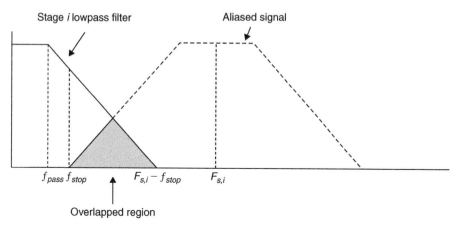

Stage *i* lowpass filter   Aliased signal

f_{pass} f_{stop} $F_{s,i} - f_{stop}$ $F_{s,i}$

Overlapped region

Figure 10.8 Frequency response of *i*th stage lowpass filter

At this point it is instructive to discuss the passband and stopband ripples of each stage. It is desired that the overall passband ripple of the composite *K*-stage filter be within $1 + \delta_{pass}$. This imposes even tighter constraints on the tolerable passband ripple in individual cascade stages. A simple, but not unique, way of specifying the passband is to require that each stage passband ripple $1 + \delta_{pass,i}$ be specified such that $\delta_{pass,i} = \delta_{pass}/K$. Note that this does not guarantee that the peaks and valleys of the overall response will align such that a precise ripple characteristic is produced. Similarly, the stopband ripple δ_{stop} specifies the overall multistage stopband ripple, or the cumulative multistage stopband filter ripple. The stopband ripple is imposed on all the individual stages of the cascade, that is $\delta_{stop,i} \cong \delta_{stop}$. This is particularly important since aliasing does not necessarily add up coherently and sufficient attenuation is required to avoid aliasing in each stage. However, it is up to the designer and the nature of the design to specify this parameter. The interesting question then becomes: what is the optimum number of decimation stages and amount by which to decimate in each stage in order to achieve maximum efficiency? The answer lies in minimizing the computational complexity of the overall filter, as will be shown herein.

In order to estimate the required order of an optimal lowpass FIR filter, define the transition band, say for a given decimation stage Δf whose width is normalized by the sampling frequency at that particular stage, that is

$$\Delta f_i = \frac{(F_{s,i} - f_{stop}) - f_{pass}}{F_{s,i-1}} \tag{10.27}$$

where f_{stop} is the stopband frequency, f_{pass} is the passband frequency, $F_{s,i}$ is the sampling rate at the output of the ith stage, and $F_{s,i-1}$ is the sampling rate frequency at the input to the ith stage. Then the optimum number of taps is given by [5] and [6]:

$$N_{i,optimum} \cong \frac{D_{\infty}(\delta_{pass,i}\delta_{stop,i})}{\Delta f_i} - f(\delta_{pass,i}\delta_{stop,i})\Delta f_i + 1 \qquad (10.28)$$

where

$$\begin{aligned} D_{\infty}(\delta_{pass,i}\delta_{stop,i}) = &[5.309 \times 10^{-3}(\log_{10}\delta_{pass,i})^2 \\ &+ 7.114 \times 10^{-2}(\log_{10}\delta_{pass,i}) - 0.4761] \times \log_{10}\delta_{stop,i} \\ &- [2.66 \times 10^{-3}(\log_{10}\delta_{pass,i})^2 + 0.5941(\log_{10}\delta_{pass,i}) \\ &+ 0.4278] \end{aligned} \qquad (10.29)$$

and

$$f(\delta_{pass,i}\delta_{stop,i}) = 0.51244\log_{10}\left(\frac{\delta_{pass,i}}{\delta_{stop,i}}\right) + 11.01217 \qquad (10.30)$$

$\delta_{pass,i}$ = Linear peak passband ripple at ith stage
$\delta_{stop,i}$ = Linear peak stopband ripple at ith stage

Note that, for a large overall decimation ratio where the transition band is small, the relation in (10.28) can be simplified as

$$N_{i,optimum} \approx \frac{D_{\infty}(\delta_{pass,i}\delta_{stop,i})}{\Delta f_i} \qquad (10.31)$$

The objective of a multistage decimator will be to minimize the total number of MADS R_{total}. Let R_i be the number of MADS in the ith decimation stage; then R_{total} is given as

$$R_{total} = \sum_{i=1}^{K} R_i \qquad (10.32)$$

Minimizing R_{total} minimizes the overall amount of computation required. Given the filter length $N_{optimum,i}$ for the ith stage,[1] then R_i can be expressed as the product of $N_{optimum,i}$ times the sampling rate of the filter $F_{s,i}$ divided by the decimation factor M_i:

$$R_i = \frac{N_{optimum,i}F_{s,i}}{M_i} \qquad (10.33)$$

[1]Assume that all FIR filters employed in this section are implemented as direct-form FIR filters.

From (10.24), (10.31), and (10.33), it was found in [6] that (10.33) can be further expressed as

$$R_i \cong D_\infty \left(\frac{\delta_{pass,i}}{K}, \delta_{stop} \right) \frac{M_i F_{s,i}^2}{F_{s,i} - f_{stop} - f_{pass}} \tag{10.34}$$

and the total computation for all stages can also be expressed in the following form:

$$
\begin{aligned}
R_{total} &\cong D_\infty \left(\frac{\delta_{pass,i}}{K}, \delta_{stop} \right) \sum_{i=1}^{K} \frac{M_i F_{s,i}^2}{F_{s,i} - f_{stop} - f_{pass}} \text{ MADS} \\
&= D_\infty \left(\frac{\delta_{pass,i}}{K}, \delta_{stop} \right) F_s \sum_{i=1}^{K} \frac{M_i}{\left(\prod_{j=1}^{i} M_j \right) \left(1 - \frac{f_{stop} + f_{pass}}{F_s} \prod_{j=1}^{i} M_j \right)} \text{ MADS}
\end{aligned}
\tag{10.35}
$$

Recall that F_s in (10.35) is the initial sampling rate. To further simplify the relationship in (10.35), let ΔF be defined as the transition-band ratio

$$\Delta F = \frac{f_{stop} - f_{pass}}{f_{stop}} \tag{10.36}$$

and rewrite (10.35) as

$$R_{total} \cong D_\infty \left(\frac{\delta_{pass}}{K}, \delta_{stop} \right) F_s \Psi \tag{10.37}$$

where

$$\Psi = \frac{2}{\left(\Delta F \prod_{j=1}^{K-1} M_j \right)} + \sum_{i=1}^{K-1} \frac{M_i}{\left(\prod_{j=1}^{i} M_j \right) \left(1 - \frac{2 - \Delta F}{2M} \prod_{j=1}^{i} M_i \right)} \tag{10.38}$$

Note that for a two-stage cascaded decimator where $K = 2$ and $M > 10$, the authors in [4] have provided an optimal solution for choosing the decimation factors M_1 and M_2. This is done by minimizing R_{total} in (10.37) for a given K. Recall that, in this case, the total decimation factor is $M = M_1 \times M_2$. The optimal solution yields:

$$M_{1,optimal} = \frac{2M\left(1 - \sqrt{\dfrac{M\Delta F}{2 - \Delta F}}\right)}{2 - \Delta F(M + 1)}, \quad K = 2 \tag{10.39}$$

and

$$M_{2,optimal} = M/M_{1,optimal}, \quad K = 2 \tag{10.40}$$

Note that the relation in (10.39) will result in most cases with a noninteger number. The choice for $M_{1,optimal}$ then would be rounded to the nearest integer such that:

$$M = M_{1,optimal} \times M_{2,optimal} \tag{10.41}$$

From (10.37), it can be observed that the choice of the optimum number of stages K is strongly influenced by ΔF and M, much more so than the passband and stopband ripples.

Example 10-1: Two-Stage Decimator for LTE 1.4-MHz Channel

Consider the sampling rate at the output of a $\Delta\Sigma$-converter to be 124.8 MHz. Assume that the desired sampling rate before any further fractional rate conversion is 1.418182 MHz (corresponding to decimation by an integer factor M = 88). Note that 1.4 MHz is the smallest channel bandwidth supported by UMTS-LTE. Assume that you are designing the last two stages of the decimation process. The data rate of the input signal to the two-stage decimator is 17.018018 MHz. Assume that the passband is 0.54 MHz, which is the bandwidth of the signal-occupied subcarriers. Compare the complexity of the two-stage decimator to that of a single-stage decimator. Assume that the overall linear passband ripple is 0.5 dB and that the linear stopband rejection is 30 dB. Note that in this example, the effect of blockers is ignored. In actuality, the blocker signals after filtering and data conversion must be accommodated, which could have profound effects on the design.

In order to prevent aliasing, the stopband frequency must be less than 1.418182 MHz/2. In this case, let's assume that the stopband frequency is ½ the Nyquist rate, that is $f_{stop} = 0.7$ MHz. The transition band ratio as defined in (10.36) is given as

$$\Delta F = \frac{f_{stop} - f_{pass}}{f_{stop}} = \frac{0.7\,\text{MHz} - 0.54\,\text{MHz}}{0.7\,\text{MHz}} = 0.2286 \tag{10.42}$$

Recall that the total decimation ratio required by the two-stage decimation process is 17.0181018 MHz/1.418182 MHz = 12:

$$
\begin{aligned}
M_{1,optimal} &= \frac{2M\left(1 - \sqrt{\dfrac{M\Delta F}{2 - \Delta F}}\right)}{2 - \Delta F(M + 1)} \\[2em]
&= \frac{2 \times 12\left(1 - \sqrt{\dfrac{12 \times 0.2286}{2 - 0.2286}}\right)}{2 - 0.2286(12 + 1)} \\[2em]
&= 6.0365
\end{aligned}
\tag{10.43}
$$

The nearest integer to 6.0365 is 6 and hence $M_{1,optimal} = 6$ and $M_{2,optimal} = 2$.

Next, let's determine $\delta_{pass,i}$ and $\delta_{stop,i}$. To determine the linear passband ripple and stopband attenuation, we know that

$$
\begin{aligned}
10\log\left(1 + \delta_{pass,i}/K\right) &= 0.5, \quad K = 2, \quad i = 1, 2 \Rightarrow \delta_{pass,i} = 0.061 \\
10\log\left(\delta_{stop,i}\right) &= -30 \Rightarrow \delta_{stop,i} = 0.001
\end{aligned}
\tag{10.44}
$$

For the first stage, the sampling rate changes from $F_{s,1} = 17.018018$ MHz to $F_{s,2} = 17.018018/6$ MHz = 2.8363 MHz. From (10.26), the stopband frequency of the first stage is given as $f_{stop,1} = F_{s,1} - f_{stop} = 17.018018/6 - 0.7 = 2.1363$ MHz. Therefore, from (10.27), the transition band ratio of the first stage is given as

$$
\begin{aligned}
\Delta f_1 &= \frac{(F_{s,1} - f_{stop}) - f_{pass}}{F_{s,0}} \\[1em]
&= \frac{(2.8363 - 0.7) - 0.54}{17.018018} = 0.0938
\end{aligned}
\tag{10.45}
$$

From (10.28), it turns out that the number of taps required is *ceil* {20.7116} = 21 taps. Next, we turn our attention to the second stage. The transition band ratio for the second stage is given as

$$
\begin{aligned}
\Delta f_2 &= \frac{(F_{s,2} - f_{stop}) - f_{pass}}{F_{s,1}} \\[1em]
&= \frac{(1.4182 - 0.7) - 0.54}{2.8363} = 0.0628
\end{aligned}
\tag{10.46}
$$

Again, using (10.28), results in the number of required taps to be *ceil* {31.3565} = 32 taps.

In order to compare the complexity of the two-stage decimation process to that of a single stage, apply (10.28) with passband ripple $2 \times \delta_{pass,i} = 0.1220$, we obtain the optimum number of taps to be *ceil* $\{164.5529\} = 165$ taps.

Note that, given that the filters are implemented in each case before the decimator, then the single-stage filter case with 165 taps results in $\Gamma = 165 \times 17.018018\,\text{MHz} = 2.8080e + 009$ multiplies per second. Comparing this to the two-stage approach, we obtain:

$$\Gamma = 21\,\text{taps} \times 17.018018\,\text{MHz} + 32\,\text{taps} \times \frac{17.018018}{6}\,\text{MHz}$$

$$\Gamma = 4.4814e8\,\text{multiplies per second} \tag{10.47}$$

It is evident from (10.47) that the two-stage solution has significantly less computational complexity than the single-stage solution.

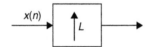

Figure 10.9 An upsampling block diagram

10.1.2 Interpolation by an Integer L

In most multirate applications, interpolation is used to change the sampling rate of a signal without changing its spectral content. In other words, increasing the sampling rate of a given signal (upsampling) increases the spectral separation between the images of the original spectrum.[2] However, this process does not add any new information to the existing signal even though the sampling rate has increased, yielding more sample points for processing. In this section, we first discuss the simplest form of signal interpolation, namely interpolation via zero insertion. The upsampling process is depicted in Figure 10.9.

Interpolation via zero insertion implies that zeros are inserted at the new rate between the original sample signals according to the relation:

$$y(n) = \begin{cases} x\left(\dfrac{n}{L}\right) & n = 0, \pm L, \pm 2L, \ldots \\ 0 & \text{otherwise} \end{cases} \tag{10.48}$$

[2]Note that, even though we are using the terms interpolation and upsampling interchangeably, upsampling does not imply interpolation in the full mathematical sense unless a filtering of the upsampled signal is involved.

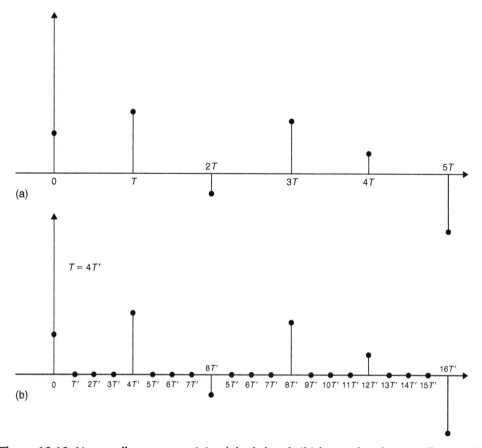

Figure 10.10 Upsampling process: (a) original signal, (b) increasing the sampling rate by inserting zeros between the original samples

This process is illustrated for an upsampling factor of 4 as shown in Figure 10.10. The relationship between $Y(z)$ and $X(z)$ is easily derived. Let

$$
Y(z) = \sum_{n=-\infty}^{\infty} y(n)z^{-n} = \sum_{\substack{k=-\infty \\ k=Ln}}^{\infty} y(k)z^{-k}
$$

$$
= \sum_{\substack{k=-\infty \\ k=Ln}}^{\infty} x\left(\frac{k}{L}\right)z^{-k} \tag{10.49}
$$

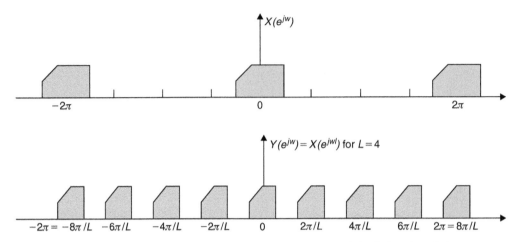

Figure 10.11 Upsampling without filtering: (a) spectrum of original signal and (b) spectrum of upsampled signal

Define $l = k/L$, then the relationship in (10.49) becomes:

$$Y(z) = \sum_{\substack{k=-\infty \\ k=Ln}}^{\infty} x(l)z^{-Ll} = X(z^L) \tag{10.50}$$

In the frequency domain, this implies that the DTFT of $Y(z)$ is

$$Y(e^{j\omega}) = X(e^{j\omega L}) \tag{10.51}$$

Pictorially, the relationship in (10.51) can be depicted as Figure 10.11 for the case where $L = 4$. The spectrum of the upsampled signal shown in Figure 10.11(b) is made up of multiple *compressed* copies of the original signal spectrum, shown in Figure 10.11(a). These compressed copies, known as images, are placed at $2\pi/L$ intervals from DC. Note that constraining the original signal to the frequency range $-\pi \leq \omega \leq \pi$ implies that the upsampled signal is now limited to the range $-\pi/L \leq \omega \leq \pi/L$. Hence, the upsampling process compresses the signal in the frequency domain by a factor of L with respect to the new sampling rate which is L times higher than the original sampling rate.

In order to suppress the images, the upsampler of Figure 10.9 is typically followed by a low pass filter, as shown in Figure 10.12. The cutoff frequency of this filter is typically set

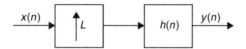

Figure 10.12 An upsampling operation followed by a lowpass filter

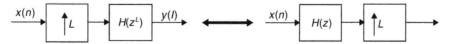

Figure 10.13 Linear time-variant property of upsampling

to π/L [1]. The resulting signal is an interpolated version of the original signal. The zero-valued samples inserted between the original samples are now replaced with nonzero samples due to the convolution operation between the upsampled signal and the impulse response of the lowpass filter $h(n)$.

Similar to the decimation case, due to the linear time-variant characteristic of interpolation, the lowpass filter can be placed either before or after the interpolator according to Figure 10.13. This property is known as the Noble identity for interpolation. Placing the filter before the upsampler, however, results in decreased computational complexity of the system.

Thus far, we have only discussed upsampling via zero insertion. Other more sophisticated and more computationally costly methods exist. One example of such involves replicating a given sample $L - 1$ times for an upsampling rate of L between any consecutive samples. That is, given two consecutive samples at the original sampling rate $x(n)$ and $x(n + 1)$, then the upsampled data is given as

$$y(m) = y(m + 1) = \cdots = y(m + L - 1) = x(n) \qquad (10.52)$$
$$\text{for samples between } x(n) \text{ and } x(n + 1)$$

This method is known as zero-order hold upsampling, illustrated for $L = 4$ in Figure 10.14.

Zero-order hold upsampling tends to degrade the desired signal beyond that which is done by zero insertion upsampling. A sinc function due to the hold operation distorts the desired signal. At the same time, the sinc response helps to attenuate distortion present in the high-frequency signal content.

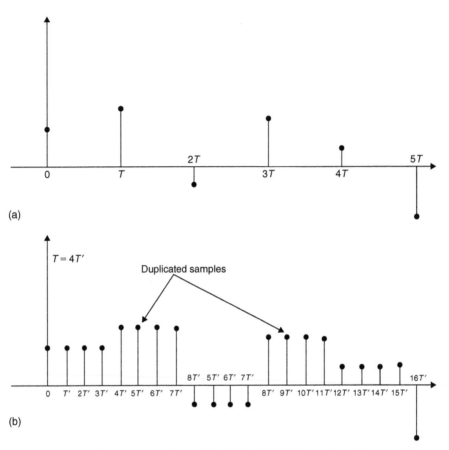

Figure 10.14 Upsampling process: (a) original signal, (b) increasing the sampling rate by inserting duplicate samples between the original samples

10.1.3 Fractional Rate Conversion

There are many sampling rate conversion scenarios where the final sampling rate cannot be obtained simply by decimation or interpolation alone. One such scenario requires the initial sampling rate and the final sampling rate to be a rational number L/M. Sampling rate conversion by a rational factor L/M can be accomplished by cascading an interpolator followed by a decimator as shown in Figure 10.15. Note that, in certain cases, it is important to perform the upsampling operation first followed by the decimation operation in order to prevent degradation due to aliasing. The frequency response of the filter

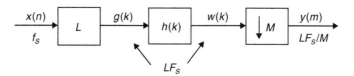

Figure 10.15 Fractional rate conversion operation by a factor of L/M

$h(k)$ has the composite transfer function of the upsampling and downsampling filters illustrated in Figure 10.4 and Figure 10.12. The ideal transfer function of $h(k)$ is

$$H(e^{j\omega_L})\Big|_{\omega_L=\frac{\omega}{L}} = \begin{cases} L & 0 \le |\omega_L| = \left|\dfrac{\omega}{L}\right| \le \min\left\{\dfrac{\pi}{M},\dfrac{\pi}{L}\right\} \\ 0 & \text{otherwise} \end{cases} \tag{10.53}$$

where $\omega_L = \omega/L = 2\pi F/LF_s$. The input to the filter can be expressed in the time domain as the upsampled sequence:

$$g(k) = \begin{cases} x\left(\dfrac{k}{L}\right) & k = 0, \pm L, \pm 2L,\dots \\ 0 & \text{otherwise} \end{cases} \tag{10.54}$$

Similarly, the output of the filter is

$$w(k) = \sum_{l=-\infty}^{\infty} h(k-l)g(l)$$

$$= \sum_{l=-\infty}^{\infty} h(k-lL)x(l) \tag{10.55}$$

In the frequency domain, the sequence $\{w(k)\}$ can be characterized as

$$W(e^{j\omega_L}) = H(e^{j\omega_L})G(e^{j\omega_L})$$
$$= H(e^{j\omega_L})X(e^{jL\omega_L})$$
$$= H\left(e^{j\frac{\omega}{L}}\right)X(e^{j\omega}) \tag{10.56}$$

The output of the downsampler in Figure 10.15 then becomes the decimated sequence $\{y(m)\}$:

$$y(m) = w(Mm) = \sum_{l=-\infty}^{\infty} h(mM-lL)x(l) \tag{10.57}$$

The relation in (10.57) can be expressed in the frequency domain as

$$Y(e^{j\omega_{L,M}})\bigg|_{\omega_{L,M}=\frac{M}{L}\omega} = \frac{1}{M}\sum_{m=0}^{M-1} W\left(e^{j\frac{(\omega_{L,M}-2\pi m)}{M}}\right) \tag{10.58}$$

It can be easily determined from (10.53) and (10.56) that

$$W(e^{j\omega_L}) = \begin{cases} LX(e^{jL\omega_L}) & 0 \le |\omega_L| \le \min\left\{\dfrac{\pi}{M},\dfrac{\pi}{L}\right\} \\ 0 & \text{otherwise} \end{cases} \tag{10.59}$$

Providing that the lowpass filter can prevent any degradation to the desired signal, the relations in (10.58) and (10.59) imply:

$$Y(e^{j\omega_{L,M}})\bigg|_{\omega_{L,M}=\frac{M}{L}\omega} = \begin{cases} \dfrac{1}{M}X\left(e^{j\frac{\omega_{L,M}}{M}}\right) & 0 \le |\omega_{L,M}| \le \min\left\{\pi,\dfrac{M\pi}{L}\right\} \\ 0 & \text{otherwise} \end{cases} \tag{10.60}$$

It is important at this juncture to emphasize that (10.60) is valid only for rational rate conversion. Furthermore, there are limitations in terms of computational complexity that at times preclude use of this technique for practical implementation.

10.2 Filter Design and Implementation

When dealing with decimation and interpolation, it is imperative for the designer to implement cost-effective, low-power filters. In this section, we introduce several filtering schemes that meet the low-power, cost-effective criteria while achieving the desired performance.

10.2.1 Polyphase Structures

The processes of downsampling and upsampling, as described thus far, can be further simplified. In single rate signal processing, a digital filter produces a single output sample for every input sample. In decimation, for example, the filter needs to process only every *Mth* output sample. This is true since only these samples will be retained after decimation. That is, in the time domain, the samples at the output of the filter that are relevant after decimation would be

$$\widehat{d}(n) = \sum_{k=0}^{N-1} x(Mn - k)h(k) \tag{10.61}$$

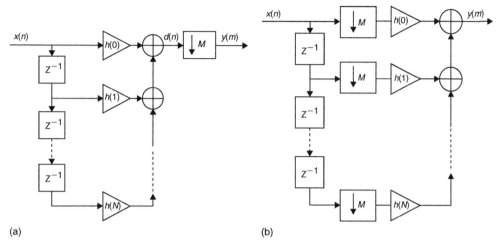

Figure 10.16 **The filtering process in decimation: (a) direct form FIR followed by decimator and (b) decimation moved into the filter**

The implication of (10.61) on the computational complexity of the filter will be discussed herein. Consider the direct-form FIR filter followed by the decimator shown in Figure 10.16(a). In this case, the filtering occurs at the high sampling rate before any rate reduction takes place. That is, the filtered sequence which constitutes the input to the decimator is the convolution computed in accordance with (10.62):

$$d(n) = \sum_{k=0}^{N-1} x(n - k)h(k) \tag{10.62}$$

In order to obtain the decimated sequence $y(m)$, the downsampling process retains only every Mth sample and the remaining $M - 1$ samples out of M samples are discarded. The pertinent question then is: do the discarded samples need to be computed at all? To answer this question, we need to look at ways of implementing (10.61). Figure 10.16(b) depicts such an implementation. This is done by moving the decimation block before each filter tap, thus reducing the number of multiplications by a factor of M. Note that the decimation process involves slowing down the clock and throwing away samples, and therefore it does not impact the complexity of the design.

The upsampling process, similar to downsampling, can afford the designer a certain savings in computational complexity, as illustrated in Figure 10.17. In zero-insertion type upsampling, $L - 1$ out of L samples are actually zero requiring that no multiplication

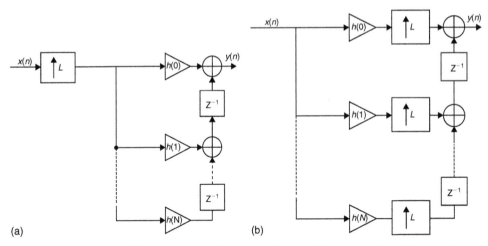

Figure 10.17 The filtering process in interpolation: (a) upsampling followed by direct form FIR, and (b) upsampling moved into the filter

need be performed. Figure 10.17(a) depicts an upsampling operation followed by a transposed direct form FIR filter $h(n)$. After each input sample $x(n)$, the interpolator inserts $L - 1$ zero samples. In this case, each tap performs only one valid multiplication operation every L samples. The remaining $L - 1$ samples are zeros and hence result in a zero output. In order to avoid these unnecessary multiplications, we can move the interpolator inside the filter at the output of each tap as shown in Figure 10.17(b). In this case, all multiplications occur at the lower rate and only at nonzero samples thus resulting in savings in both power consumption and computational complexity. Note that further reduction in complexity may be achieved by exploiting certain symmetries.

In upsampling, it is desired that the filter passes the nonzero samples of the original sequence unchanged. That is, the output of the lowpass filter in Figure 10.17 is a filtered form of:

$$y(n) = \begin{cases} x\left(\dfrac{n}{L}\right) & n = 0, \pm L, \pm 2L, \ldots \\ \text{Upsampled and filtered value} & \text{otherwise} \end{cases} \qquad (10.63)$$

Since all input samples other than $x(n/L)$, for $n =$ multiple of L, are zero, we can express the output of the filter of Figure 10.17 as

$$y(n) = \sum_{k=-p/L}^{p/L} x\left(\frac{n-k}{L}\right) h(Lk) \qquad (10.64)$$

Now in order to guarantee the condition disclosed in (10.63) concerning the output $y(n)$ at integer multiples of L, the filter $h(n)$ must be constrained such that

$$h(n) = \begin{cases} 1 & n = 0 \\ 0 & n \neq 0 \text{ and } n = \pm L, \pm 2L, \ldots \end{cases} \tag{10.65}$$

A further constraint on length of $h(n)$ can be stated as follows: The length of $h(n)$ must be chosen such that each output sample is generated using an equal number of nonzero input samples. That is, the filter length N depends on the upsampling factor L and the number of input samples I or:

$$N = \begin{cases} LI & LI \text{ is odd} \\ LI - 1 & LI \text{ is even} \end{cases} \tag{10.66}$$

where N is an odd value. In this case, the filter will always cover the same number of input samples I to produce an output sample.

10.2.1.1 Polyphase Decimation Filters

Additional insight can be gained by examining the sample redundancy in the filter. Consider by way of an illustration the decimator by 5 shown in Figure 10.18. The filter is a 15-tap FIR filter. Consider the input sequence $\{x(n)\}$ passing through the filter and decimators. Note that the decimator discards four out of every five samples as depicted in Figure 10.18. A further examination of Figure 10.18 shows a significant amount of redundancy in the input sequences to each tap. For example, the input sequence to $h(5)$, $\{x(n - 15), x(n - 10), x(n - 5)\}$, is identical to that input to $h(0)$, $\{x(n - 10), x(n - 5), x(n)\}$, delayed by 5, which incidentally is the decimation ratio. In the same manner, the input to $h(10)$ is also a delayed version of the input to $h(5)$ and $h(0)$. Next, if we examine the input sequence to $h(6)$, $\{x(n - 17), x(n - 11), x(n - 6)\}$, it is a delayed version of the input to $h(2)$ namely $\{x(n - 11), x(n - 6), x(n)\}$. Similarly, the input to $h(11)$ is also a delayed-by-five version of the input to $h(6)$, and a delayed-by-10 version of the input to $h(2)$. These observations indicate that further consolidation of the filter resources is possible. This is illustrated in Figure 10.19.

In Figure 10.19, the taps with the same but delayed sample inputs are grouped together resulting in $M = 5$ different groups. The outer delays are reduced to $M - 1 = 4$ elements. Note that the multipliers are now operating at the low sampling rate and therefore their

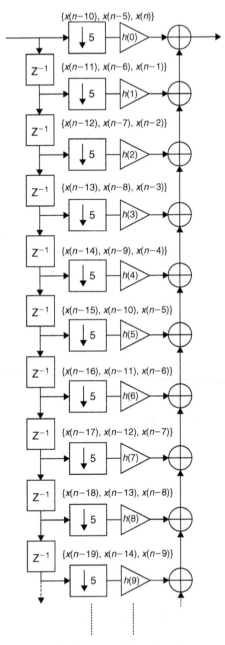

Figure 10.18 A 15-tap decimator with redundant samples for $M = 5$

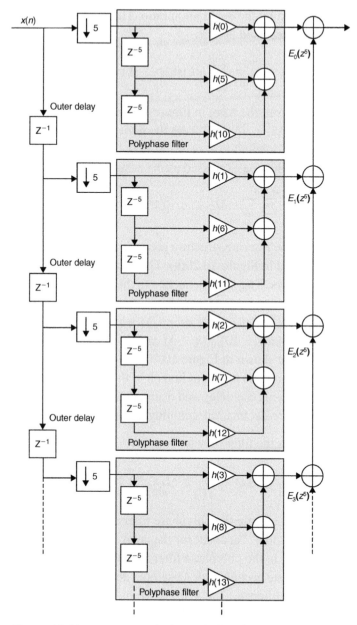

Figure 10.19 A 15-tap polyphase decimation filter for $M = 5$

clock has been slowed down by the decimation ratio. Therefore, for a decimation factor of M, shown in (10.5) and repeated here as

$$y(n) = \sum_{k=0}^{N-1} h(k)x(Mn - k) \tag{10.67}$$

can be further expressed to suit the form in Figure 10.19. That is, let $k = Mq + p$ for $p = 0,1,\ldots, N-1$ and $q = 0,1,\ldots, \lfloor N/M \rfloor$, where the floor function of a real number r is the largest integer less than or equal to r, or $\lfloor r \rfloor = \max \{n \in \mathbb{Z} | n \le r\}, r \in \mathbb{R}$. The relationship in (10.67) can then be expressed as the nested sum:

$$y(n) = \sum_{p=0}^{M-1} \sum_{q} h(Mq + p)x[(M(n - q) - p)] \tag{10.68}$$

To gain more insight into the filtering structure presented in (10.68), consider the delay and decimate chain depicted in Figure 10.20(a). This structure is made up of $M - 1 = 4$ delays and $M = 5$ decimators. The input sequence to the ith decimator is the delayed version of the input sequence to the $(i - 1)th$ decimator. For example, the input to the second decimator is $\{x(n - 11), x(n - 6), x(n - 1)\}$ which is a delayed version of the input to the first decimator $\{x(n - 10), x(n - 5), x(n)\}$. This can be perfectly represented by the commutator structure shown in Figure 10.20(b) [7]. At every sampling instant, the commutator, moving from one tap input branch to the next, delivers a data sample to the filters thus in effect creating the delay and decimate effect. The commutator is then synchronized to the high, pre-decimation sampling clock.

In summary, we can express the filter transfer function

$$H(z) = \sum_{k=0}^{M-1} z^{-k} E_k(z^M) \tag{10.69}$$

where $E_k(z^M), k = 0,\ldots, M - 1$ are the polyphase filters that form the basis for the polyphase transformation, and z^{-k} account for the delays preceding the decimators. Recall that the argument z^M in the polyphase filter accounts for the fact that a one sample delay at the low sampling rate accounts for M-sample delay at the original high sampling rate. Each of the polyphase filters $e_k(.), k = 0,1,\ldots M$, has an impulse response that corresponds to a decimated version of the original filter's impulse response $h(n)$ as shown in Figure 10.21. It is interesting to note that when examining the polyphase filters as stand alone filters, we observe that these filters are, in general, not linear phase. Furthermore,

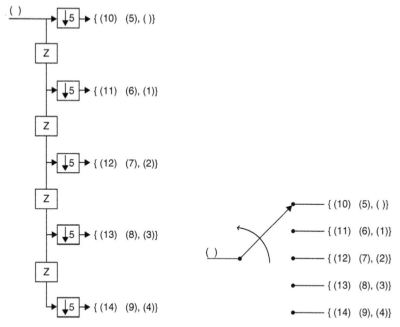

Figure 10.20 Polyphase decomposition: (a) delay-and-decimate chain corresponding to Figure 10.19 and, (b) commutator structure equivalent to decimator in (a)

at the low sampling rate, when we rescale the frequency axis, the ideal polyphase filters have an allpass frequency response.

10.2.1.2 Polyphase Interpolation Filters

A similar set of transformations can be performed for polyphase interpolation starting with Figure 10.17(b). Again consider the lowpass filter depicted in Figure 10.12:

$$H(z) = \sum_{\text{over } n} h(n)z^{-n} \tag{10.70}$$

Next express the coefficients of $H(z)$ in terms of the polyphase components following the structure depicted in Figure 10.22—that is, define the new variable k such that $n = kL + p, p = 0,...,L - 1$ and $k = \lfloor n/L \rfloor$—then the relation in (10.70) becomes:

$$H(z) = \sum_{\text{over } n} \left\{ \sum_{p=0}^{L-1} h(kL + p)z^{-(-kL+p)} \right\} \tag{10.71}$$

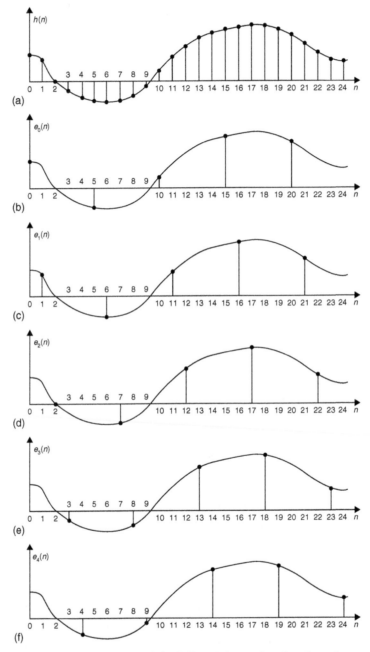

Figure 10.21 Impulse response of original filter (a) as related to impulse responses of polyphase filters (b) through (f)

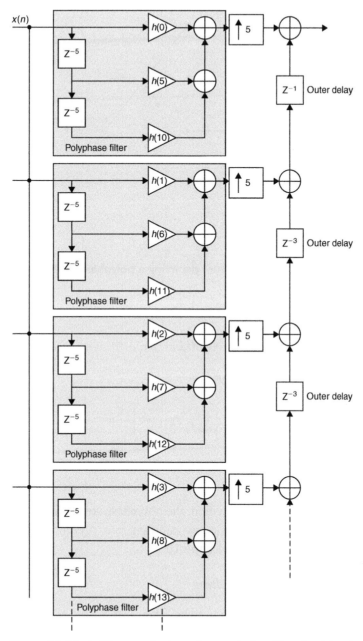

Figure 10.22 A 15-tap polyphase interpolation filter for $M = 5$

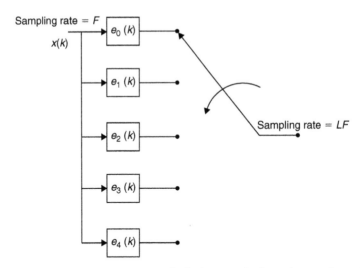

Figure 10.23 A commutator structure depicting a polyphase interpolator for L = 5

Define the *kth* polyphase filter as

$$e_p(k) = h(kL + p) \tag{10.72}$$

Substituting (10.72) into (10.71), we obtain:

$$H(z) = \sum_{p=0}^{L-1} \left\{ \sum_k e_p(k) z^{-kL} \right\} z^{-p}$$

$$= \sum_{p=0}^{L-1} E_p(z^L) z^{-p} \tag{10.73}$$

Again, as mentioned in the previous section, the polyphase structure has the advantage of performing the filtering process at the low sampling rate. A commutator structure depicting this interpolator is shown in Figure 10.23.

10.2.2 Half-Band and Kth-Band Filters

Half-band filters play a prominent role in SDR. Their use as cost-efficient, easy to implement filters has contributed greatly to their popularity. A half-band filter is a filter that possesses a certain symmetry property such that the time domain impulse response

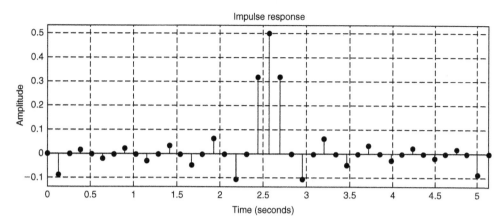

Figure 10.24 Halfband 39-tap filter with $F_s = 7.8\,\text{MHz}$ and $F_{pass} = 1.9\,\text{MHz}$

has every other coefficient equal to zero except at the center, as depicted in Figure 10.24 [8],[9]. This translates into reduced complexity in terms of the number of multiplies required for this type of filter. For an N-tap half-band filter only $(N + 1)/2 + 1$ multiplies are required to generate a single output sample. In this case the number of multiplies has been reduced by approximately one half. In the frequency domain, this implies that its frequency response is symmetrical around ¼ the sampling frequency—that is, symmetrical around the 3-dB attenuation point $F_s/4$ as shown in Figure 10.25. Note that the symmetry in this case is preserved, since the passband ripple δ_{pass} is equal to the stopband ripple δ_{stop}.

To simplify the understanding of half-band FIR filters, consider first the *noncausal* zero-phase impulse response

$$h(n) = h(-n) \tag{10.74}$$

In the frequency domain, this simply implies:

$$H(e^{j\omega}) = H(e^{-j\omega}) \tag{10.75}$$

The transfer function $H(e^{j\omega})$ is real and even due to the even characteristic of the impulse response in (10.74). The transfer function $H(z)$ can be expressed in terms of two polyphase components

$$H(z) = E_0(z^2) + z^{-1}E_1(z^2) \tag{10.76}$$

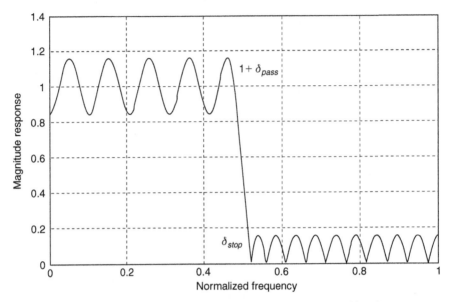

Figure 10.25 Frequency response of half-band filter depicted in Figure 10.24

For half-band filters, the first polyphase component is usually a constant defined as

$$E_0(z^2) = \frac{1}{2} \tag{10.77}$$

The relations in (10.75) and (10.76) imply that the number of coefficients N is odd. Furthermore, we can now state that, except for $n = 0$, all the even coefficients are zeros, that is $h(2n) = 0$ for $n \neq 0$. This in turn implies that $N + 1$ must be an integer multiple of 4, or

$$N = 4m - 1, \quad m \in \mathbb{N} \tag{10.78}$$

which implies that $(N - 1)/2$ is always odd.

Substituting (10.77) into (10.76), we obtain an interesting property of half-band filters:

$$H(z) + H(-z) = 1 \tag{10.79}$$

In the frequency domain, the relation in (10.79) implies:

$$H(e^{j\omega}) + H(e^{j(\omega - \pi)}) = 1 \tag{10.80}$$

From (10.80), the frequency response of the half-band filter has a point of symmetry at $\omega = \pi/2$ where the passband frequency ω_{pass} is situated symmetrically from the stopband frequency ω_{stop} according to the relationship:

$$\omega_{pass} = \pi - \omega_{stop} \tag{10.81}$$

Next we turn our attention to causal half-band filters. In this case, the noncausal filter discussed thus far can be made causal by shifting the impulse response by $(N - 1)/2$. Mathematically, this means that the transfer function in (10.76) becomes:

$$\begin{aligned} H_{causal}(z) &= z^{-(N-1)/2} H(z) \\ &= z^{-(N-1)/2} E_0(z^2) + z^{-[(N-1)/2-1]} E_1(z^2) \\ &= \frac{1}{2} z^{-(N-1)/2} + z^{-[(N-1)/2-1]} E_1(z^2) \end{aligned} \tag{10.82}$$

where again $(N - 1)/2$ is odd. The relation in (10.82) can be further written in a conventional polyphase form. Let

$$\begin{aligned} E_{causal,0}(z) &= z^{-[(N-1)/2-1]} E_1(z^2) \\ z^{-1} E_{causal,1}(z) &= z^{-(N-1)/2} E_0(z^2) \end{aligned} \tag{10.83}$$

Then

$$H_{causal}(z) = E_{causal,0}(z^2) + z^{-1} E_{causal,1}(z) \tag{10.84}$$

Next we turn our attention to the more general form of this type of filter, namely the Kth-band lowpass filters, or simply the Kth-band filters, where K is an integer. In this case, the transfer function of the Mth-band filter is

$$H(z) = \sum_{q=0}^{K-1} z^{-q} E_q(z^K) \tag{10.85}$$

Similar to the half-band case, the first polyphase component of (10.85) corresponding to $q = 0$ is simply a constant, that is:

$$E_0(z^K) = \frac{1}{K} \tag{10.86}$$

The impulse response of the Kth-band filter is such that:

$$h(n) = 0 \quad \text{for } n = \pm lK, \text{ and } l = 1, 2, \ldots \tag{10.87}$$

The *Kth*-band filter also retains the spectral property:

$$\sum_{k=0}^{K-1} H(e^{j(\omega - 2\pi k/K)}) = 1 \tag{10.88}$$

The relationship in (10.88) indicates that the *K*-shifted duplicates of the frequency response of the *Kth*-band filter, each shifted successively by $2\pi/K$, result in the sum of the responses being unity. The passband and stopband ripples $H(z)$ are bounded such that [9]:

$$\delta_{pass}^{(K)} \leq \frac{K-1}{K}\delta_{pass} + \frac{K-1}{K}\delta_{stop} \tag{10.89}$$

and

$$\delta_{stop}^{(K)} \leq \frac{1}{K}\delta_{pass} + \frac{2K-3}{K}\delta_{stop} \tag{10.90}$$

Both (10.89) and (10.90) imply that

$$\delta_{pass}^{(K)}, \delta_{stop}^{(K)} \leq \frac{2(K-1)}{K}\max\{\delta_{pass}, \delta_{stop}\} \tag{10.91}$$

Thus, one may claim that a *Kth*-band filter could provide near optimal performance for $1/K$th the computational complexity.

In [10], the authors provide a design technique for the design of half-band filters that relies on the Parks-McClellan filter design method [11]. First, consider an FIR filter with $g(n)$ with $(N+1)/2$ coefficients. Again, *N* is the number of coefficients for the half-band filter defined in (10.78). Define the cutoff frequency of the filter $G(z)$ as $\omega_{g(n),pass} = 2\omega_{pass}$ and the stopband frequency of $G(z)$ as $\omega_{g(n),stop} = \pi$, which corresponds to the normalized frequency of $f = 1/2$. This filter has only a passband and a stop band but no transition band as shown by way of the example in Figure 10.26. The half-band filter is then obtained by upsampling the response of $g(n)$ by 2 and setting the center coefficient to ½, that is

$$H(z) = \frac{G(z^2) + z^{-\frac{(N-1)}{2}}}{2} \tag{10.92}$$

10.2.3 Cascaded Integrator-Comb Filters

Cascaded integrator comb (CIC) filters, first introduced by Hogenauer [12], provide cost efficient means of performing decimation and interpolation. These filters are particularly

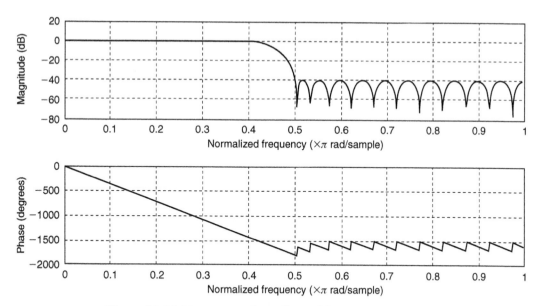

Figure 10.26 Frequency of FIR filter with no transition band

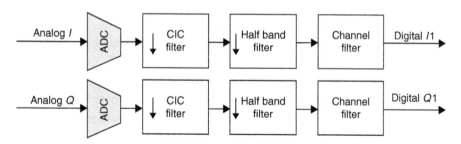

Figure 10.27 CIC filters perform the initial decimation

attractive when the sampling rate change is very large. For example, in SDR, it is fairly common to decimate a signal sampled, say, at 20 MHz to a bandwidth of 100 kHz. In this case, the CIC filter despite its performance limitation can provide a cost and computationally efficient filtering and decimation scheme to bring the sampled signal close to its desired final rate. The last decimation stage is done with a filter with sharper response than a CIC, such as a half-band filter for example, to preserve the response of the desired signal. This is illustrated in Figure 10.27. In this case, the sampling rate is much higher than the signal rate. To reduce the rate of the signal to the desired

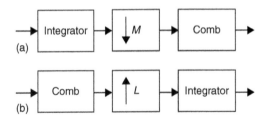

Figure 10.28 CIC filter is comprised of three basic blocks: (a) decimator and (b) interpolator

bandwidth, the data rate is first decimated via the CIC block to significantly reduce the bandwidth as much as possible without degrading the desired signal's spectral response. The final decimation stage is done via a half-band filter. If necessary, the signal is finally filtered by a channel filter in order to band-limit the signal to the desired bandwidth. This will become more apparent in the ensuing discussion concerning the performance of CICs. Finally, CIC filters are attractive in terms of implementation complexity (no multipliers) and are ideal for large decimation and interpolation ratios because of their simplicity and ease of implementation.

10.2.3.1 Signal Processing of CIC Filters

A CIC filter is comprised of three major blocks: an integrator section, a decimation or interpolation section, and a comb section as shown in Figure 10.28. Note that the rate change block, that is the decimation or interpolation block, is always situated between the integrator and comb sections. The particular arrangement of these blocks depends on the operation, be it upsampling or downsampling. To perform downsampling, the first block is the integrator block followed by the downsampling operation, which in turn is followed by the comb section. To perform upsampling, the comb section is followed by the upsampling operation which is then followed by the integrator section. The function of these various blocks will be discussed next.

The integrator block is comprised of one or more basic integrators connected in series. A basic integrator is shown in Figure 10.29(a). The integrator block is then simply a single-pole IIR filter with difference equation:

$$y(n) = x(n) + y(n - 1) \tag{10.93}$$

where $x(n)$ is the input to the integrator and $y(n)$ is the output of the integrator. The integrator frequency response is then simply $H_I(z) = 1/(1 - z^{-1})$. In the frequency

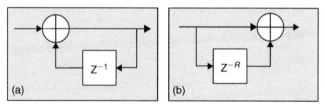

Figure 10.29 Basic building blocks: (a) integrator and (b) comb

domain, this implies that for a sampling rate, say T_s, the frequency response of the integrator can be expressed as

$$H_I(e^{j\omega T_s}) = \frac{1}{1 - e^{j\omega T_s}} = \frac{1}{2je^{-j\frac{\omega T_s}{2}}\sin\left(\frac{\omega T_s}{2}\right)} \tag{10.94}$$

The magnitude squared (power) response of (10.94) is

$$|H_I(e^{j\omega T_s})|^2 = \frac{1}{4\sin^2\left(\frac{\omega T_s}{2}\right)}$$

$$= \frac{1}{2(1 - \cos(\omega T_s))} \tag{10.95}$$

This is the power response of a lowpass filter with a 20-dB per decade (or 6 dB per octave) rolloff. The gain at DC for $\omega = 0$ is $|H_I(e^{j\omega T_s})|^2_{\omega=0} = \infty$. This is true since the pole of $H_I(z)$ is $z^{-1} = 1$. Recall, that a single integrator on its own is not necessarily stable. That is for a bounded input the output is not necessarily bounded. Figure 10.30 displays the magnitude response of $H_I(z)$.

Next, consider the difference equation governing the comb section:

$$y(n) = x(n) - x(n - R) \tag{10.96}$$

where R is an integer denoted as the differential delay. The transfer function corresponding to (10.96) is $H_c(z) = 1 - z^{-R}$. A typical value for R could range between 1 and 3. The magnitude and phase response of $H_c(z)$ corresponding to $R = 1$ are shown in Figure 10.31. Obviously, the comb filter in this case, with a null placed at DC, acts as a highpass filter. Next, for illustration's sake, consider the amplitude and phase response

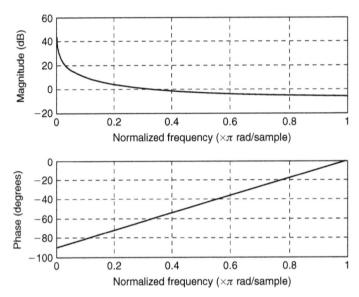

Figure 10.30 Magnitude response of single pole IIR integrator $H_I(z) = 1/(1 - z^{-1})$

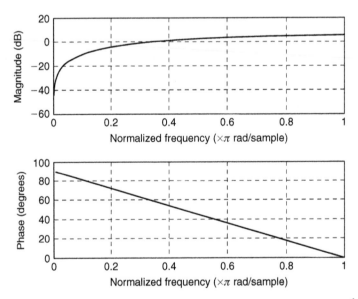

Figure 10.31 Magnitude and phase response of $H_c(z) = 1 - z^{-1}$

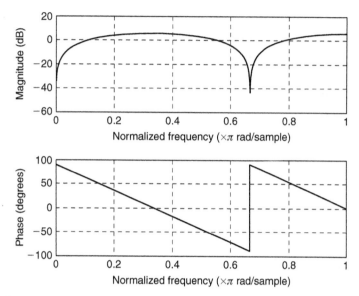

Figure 10.32 Magnitude and phase response of $H_c(z) = 1 - z^{-3}$

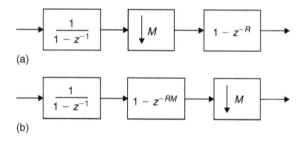

Figure 10.33 Single-stage CIC decimator filter (a) integrator, followed by decimator, followed by comb, and (b) using Noble identity the decimator and comb are swapped

corresponding to $R = 3$ as shown in Figure 10.32. Note the presence of the additional null in the magnitude response. We will elaborate on the effect of these nulls in the context of a CIC decimator next.

To further understand the behavior of a CIC decimator, consider a single-stage decimation filter as depicted in Figure 10.33(a). Consider the comb section

$H_c(z) = 1 - z^{-R}$; using the Noble identity for decimation, we can simply move the comb block closer to the integrator as shown in Figure 10.33(b). This implies that the comb section is now operating at the sampling rate of F_s instead of F_s/M. The comb transfer function then becomes $H_c(z) = 1 - z^{-RM}$. The transfer function of the CIC filter before decimation can then be expressed as

$$H(z) = H_I(z)H_C(z^M) = \frac{1 - z^{-RM}}{1 - z^{-1}} \tag{10.97}$$

Note that the relationship expressed in (10.97) is none other than the FIR filter $H(z) = 1 + z^{-1} + \cdots z^{-RM+1}$. This is true since $z^{-1}H(z) = z^{-1} + z^{-2} \ldots z^{-RM}$ and

$$\{1 - z^{-1}\}H(z) = 1 - z^{-RM} \Rightarrow H(z) = \frac{1 - z^{-RM}}{1 - z^{-1}} \tag{10.98}$$

First let's examine the comb section's frequency response, which is the numerator of (10.97):

$$H_C(e^{j\omega T_s M}) = 1 - e^{-j\omega T_s RM}$$

$$= 2je^{-j\frac{\omega T_s RM}{2}} \left[\frac{e^{j\frac{\omega T_s RM}{2}} - e^{-j\frac{\omega T_s RM}{2}}}{2j} \right]$$

$$= 2je^{-j\frac{\omega T_s RM}{2}} \sin\left(\frac{\omega T_s RM}{2}\right) \tag{10.99}$$

The magnitude-squared response of $H_c(z) = 1 - z^{-RM}$ can then be derived from (10.99) as

$$|H_C(e^{j\omega T_s M})|^2 = 4\sin^2\left(\frac{\omega T_s RM}{2}\right)$$

$$= 2\{1 - \cos(\omega T_s RM)\} \tag{10.100}$$

The nulls in (10.100) are found at the frequencies $\{\omega = \pm(2\pi/RMT_s)k\}$ for $k \in \mathbb{N}$. In all cases, the first null is always at DC. Furthermore, for $k = R$, the null is at $\omega = 2\pi/MT_s$, which occurs at the center frequency of the aliased signal. This is a very attractive feature of CIC filters, since the null occurs automatically over the aliasing signal, thus minimizing degradation.

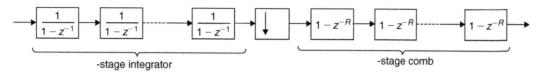

-stage integrator   -stage comb

Figure 10.34 N-stage CIC decimator filter

Next consider the N-stage CIC filter shown in Figure 10.34. Using the relationship in (10.97) and (10.98) we can express the transfer function as

$$H(z) = H_I^N(z)H_C^N(z^M) = \left(\frac{1 - z^{-RM}}{1 - z^{-1}}\right)^N \tag{10.101}$$

In the frequency domain, the magnitude response of (10.101) expressed at the sampling rate F_s is

$$H(e^{j\omega T_s}) = \left|\frac{\sin\left(\dfrac{\omega T_s RM}{2}\right)}{\sin\left(\dfrac{\omega T_s}{2}\right)}\right|^N \tag{10.102}$$

The resulting filter in (10.102) specifies a linear-phase lowpass filter indicative of a time-domain rectangular window. By the same token, the transfer function in (10.102) can be expressed in terms of the low sampling rate; that is, at F_s/M the frequency becomes:

$$H(e^{j\omega_M T_s})\Big|_{\omega_M = \frac{\omega}{M}} = \left|\frac{\sin\left(\dfrac{\omega_M T_s R}{2}\right)}{\sin\left(\dfrac{\omega_M T_s}{2M}\right)}\right|^N \tag{10.103}$$

Recall that for small α, the sinusoid can be approximated as $\sin(\alpha) \approx \alpha$. Then for large M, the relationship in (10.103) becomes:

$$H(e^{j\omega_M T_s})\Big|_{\omega_M = \frac{\omega}{M}} \approx R^N M^N \left|\frac{\sin\left(\dfrac{\omega_M T_s R}{2}\right)}{\dfrac{\omega_M T_s R}{2}}\right|^N \tag{10.104}$$

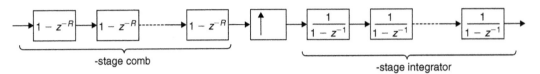

Figure 10.35 *N*-stage CIC interpolator filter

From (10.104), note that increasing the number of stages serves to improve the aliasing or image rejection of the filter at the expense of degrading the passband response which tends to droop more with increasing values of *N*. Furthermore, the DC gain as a function of *R*, *M*, and *N* can become large, thus requiring very large register bit widths.

CIC filters can also be used in interpolation. An *N*-stage interpolation CIC filter is depicted in Figure 10.35. In this case, the input is fed to the comb section. The comb section is followed by the upsampler, which is then followed by the integrator section. Note that the signal processing discussed thus far applies to both. Like CIC decimators, CIC interpolators automatically suppress images due to the zero-insertion process performed by the upsampler.

10.2.3.2 Computational Complexity and Finite Word-Length Effects

CIC filters are based on the premise that perfect pole and zero cancellation is possible. To do so, the designer must strive to minimize errors due to fixed point implementations. This is possible using 2's complement arithmetic as well as the residue number system (RNS) and carry-save adders (CSA). This is particularly important when implementing a CIC decimator to curb any arithmetic overflow due to the integrator section. The required bit-widths for a CIC filter that ensures adequate performance can be obtained by examining the DC gain of the filter:

$$gain = (RM)^N \tag{10.105}$$

This gain designates the maximum growth at the output of the filter. In terms of total number of bits, if b_{in} is the number of input bits, then the number of output bits b_{out} is

$$
\begin{aligned}
b_{out} &= b_{in} + \lceil \log_2 (gain) \rceil \\
&= b_{in} + \lceil N \log_2 (RM) \rceil \\
&= b_{in} + b_{total, growth}
\end{aligned}
\tag{10.106}
$$

This is the minimum number of output bits required to avoid any degradation to the signal due to overflow. The operator $\lceil \cdot \rceil$ indicates the ceiling function, which determines the total bit-growth, $b_{total,growth}$. The individual stages themselves are subject to truncation or rounding with bit-growth that can be determined for each stage based on the relation in (10.106):

$$b_{out,k} = b_{in} + b_{growth} \tag{10.107}$$

where $b_{out,k}$ is the internal output of the various stages. Signal truncation in 2's complement format, unlike rounding, introduces DC errors regardless of whether the input signal is DC-free or not. Note that in many practical systems employing oversampled $\Sigma\Delta$ converters, for example, the number of bits could become quite large. In this case additional rounding or truncating at each stage may become necessary. For an N-stage CIC filter, there are $2N + 1$ error sources where the first $2N$ sources are due to the adders and subtractors being rounded in the various filter stages. The last rounding occurs at the output register. Each error source is assumed to be uncorrelated with the input and other error sources. However, this is true only if the input signal does not exhibit periodicity. Assuming that the errors are uniformly distributed, then the error at the jth-source is given as

$$e_j = \begin{cases} 2^{b_j} & \text{with rounding} \\ 0 & \text{with no rounding} \end{cases} \tag{10.108}$$

where b_j is the number of LSBs that were pruned at the jth-source. The variance of this error is simply

$$\sigma_j^2 = \frac{1}{12} e_j^2 = \frac{1}{12} 2^{b_j} \tag{10.109}$$

Given the system transfer function in FIR form starting from the jth-stage and up to the last stage, we have

$$H_j(z) = \begin{cases} H_I^{N-j+1}(z)H_C^N(z) = \displaystyle\sum_{n=0}^{(RM-1)N+j-1} h_j(n)z^{-n}, & j = 1,\ldots,N \\ H_C^{j-N}(z) = \displaystyle\sum_{n=0}^{2N+1-j} h_j(n)z^{-nRM}, & j = N+1,\ldots,2N \end{cases} \tag{10.110}$$

where the coefficients:

$$h_j(n) = \begin{cases} \displaystyle\sum_{l=0}^{\lfloor n/RM \rfloor} (-1)^l \binom{N}{l} \binom{N-j+n-RMl}{n-RMl}, & j = 1,\ldots,N \\ (-1)^n \binom{2N+1-j}{n}, & j = N+1,\ldots,2N \end{cases} \tag{10.111}$$

Note that the maximum DC value may be obtained as

$$\sum_n |h_1(n)|^2 = (RM)^N \tag{10.112}$$

The total error variance up to the *jth*-stage is

$$\sigma_{T,j}^2 = \sigma_j^2 \sum_n h_j^2(n) \tag{10.113}$$

and the total error variance due to rounding as shown in [12] is simply the sum of all the error-variances produced at the various independent sources, that is:

$$\sigma_T^2 = \sigma_j^2 \sum_{j=1}^{2N+1} \sigma_{T,j}^2 \tag{10.114}$$

10.3 Arbitrary Sampling Rate Conversion

The sampling rate conversion techniques that we have discussed thus far assume that the ratio of input to output sampling rates is an integer or a ratio of integers. However, in certain cases where the ratio is an irrational number or a ratio of integers requiring high upsampling and downsampling rates, it becomes more practical to implement the rate change operation with the help of a fractional delay filter. One such example is to change the sampling rate, say by the ratio of 64/65 = 0.984. Obviously, polyphase filter structure would be impractical. In this section, we will discuss techniques that may be used to perform such sampling-rate conversion with the aid of fractional delay filters.

10.3.1 The Ideal Fractional Delay Filter

An ideal fractional delay filter is a filter through which the signal is simply delayed by an arbitrary amount

$$y(n) = h_{ideal}(n) * x(n) = x(n - D) \tag{10.115}$$

where $x(n)$ and $y(n)$ are the input and output signals respectively. In the frequency domain, this simply implies that $H_{ideal}(z) = z^{-D}$. In the frequency domain, the filter response is given as

$$H(e^{j\omega}) = e^{-j\omega D} \quad \Leftrightarrow \quad \begin{array}{l} |H(e^{j\omega})| = 1, \forall \omega \\ \arg\left\{H(e^{j\omega})\right\} = \varphi(\omega) = -\omega D \end{array} \tag{10.116}$$

The group delay can be derived from the phase as

$$\tau_g(\omega) = -\frac{\partial \varphi(\omega)}{\partial \omega} = D \tag{10.117}$$

Note that the implication of (10.115) in the z-domain is that

$$Z\{x(n - D)\} = z^{-D}X(z) \tag{10.118}$$

This in essence implies that D *must be* an integer. In reality z^{-D} cannot be simply realized for non-integer values of D. Therefore, an approximation of this all pass function is then in order. From (10.116) the ideal solution in the time domain can be obtained as

$$
\begin{aligned}
h_{ideal}(n) &= \frac{1}{2\pi} \int_{-\pi}^{\pi} H(e^{j\omega})e^{j\omega n}d\omega \\
&= \frac{\sin\left[\pi(n - D)\right]}{\pi(n - D)} = \mathrm{sinc}(n - D)
\end{aligned}
\tag{10.119}
$$

When the delay D is an integer, the relationship in (10.119) is none other than the delta function $\delta(n - D)$ which is the pulse function. This is illustrated in Figure 10.36(a) for $D = 2$. However, for noninteger values of D, the ideal delay is represented by a delayed *sinc* response. This discrete ideal response becomes infinitely long and noncausal, as shown in Figure 10.36(b). Therefore, it becomes evident that in order to approximate a fractional delay, a proper approximation of (10.119) is needed [13].

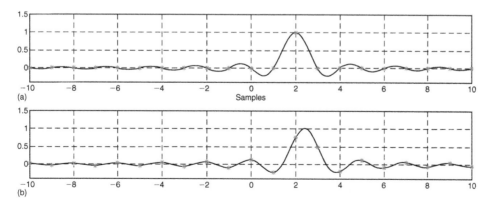

Figure 10.36 Ideal sinc response of filter for (a) delay $D = 2$, and (b) delay $D = 2.4$

10.3.2 Fractional Delay Filter via Direct Least Squared Design

Let the FIR filter $H(z)$ be the approximation of the ideal fractional delay D:

$$H(z) = \sum_{n=0}^{N} h(n) z^{-n} \tag{10.120}$$

The N-th order filter coefficients may be determined such that the L_2-norm of the error function is minimized. In other words, define the frequency domain error signal:

$$E(e^{j\omega}) = H(e^{j\omega}) - H_{ideal}(e^{j\omega}) \tag{10.121}$$

The L_2-norm of the error function:

$$
\begin{aligned}
|E| &= \frac{1}{\pi} \int_0^\pi |E(e^{j\omega})|^2 d\omega \\
&= \frac{1}{\pi} \int_0^\pi |H(e^{j\omega}) - H_{ideal}(e^{j\omega})|^2 d\omega
\end{aligned}
\tag{10.122}
$$

Using Parseval's relation, the L_2-norm can also be expressed as

$$|E| = \sum_{n=-\infty}^{\infty} |h(n) - h_{ideal}(n)|^2 \tag{10.123}$$

It turns out that in order to minimize the L_2-norm of (10.123) the optimal FIR filter can simply be obtained by truncating the ideal *sinc* function in (10.119) to $N + 1$ terms; that is,

$$h(n) = \begin{cases} \text{sinc}(n - D) & 0 \le n \le N \\ 0 & \text{otherwise} \end{cases} \tag{10.124}$$

In the event where the overall delay D is large, part of the filter is realized as a simple delay line and the rest as a fractional delay filter. Furthermore, as is obvious from (10.123) the error becomes smaller and smaller as we increase N. This method suffers from the Gibbs phenomenon and its performance is generally unacceptable in practice. In order to improve on the performance of the filter and reduce the degradation, a windowing version of (10.124) is used:

$$h(n) = \begin{cases} w(n - D) \, \text{sinc}(n - D) & 0 \le n \le N \\ 0 & \text{otherwise} \end{cases} \tag{10.125}$$

The window function provides some weighting to the coefficients of $h(n)$ thus emphasizing the values near the filter's center of gravity, alleviating Gibbs phenomenon.

10.3.3 Maximally Flat Filter Design via Lagrange Interpolation

The maximally flat FIR fractional delay filter ensures that the error function is maximally flat at a certain frequency, say at DC, for example. This implies that the filter approximation is at its best near $\omega = \omega_0$. This can be done by ensuring that the derivatives of the error function at ω_0 are equal to zero, or:

$$\left. \frac{d^k E(e^{j\omega})}{d\omega^k} \right|_{\omega=\omega_0} = 0, \quad \text{for } k = 0, 1, \ldots, N \tag{10.126}$$

where the complex frequency domain error function is as defined in (10.121). The relationship in (10.126) can be explicitly written as

$$\left. \frac{d^k}{d\omega^k} \left\{ \sum_{n=0}^{N} h(n, D)e^{-j\omega n} - e^{-j\omega D} \right\} \right|_{\omega=\omega_0} = 0, \quad \text{for } k = 0, 1, \ldots, N \tag{10.127}$$

Note that D in the coefficient $h(n, D)$ indicates that the computed coefficient is pertinent to the fractional delay D. For $k = 0$ and 1, the relationship in (10.127) at DC is

$$\text{For } k = 0, \ \sum_{n=0}^{N} h(n, D) - 1 = 0 \Rightarrow \sum_{n=0}^{N} h(n, D) = 1$$

$$\text{For } k = 1, \ \frac{d}{d\omega} \left\{ \sum_{n=0}^{N} h(n, D)e^{-j\omega n} - e^{-j\omega D} \right\} = -\sum_{n=0}^{N} jnh(n, D) + jD = 0$$

$$\Rightarrow \sum_{n=0}^{N} nh(n, D) = D \tag{10.128}$$

Note that $\sum_{n=0}^{N} h(n, D) = 1$ for $k = 0$ in (10.128) implies that the magnitude response of the filter at DC is 1.

In general, minimizing the error function at DC, that is, for $\omega_0 = 0$, results in the set of $(N + 1)$ linear equations defined as [14]:

$$\sum_{n=0}^{N} n^k h(n, D) = D^k, \quad \text{for } k = 0, 1, \ldots, N \tag{10.129}$$

In matrix notation, the relation expressed in (10.129) can be written in the form

$$\Psi \vec{h} = \vec{\Delta} \tag{10.130}$$

where $\vec{h}_D = [h(0, D), h(1, D), \ldots, h(N, D)]^T$, $\vec{\Delta} = [1, D, D^2, \ldots, D^N]^T$, and Ψ is the Vandermonde $(N + 1)$ by $(N + 1)$ *nonsingular* matrix:

$$\Psi = \begin{bmatrix} 1 & 1 & 1 & \cdots & 1 \\ 0 & 1 & 2 & \cdots & N \\ 0 & 1 & 2^2 & \cdots & N^2 \\ \vdots & \vdots & \vdots & \vdots & \vdots \\ 0 & 1 & 2^N & \cdots & N^N \end{bmatrix} \tag{10.131}$$

The nonsingularity of (10.131) implies that Ψ has an inverse such that $\Psi\,\Psi^{-1} = I$. The solution to (10.130) can be readily obtained via Lagrange interpolation. Lagrange interpolation uses polynomial curve fitting based on a given finite number of points to predict the value of the missing samples. In other words, the filter coefficients may be obtained as

$$\begin{aligned} h(n, D) &= \prod_{k=0, k \neq n}^{N} \frac{D - k}{n - k}, \quad \text{for } n = 0, 1, 2 \cdots N \\ &= (-1)^{N-n} \binom{D}{n}\binom{D - n - 1}{N - n} \\ &= \frac{D}{n} \times \frac{D - 1}{n - 1} \times \cdots \times \frac{D - n + 1}{1} \times \frac{D - n - 1}{-1} \times \cdots \times \frac{D - N}{n - N} \end{aligned} \tag{10.132}$$

The case for $N = 1$ is none other than the linear interpolation case. The coefficients for this case can be obtained from (10.132):

$$\begin{aligned} h(n, D) &= \prod_{k=0, k \neq n}^{1} \frac{D - k}{n - k}, \quad \text{for } n = 0, 1 \\ h(0, D) &= 1 - D \\ h(1, D) &= D \end{aligned} \tag{10.133}$$

The coefficients for the case where $N = 3$ are

$$h(n, D) = \prod_{k=0, k \neq n}^{3} \frac{D - k}{n - k}, \quad \text{for } n = 0, 1, 2, 3 \tag{10.134}$$

$$h(0, D) = \frac{(1 - D)(D - 2)(D - 3)}{6}$$

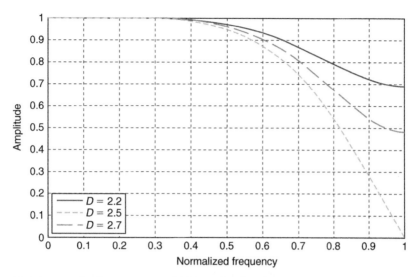

Figure 10.37 Magnitude response of fifth order fractional delay filter for D = 2.2, 2.5, and 2.7 samples

$$h(1, D) = \frac{D(D - 2)(D - 3)}{2}$$

$$h(2, D) = \frac{-D(D - 1)(D - 3)}{2}$$

$$h(3, D) = \frac{D(D - 1)(D - 2)}{6}$$

The magnitude and phase delay responses for a fifth order Lagrange fractional delay filter for various delays are shown in Figure 10.37 and Figure 10.38, respectively. It can be easily shown from (10.132) that the Lagrange coefficients possess certain interesting symmetry. The coefficients for the fractional delay $N - D$ are the same as the coefficients for the fractional delay D except in reverse order, that is

$$h(N - n, N - D)\Big|_{\text{Fractional delay } = N - D} = \prod_{k=0, k \neq n}^{N} \frac{N - D - k}{N - n - k}, \quad \text{for } n = 0, 1, 2 \ldots N$$

$$= h(n, D)\Big|_{\text{Fractional delay } = D}$$

$$(10.135)$$

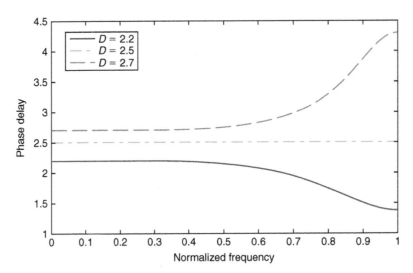

Figure 10.38 Phase delay of fifth order fractional delay filter for $D = 2.2$, 2.5, and 2.7 samples

In [15], a generalized expression for computing the coefficients for $\omega_0 \neq 0$ was found to be:

$$h(n, D) = e^{j\omega_0(n-D)} \prod_{k=0, k \neq n}^{N} \frac{D - k}{n - k}, \quad \text{for } n = 0, 1, 2 \cdots N, \omega_0 \in [0, \pi] \qquad (10.136)$$

Note that the relation in (10.132) is a special case of (10.136). For $0 < \omega_0 < \pi$, the coefficients are complex valued. For $\omega_0 = \pi$, the coefficients are real and identical to the Lagrange coefficients except that the sign of every other coefficient is inverted.

Given the $(N + 1)$ samples of $x(t)$ at certain known sampling instances, then using the coefficients computed using the Lagrange polynomial, we can find the missing sample $x(n - D)$ according to the relation

$$x(n - D) = \sum_{n=0}^{N} h(n, D)x(n), D \text{ is real in } [0, N] \qquad (10.137)$$

To compute the coefficients of the Lagrange fractional delay filter using the method above requires $N(N + 1)$ additions and $N(N + 1)$ multiplies. In practice, for variable D, the coefficients are retrieved via table look-up. It is important to note from (10.137) given the order of the Lagrange interpolator N, that the order number of samples used in the interpolation is equal to $N + 1$. This is a different scenario from the Farrow structure discussed below.

At this point, it is instructive to examine the approximation error for various values D. For integer values of D, that is $D \in \mathbb{N}$, the relationship in (10.132) reduces to the Kronecker delta function thus representing a pure delay. That is, in this case $h(n) = \delta$ $(n - D)$ for $n = 1,2, \dots ,N$. The approximation error itself reduces to zero, which in turn implies that the interpolation polynomial passes through the known signal samples. The worst case approximation error occurs when the fractional portion of D is 0.5 (e.g., $D = 2.5$ samples). For this worst case, we present two different scenarios, namely the odd order case and the even order case:

- The odd order case. The impulse response of the interpolator in this case is said to be even symmetric, that is $h(n) = h(N - n)$ for $n = 1,2, \dots ,N$. This property ensures that the interpolating function is linear phase with phase response $D\omega$, as can be inferred from the phase-delay response shown in Figure 10.38. The magnitude response, however, possesses a zero at the Nyquist frequency, thus causing a large error deviation at high frequencies.

- The even case. There is no particular symmetry associated with this case, however the overall approximation error, considering both amplitude and phase responses, is the worst. A quick comparison between the fifth order magnitude response and the sixth order magnitude response, depicted in Figure 10.39, shows that the latter does not possess a null at the Nyquist frequency. Consequently, the magnitude degradation by itself is not as severe for the even case as it is for the odd case. Note, however, that for either order the Lagrange method produces accurate results at low frequencies.

10.3.4 The Farrow Structure for Fractional Delay FIR Filters

The Farrow structure, first proposed in [16], offers a new structure in which fractional delay filters can be implemented using Lagrange polynomials. The Farrow filter relies on a filter bank structure, whereby each filter coefficient is approximated as an Nth order polynomial in D:

$$h(n, d) = \sum_{k=0}^{P} c_k(n)d^k, \quad n = 0,1,\dots,N, 0 \le d \le 1 \tag{10.138}$$

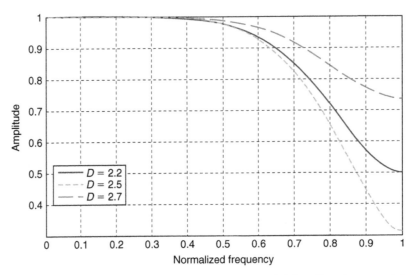

Figure 10.39 Magnitude response of sixth order fractional delay filter for $D = 2.2$, 2.5, and 2.7 samples

where the set of real-valued coefficients $\{c_k(.)\}$ is obtained by solving a set of $N + 1$ equations as will be shown promptly. Note that the coefficients presented in (10.138) are expressed in terms of the fractional delay d, that is $0 \leq d \leq 1$. The Farrow structure is shown in Figure 10.40. In the z-domain, the filter transfer function can be expressed as

$$H_d(z) = \sum_{n=0}^{N} h(n,d)z^{-n} = \sum_{n=0}^{N} \left\{ \sum_{k=0}^{P} c_k(n)d^k \right\} z^{-n} \qquad (10.139)$$

Define $C_k(z) = \sum_{n=0}^{N} c_k(n)z^{-n}$; then the relation in (10.139) can be rewritten in the form:

$$H_d(z) = \sum_{n=0}^{N} \left\{ \sum_{k=0}^{P} c_k(n)d^k \right\} z^{-n} = \sum_{k=0}^{P} \left\{ \sum_{n=0}^{N} c_k(n)z^{-n} \right\} d^k$$

$$= \sum_{k=0}^{P} C_k(z)d^k \qquad (10.140)$$

Again, the relation in (10.140) suggests that the filter structure depicted in Figure 10.40 is made up of a bank of fixed filters weighted by the fractional delay d and summed at the

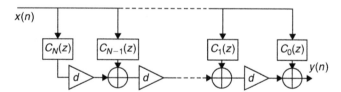

Figure 10.40 Implementation of polynomial approximation of filter coefficients for Farrow structure

output of every tap. Unlike the Lagrange interpolator discussed above, the degree P of the polynomial is independent from the number of samples used in the interpolation. Furthermore, unlike the Lagrange interpolator case, where the same polynomial is utilized for all the samples used in the interpolation, the Farrow filter uses a unique polynomial in the implementation of each tap.

Example 10-2: Realizing a Farrow Filter Using Lagrange Interpolation

Realize a Farrow filter using a second order Lagrange polynomial for an arbitrary delay d.

Let's start by reexamining the relationship in (10.132) for $N = 2$ and a fractional delay d, that is:

$$h(n, d) = \prod_{k=0, k \neq n}^{2} \frac{d - k}{n - k}, \quad \text{for } n = 0, 1, 2 \tag{10.141}$$

The various coefficients may be obtained then using (10.141):

$$h(0, d) = \prod_{k=1, k \neq 0}^{2} \frac{d - k}{0 - k} = \left(\frac{d - 1}{-1}\right)\left(\frac{d - 2}{-2}\right) = \frac{1}{2}(d^2 - 3d + 2)$$

$$h(1, d) = \prod_{k=0, k \neq 1}^{2} \frac{d - k}{1 - k} = \left(\frac{d}{1}\right)\left(\frac{d - 2}{1 - 2}\right) = -d^2 + 2d$$

$$h(2, d) = \prod_{k=0, k \neq 2}^{1} \frac{d - k}{2 - k} = \left(\frac{d}{2}\right)\left(\frac{d - 1}{1}\right) = \frac{1}{2}(d^2 - d) \tag{10.142}$$

Recall that the filter in (10.139) is of the form:

$$H_d(z) = \sum_{n=0}^{2} h(n,d)z^{-n} = h(0,d) + h(1,d)z^{-1} + h(2,d)z^{-2}$$

$$= \frac{1}{2}(d^2 - 3d + 2) + (-d^2 + 2d)z^{-1} + \frac{1}{2}(d^2 - d)z^{-2} \qquad (10.143)$$

Rearrange (10.143) in the form of (10.140); that is,

$$C_0(z) = 1$$
$$C_1(z) = -\frac{3}{2} + 2z^{-1} - \frac{1}{2}z^{-2}$$
$$C_2(z) = \frac{1}{2} - z^{-1} + \frac{1}{2}z^{-2} \qquad (10.144)$$

A conceptual implementation of this filter is shown in Figure 10.41. Again it is conceptual because in reality the implementation will share elements such as unit delays, resulting in a more efficient structure. Note that the filters in (10.144) can be further expressed in matrix form as

$$\vec{C}(z) = \Phi^T \vec{z} \qquad (10.145)$$

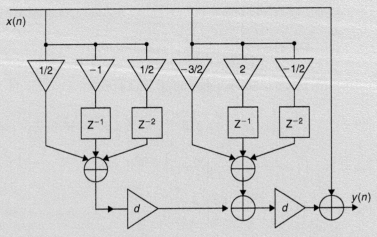

Figure 10.41 Farrow filter implemented using a second order Lagrange polynomial

where

$$\bar{C}(z) = \begin{bmatrix} C_0(z) \\ C_1(z) \\ C_2(z) \end{bmatrix}$$

$$\Phi^T = \begin{bmatrix} 1 & 0 & 0 \\ -3/2 & 2 & -1/2 \\ 1/2 & -1 & 1/2 \end{bmatrix}$$

$$\bar{z} = \begin{bmatrix} 1 & z^{-1} & z^{-2} \end{bmatrix}^T \tag{10.146}$$

Note that the transposed matrix Φ^T in (10.146) is none other than the inverse Vandermonde matrix shown in (10.131); that is, $\Psi^{-1} = \Phi^T$. Therefore, it can be easily shown from the analysis presented thus far that the relationship in (10.145) is true for any size matrix. And furthermore, we note that

$$\sum_{k=0}^{P} C_k(z)d^k = z^{-d}\Big|_{d=0} \Rightarrow C_0(z) = 1 \tag{10.147}$$

is for any order Farrow filter. All other transfer functions $\{C_k(z), k = 1, \ldots, N\}$ are of order N. These transfer functions are fixed for a given N. It is important to realize that a Farrow structure is suitable for applications where the fractional delay may change frequently from one SDR application to the next. This allows for a great reuse of hardware at minimal cost. However, if the fractional delay is fixed, the Lagrange interpolator method discussed in the previous section is more suitable, since it is less computationally intensive.

References

[1] Crochiere R, Rabiner L. Interpolation and decimation of digital signals—a tutorial review. Proceedings of the IEEE March 1981;69(3):300–31.
[2] Vaidyanathan PP. Multirate digital filters, filter banks, polyphase networks, and applications: a tutorial. Proceedings of the IEEE January 1990;78(1):56–93.
[3] Vaidyanathan PP. A tutorial on multirate filter banks. IEEE International Symposium on Circuits and Systems June 1998;3:2241–8.

[4] Crochiere R, Rabiner L. Optimum FIR Digital Implementations for Decimation, Interpolation, and Narrow-band Filtering. IEEE Trans. on Acoust. Speech, and Signal Proc. October 1975;23(5).

[5] Hermann O, Rabiner LR, Chan DSK. Practical design rules for optimum finite impulse response low-pass digital filters. Bell Syst. Tech. Journal July–August 1973;52:769–99.

[6] Rabiner LR. Approximate design relationships for low-pass FIR digital filters. IEEE Trans. Audio Electroacoust. October 1973;AU-21:456–60.

[7] Crochiere R, Rabiner L. Multirate Digital Signal Processing. Englewood Cliffs, N.J.: Prentice Hall; 1996.

[8] Bellanger M, Daguet JL, Lepagnol GP. Interpolation, extrapolation, and reduction of computation speed in digital filters. IEEE Trans. on Acoustics, Speech, and Signal Processing August 1974;22:231–5.

[9] Mintzer F. On half-band, third-band, and Nth-band FIR filters and their design. IEEE Trans. on Acoustics, Speech, and Signal Processing October 1982;30(5):734–8.

[10] Vaidyanathan PP, Nguyen. A trick for the design of FIR half-band filters. IEEE Trans. on Circuits and Systems March 1987;34:297–300.

[11] McClellan JH, Parks TW, Rabiner LR. A computer program for designing optimum FIR linear phase digital filters. IEEE Trans. on Audio Electroacoust., Speech, and Signal Processing Dec. 1973;21:506–26.

[12] Hogenauer EB. An economical class of digital filters for decimation and interpolation. IEEE Trans. on Acoustics, Speech, and Signal Processing 1997;29:457–67.

[13] Laasko T, et al. Splitting the unit delay. IEEE Signal Processing Magazine January 1996:30–60.

[14] Oetken G. A new approach for the design of digital interpolation filters. IEEE Trans. on Acoustics, Speech, and Signal Processing Dec. 1979;26(6):637–43.

[15] Hermanowicz E. Explicit formulas for weighting coefficients of maximally flat tunable FIR delayers. Electronic Letters 1992;28:1936–7.

[16] Farrow CW. A continuously variable digital delay element. Proc. IEEE Int. Symp. on Circuits and Systems (ISCAS'88) 1988;3:2641–5.

Index